BASIC BIOCHEMICAL METHODS

BASIC BIOCHEMICAL METHODS

Second Edition

Renee R. Alexander and Joan M. Griffiths

Division of Biological Sciences
Section of Biochemistry, Molecular and Cell Biology
Cornell University, Ithaca, New York

WILEY-LISS

A JOHN WILEY & SONS, INC., PUBLICATION
New York • Chichester • Brisbane • Toronto • Singapore

Address All Inquiries to the Publisher
Wiley-Liss, Inc., 605 Third Avenue, New York, NY 10158-0012

Printed in the United States of America

Library of Congress Cataloging-in-Publication Data
Alexander, Renee R., 1932-
Basic biochemical methods / Renee R. Alexander, Joan M. Griffiths. - 2nd ed.
p. cm.
Includes bibliographical references and index.
ISBN 0-471-56153-3
1. Biochemistry—Laboratory manuals. I. Griffiths, Joan M., 1935- . II. Title.
[DNLM: 1. Biochemistry—laboratory manuals. QU 25 A377b]
QP519.A44 1992
574.19'2'078—dc20
DNLM/DLC
for Library of Congress 92-5540
CIP

The text of this book is printed on acid-free paper.

Contents

Preface

Laboratory experience in biochemistry is important to students and researchers in the biological sciences. This text is designed to introduce the student to methods used in the isolation and quantitation of various cell fractions or compounds having biological significance. It is intended for a laboratory course for advanced undergraduate and beginning graduate students in the biological sciences. Our students are majors in plant and animal sciences, genetics, microbiology, neurobiology and nutrition. Therefore, the experiments were designed for a broad spectrum of interest areas to teach the use of biochemical methods in these related fields. The students using this manual should have completed basic courses in biology and inorganic and organic chemistry. They should also be familiar with the material covered in an introductory lecture course in biochemistry. The principles discussed in conjunction with the experiments presented are addressed mostly to the methods themselves, and it is assumed that the reader has an understanding of the theory or concepts involved. For further background material, references to relevant texts and literature citations are given in each of the chapters.

Many of the techniques described here are introduced as part of a larger study rather than presented as individual, unrelated procedures. The experiments are therefore grouped in modules. Of these modules, "Proteins and Buffers" and "Enzymology" are designed for all students, as they include basic methods. The other modules are scheduled for three-week periods, and students should select those that fit into the time allotted and suit their interests. Although some of these experiments require the use of various instruments that may not be available in all student laboratories, experiments may be selected from each module that require essentially only pipettes, a centrifuge, and a spectrophotometer.

Appendix I is a compilation of equipment and supplies needed for each experiment and is also a guide to the preparation of reagents and media. Many commonly used biochemical procedures appear as self-contained units; in this manner the manual can serve as a guide to researchers as well as a classroom text.

The experiments in this manual are the result of a number of years of teaching an introductory course in biochemistry in the Department of Biochemistry, Molecular and Cell Biology at Cornell University. This revised edition includes updated methods and describes new techniques for the study of proteins, nucleic acids, and lipids that are of major importance in biochemical research, and the book introduces new protocols for application to studies in molecular biology.

Many colleagues who have taught with us over the years have contributed toward this effort. We are particularly indebted to our graduate teaching assistants and to our many students who have helped us to design and test these experiments. In particular, we acknowledge the contributions to the second edition made by James Blankenship, Chris Kroupis, Eric Rasmussen, and Xenia Young. We appreciate the valuable editorial comments of Martin Alexander and his patience with one of us (R.R.A.) during the preparation of this book.

We also thank Virginia Scarpino for help in assembling Appendix I, Monica Howland for providing many of the illustrations, and Virginia Slator for assistance in preparing the manuscript.

Renee R. Alexander
Joan M. Griffiths
Ithaca, New York
October 1992

CHAPTER 1

The Laboratory Notebook

REFERENCES: American Public Health Association. *Standard Methods for the Examination of Water and Wastewater*, 17th ed. APHA, Washington, DC, pp. 28–30 (1989).

Casella, G., and Berger R.L. *Statistical Inference*. Wadsworth and Brooks, Pacific Grove, CA (1990).

Day, R.A. *How to Write and Publish a Scientific Paper*. Oryx Press, Phoenix, AZ (1988).

McMillan, V.A. *Writing Papers in the Biological Sciences*. St. Martin's Press, New York (1988).

Skoog, D., and White, D. *Fundamentals of Analytical Chemistry*, 4th ed. Saunders College Publishing Co., Philadelphia pp. 80–82 (1982).

Steel, R.G.D., and Torrie, J.H. *Principles and Procedures of Statistics*. McGraw Hill, New York pp. 9–56 (1960).

Wharton, D., and McCarty, R. *Experiments and Methods in Biochemistry*. Macmillan, New York pp. 5–20 (1972).

Communication is an essential aspect of conducting laboratory experiments. Much of the value of collecting experimental results is lost if proper records are not kept and if the data are not written up clearly so that the information can be transmitted effectively to one's colleagues. It is important to record all experimental findings in a notebook *at the time that the mea-*

surements and observations are made. It may be convenient to record information on some scrap paper with the intention of copying the data into a notebook at a later time, but such loose sheets are easily misplaced. Memory also becomes faulty, even within a relatively short time period.

The importance of maintaining proper records was reinforced recently when an investigation by the U.S. Secret Service was initiated by the Chairman of the House Subcommittee on Oversight and Investigations. At issue was an article in *Cell* written by a prominent scientist and an associate whose notebooks showed that data had been transcribed as much as 2 years after the experiments were done. It was shown that some of the entry dates had been altered, and some of the published results did not appear to match the original records. An internal probe conducted previously for NIH by a panel of scientists had found no evidence of wrong doing. The result of the 9-month investigation by the Secret Service likewise led to the conclusion that there had been no intent to defraud. However, the parties involved were responsible for maintaining sloppy notebooks and for carelessness in their record keeping (*Science*, 244:643–646, May 1989).

This costly incident and the adverse publicity which ensued are regrettable. Scientists have traditionally monitored their own activities through the peer review process. This incident points to the need for vigilance and to the value of keeping good records in the laboratory. Had these practices been observed, the alleged fraudulence could have been easily refuted. One of the goals of any laboratory course is to instruct students on how to keep clear and accurate records at the beginning of their careers.

The importance of communication to the scientist cannot be overemphasized. For the science teacher, the need to guide students in writing is essential. *Writing Papers in the Biological Sciences* is a valuable text designed to help biologists write more effectively. In her book, Professor McMillan concisely develops an orderly strategy for writing a research paper. In this chapter, a format suitable for keeping a biochemistry laboratory notebook and for writing research reports is briefly outlined.

A. The Notebook

A bound quadrille ruled notebook with numbered pages is suitable. Several blank pages in the front should be reserved for a table of contents.

B. Organization of the Subject Matter for Each Experiment

1. Brief descriptive title.
2. Date on which the experiment is performed.
3. Purpose of the experiment.
4. Procedure or methods section. This section may include an outline of the steps taken to perform the experiment or, if a published text is used, a reference to the published procedure can be made.
5. Presentation of data.
   a. *Flow charts.* If an isolation scheme is to be followed, a flow chart provides a convenient diagram to follow. A procedure is given, and space is made available to record data, such as volume of the fractions, additions made, and information about the time and speed of centrifugation.

Example: Partial Purification of α-Amylase

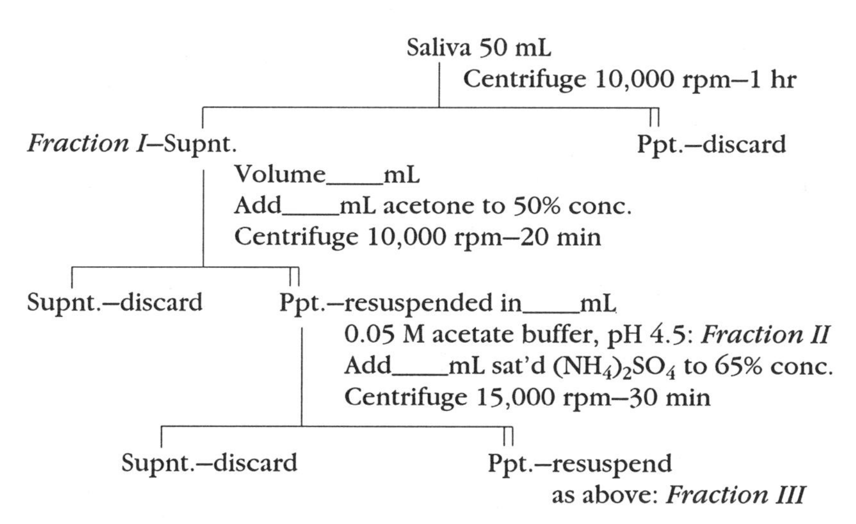

   b. *Protocols.* A protocol consists of a chart that provides places to plan a series of tests and to record the results obtained. An example is shown in Table 1.1 for establishing a standard curve. It is customary to show the tube numbers horizontally and to list the additions made to each tube vertically and in the order in which the additions are made. The variable is added first, and water is added to bring the volume to a constant, 1.5 mL.

c. *Tables*. Tables should be numbered and have titles (see Table 1.1). Headings for each column should be clearly marked and include units (*i.e.*, mg/mL, percent), if applicable. It is advisable to list variables down rather than across, and the values or properties referring to the variable should be given horizontally. Tables should be compact and should not contain procedural detail that can be written into the text.

TABLE 1.1. Protein Determination by the Biuret Method

Additions (mL)	Tube no. 1	2	3	4	5	6
Bovine serum albumin (2 mg/mL)	0	0.1	0.2	0.4	0.8	1.0
H_2O	1.5	1.4	1.3	1.1	0.7	0.5
Biuret reagent	1.5	→	→	→	→	→
$A_{540\ nm}$	0	0.025	0.050	0.105	0.210	0.265
mg/aliquot	0	0.2	0.4	0.8	1.6	2.0

d. *Graphs*. The data obtained can now be plotted to give a standard curve (see Figure 1.1). The variable, bovine serum albumin in this case, is given on the abscissa, and absorbance is recorded on the ordinate. The wavelength at which the measurements are made is indicated as a subscript of A, and the units of protein (mg) are also included. A title should clearly indicate what is represented on the graph, and a legend, if required, should be included. Several curves can be accommodated if the same units are involved, but graphs should not be cluttered or they become difficult to interpret.

C. Calculations and Analysis of Data

When measurements are made, it is important to take the *sensitivity* of the method and the instrument into account. *Sensitivity* refers to the minimum differences or changes in

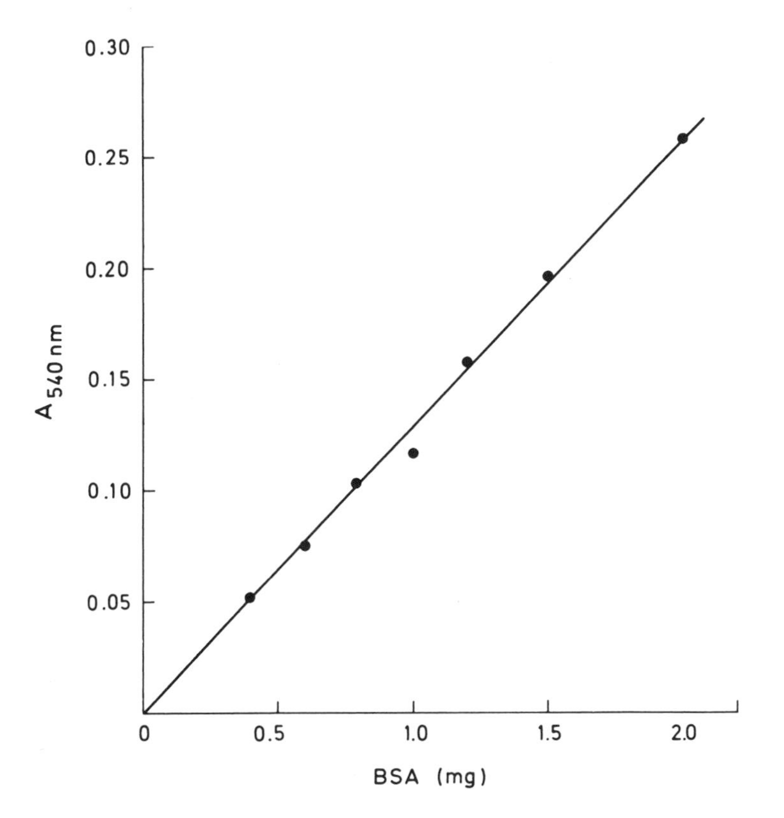

Fig. 1.1. Standard curve for protein determination by the Biuret Method.

measurement possible for the method or the instrument, *i.e.*, how little can be detected.

Accuracy is achieved when the observed value is close to the actual value. It is dependent on the quality of the reagents used and on the care exercised in performing the experiment. *Precision* refers to how well experimental values agree with each other. The two terms are therefore not synonymous and should not be used interchangeably. For instance, in a game of darts, the dart hitting the center was thrown accurately. However, when several darts land in close proximity to one another, the shots were precise, but unless located in the center of the dart board, they were not accurate.

It is important to report *significant figures* correctly. A problem frequently arises because a series of mathematical manipulations is carried out on a calculator that provides an answer with a large group of numbers. The valid value should contain only the digits known with certainty plus the first uncertain one. For example, if four protein determinations give values of 3.6, 3.7, 3.7 and 3.5 mg, the mathematical average is 3.63 mg and the standard deviation of the sum is $\pm$ 0.11. The standard deviation,

S.D. = $\sqrt{S^2/n}$, where S is the difference between the individual observation and the mean, and n is the number of observations included in determining the mean. In this example, one might be inclined to record the protein content of the sample as 3.63 ± 0.11 mg, indicating a possible range from 3.52 to 3.74 mg. However, only the first decimal place is certain, and accordingly the average value for the protein content should be reported as 3.6 ± 0.1 mg.

When the decision to *round off* a value has been made because a calculation results in digits that are not significant, the following procedure is used. If the number 0, 1, 2, 3, or 4 is dropped, the preceding digit is not altered. If the number 6, 7, 8, or 9 is dropped, the preceding digit is increased by one. If the number 5 is dropped, and the preceding digit is even, the preceding digit is not changed; if the preceding digit is odd, it is increased by one.

D. Discussion and Interpretation of Results

In this section, the results obtained should be interpreted. What do the facts suggest about the system studied? Can a general conclusion be reached? How do the results and their interpretation agree with published data?

The language of the discussion should be clear, and the statements should be concise and to the point. The writing is generally impersonal. The discussion should end with a brief summary or a conclusion stating the significance of the study. If the laboratory findings are to be submitted for publication, the journal for which the scientific paper is written should be consulted for the style and format to use.

CHAPTER 2

Water, pH, and Buffers

BUFFERS, THIN-LAYER CHROMATOGRAPHY, AND PROTEINS

An Introductory Module

Day 1 (Chapter 3)

1. Biuret method.
2. Warburg-Christian method.

Day 2 (Chapter 3)

1. Lowry method.
2. Protein dye-binding assay.

Day 3 (Chapter 2)

1. Use and care of the pH meter.
2. Titration of an amino acid.

Day 4 (Chapters 2 and 4B)

1. Preparation and titration of a buffer.
2. TLC analysis of a mixture of amino acids.

REFERENCES: Cooper, T.G. *Tools of Biochemistry.* John Wiley & Sons, New York, Ch. 1 (1977).

Rawn, J.D. *Biochemistry.* Neil Patterson, Burlington, NC, pp. 27–47 (1989).

Robyt, J.F., and White, B.J. *Biochemical Techniques—Theory and Practice.* Waveland Press, Prospect Heights, IL, pp. 29–39 (1990).

Segel, I.H. *Biochemical Calculations,* 2nd ed. John Wiley & Sons, New York (1976).

A. Properties and Importance of Water to Biological Systems

All living organisms are dependent on water. It is the solvent for polar and ionic substances, and it is present in all cells, where it participates in many of their metabolic reactions. Because of its ionization properties, water takes part in acid–base reactions, thus affecting the pH value. The reactions studied by biochemists occur mostly in solutions, which must therefore be buffered to simulate as much as possible the natural surroundings encountered in the cell.

Preparation of Solutions

In preparing solutions in which reactions take place, several units of measurement may be used.

1. Concentration in percent:
   a. Percent by weight (w/w; g "solute" /100 g).
   b. Percent by volume (v/v; mL/100 mL).
   c. Percent by weight per volume (w/v; g "solute" /100 mL).
2. Concentration in molarity:

$$\text{Molarity} = \frac{\text{number of moles of solute}}{\text{L solution}}$$

   To make a solution of a given molarity, it is necessary to know the molecular weight of the solute. The unit of molarity is indicated by a capital letter M.
3. Concentration in normality:

$$\text{Normality} = \frac{\text{number of equivalents of solute}}{\text{L solution}}$$

   The unit of normality is indicated by a capital letter N. It is commonly used when working with acids and bases.

Hydrogen Ion Concentration in Aqueous Solutions

In physiological solutions, the expression of acidity or alkalinity is in pH units. *pH* is defined as the logarithm of the reciprocal of $[H^+]$, the hydrogen ion concentration, or $pH = -\log[H^+]$. The pH scale has been introduced as a matter of convenience because $[H^+]$ values are very small and unwieldy numbers.

B. Acids and Bases

Dissociation of Weak Acids and Bases

To neutralize the acidic or basic products of biological reactions *in vitro* and maintain the pH near the physiological range, buffers are used. A buffer is a solution that resists pH change upon addition of acid or base to the solution. Commonly used buffers consist of a mixture of a weak acid and its salt. Acetic acid is a good example of such a substance. In water it dissociates only partially:

$$CH_3COOH \rightleftharpoons CH_3COO^- + H^+$$

Its dissociation constant, K_a, can be represented as

$$K_a = \frac{[CH_3COO^-]\ [H^+]}{[CH_3COOH]}$$

If the reciprocal of both sides of the equation is taken, the equation becomes

$$\frac{1}{K_a} = \frac{[CH_3COOH]}{[H^+][CH_3COO^-]}$$

By taking the logarithm of both sides, the equation now becomes

$$\log\frac{1}{K_a} = \log\frac{1}{[H^+]} + \log\frac{[CH_3COOH]}{[CH_3COO^-]}$$

Since, by definition, the $\log 1/K_a = pK_a$ and the $\log 1/[H^+] = pH$, then

$$pK_a = pH + \log\frac{[CH_3COOH]}{[CH_3COO^-]}$$

$$\text{or pH} = pK_a - \log \frac{[CH_3COOH]}{[CH_3COO^-]}$$

$$\text{and pH} = pK_a + \log \frac{[CH_3COO^-]}{[CH_3COOH]}$$

This equation is known as the Henderson-Hasselbalch equation, which serves as a convenient formula for making buffer solutions.

Properties of Buffers

When concentrations of acid and conjugate base are present in equal amounts, *i.e.*, $[CH_3COOH] = [CH_3COO^-]$, then pH = pK_a, and the buffer is at its optimum capacity. This can be shown graphically when a titration curve is generated. The following is an example of how this equation is used in practice.

What volume of glacial acetic acid (17.6 N) and what weight of sodium acetate (molecular weight, M.W. = 82) would be required to make 100 mL of 0.2 M buffer at pH 3.9 (the pK_a of acetic acid = 4.8)?

$$\text{Let } [CH_3COO^-] = [\text{salt}] = (0.2-x) \text{ mol/L}$$

$$[CH_3COOH] = [\text{acid}] = (x) \text{ mol/L}$$

$$pH = pK_a + \log\frac{[\text{salt}]}{[\text{acid}]}$$

$$3.9 = 4.8 + \log \frac{[\text{salt}]}{[\text{acid}]}$$

$$-0.9 = \log \frac{[\text{salt}]}{[\text{acid}]}$$

To obtain a positive value:

$$0.9 = \log \frac{[\text{acid}]}{[\text{salt}]}$$

On taking the log of both sides, the equation becomes:

$$7.95 = \frac{x}{0.2-x}$$

Solving for x:

$$x = 0.178 \text{ mol of } CH_3COOH \text{ per L}$$
$$0.2 - x = 0.022 \text{ mol of } CH_3COONa \text{ per L}$$

For 100 mL, the following must be mixed and brought to volume with water in a volumetric flask.

$$0.0022 \text{ mol} \times 82 \text{ g/mol} = 0.18 \text{ g of } CH_3COONa$$
$$\frac{0.0178 \text{ mol} \times 1000 \text{ mL/L}}{17.6 \text{ mol/L}} = 1.01 \text{ ml } CH_3COOH$$

Titration Curve for a Weak Acid

The effectiveness of a buffer system as a function of its pKa is evident when a titration curve of the weak acid is generated. If 100 mL of a 0.2 M solution of the acid is titrated with a 1 N solution of NaOH, the 0.02 mol of acid will require 20 mL of this base to titrate as shown in Figure 2.1.

The pK_a value is at the inflection point of the titration curve. This curve also shows that the buffer is useful from a pH of about 4 to 5 or ± 1 pH unit on either side of the pK_a.

Some compounds with more than one pK_a are multiprotic. Phosphoric acid is the most commonly used of these multiprotic compounds. It has three dissociable hydrogens and undergoes the following dissociations upon titration with base:

$$H_3PO_4 \rightleftharpoons H^+ + H_2PO_4^{1-}$$

$$H_2PO_4^{1-} \rightleftharpoons H^+ + HPO_4^{2-}$$

$$HPO_4^{2-} \rightleftharpoons H^+ + PO_4^{3-}$$

A titration curve of H_3PO_4 shows three inflection points: at pH values 2.0, 7.0, and 12.5. This compound therefore has three buffering regions.

Amino Acids as Buffers

Of particular interest to biochemists is another category of buffers that have two or more different functional groups. All amino acids have $[NH_3^+]$ and $[COO^-]$ groups, which have pK_a values in the regions 9 to 10 and 2 to 3, respectively. An experiment involving an amino acid titration will be performed in a subsequent section.

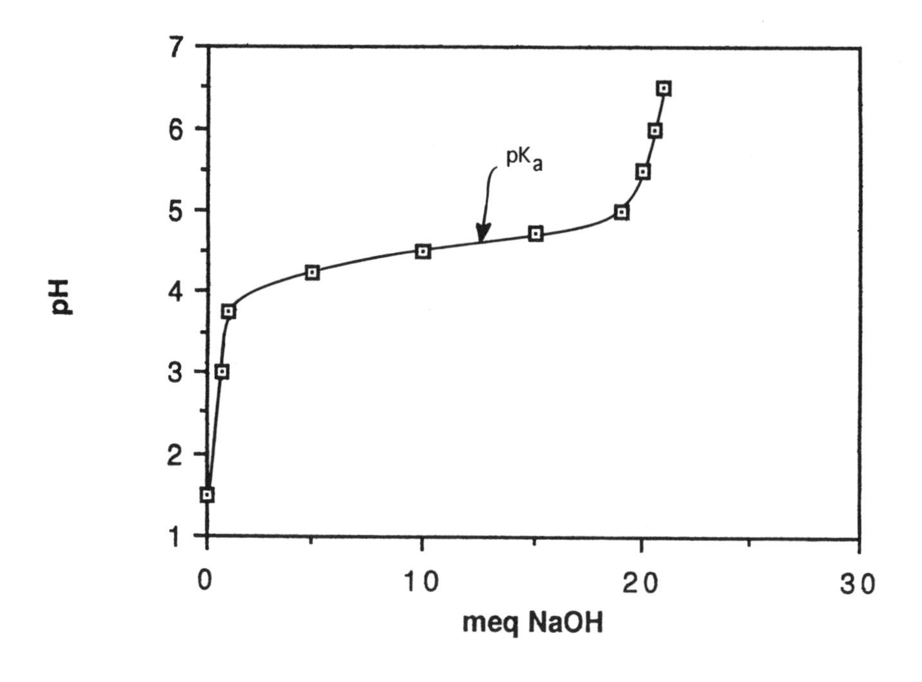

Fig. 2.1. Titration of 100 mL of 0.2 M weak acid with NaOH.

Glycine, the simplest amino acid, illustrates these characteristics. A solution of 100 mL of 0.1 M glycine is titrated with 1 N NaOH, and the resulting curve is shown in Figure 2.2. It is evident that the amino acid has two buffering regions, one at pH 2.3 ± 1.0 and the other at pH 9.8 ± 1.0.

C. Preparation and Properties of Buffers

Use and Care of pH Meters

Before using a pH meter, the electrode must be washed thoroughly with deionized water and blotted gently with a tissue. The temperature-compensation knob should be adjusted to the temperature of the test solution and the standard buffers should be used to calibrate the meter. Be aware that the pH values of the standard buffers are temperature dependent. This temperature dependence is a consequence of the temperature dependence of the dissociation constant of the buffers. For example, a common standard buffer has a pH of 7.00 + 0.02 at 25°C, but at 0°C the pH is 7.12 + 0.02.

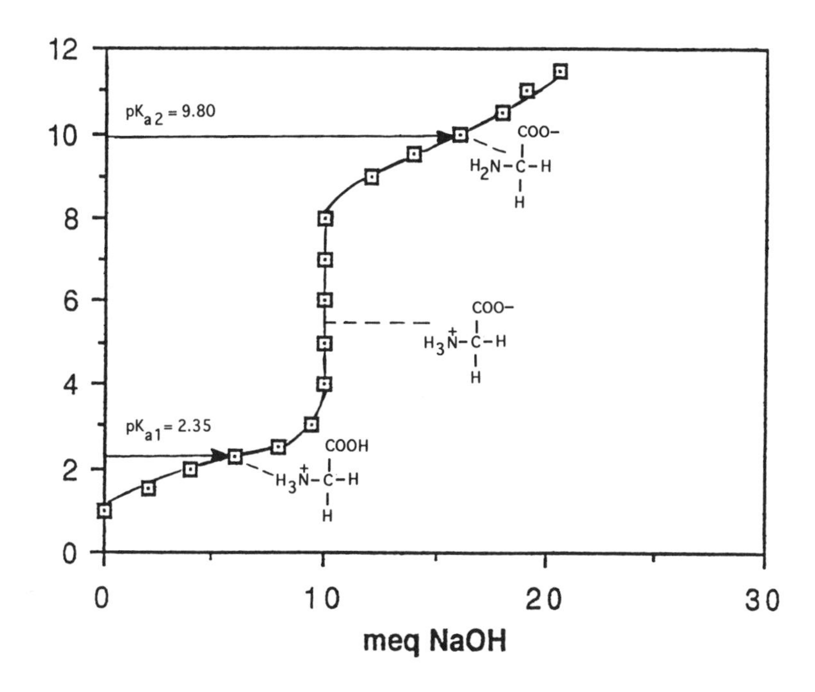

Fig. 2.2. Titration of 100 mL of 0.1 M glycine with NaOH.

Instructions for Standardizing the pH Meter

1. Check the temperature of the room and set the temperature dial accordingly. (See Figure 2.3 for location of knobs). Immerse the electrode into a buffer solution of pH 7.0, and adjust the meter with the calibration knob to this pH.

2. Rinse the electrode with deionized water, wipe dry, and check the operation of the pH meter with another standard buffer, e.g., a buffer of pH 4.0. The pH of each of the pair of buffers selected to calibrate the meter should overlap the expected pH of the test solution. If the pH of the second buffer is within 0.2 pH units of the expected value, the measurements will be adequate for most biological work. Since the characteristics of the glass electrode change with time, the calibration of the meter must be checked at regular intervals.

3. Once the meter is properly calibrated, the pH of the test solution may be determined.

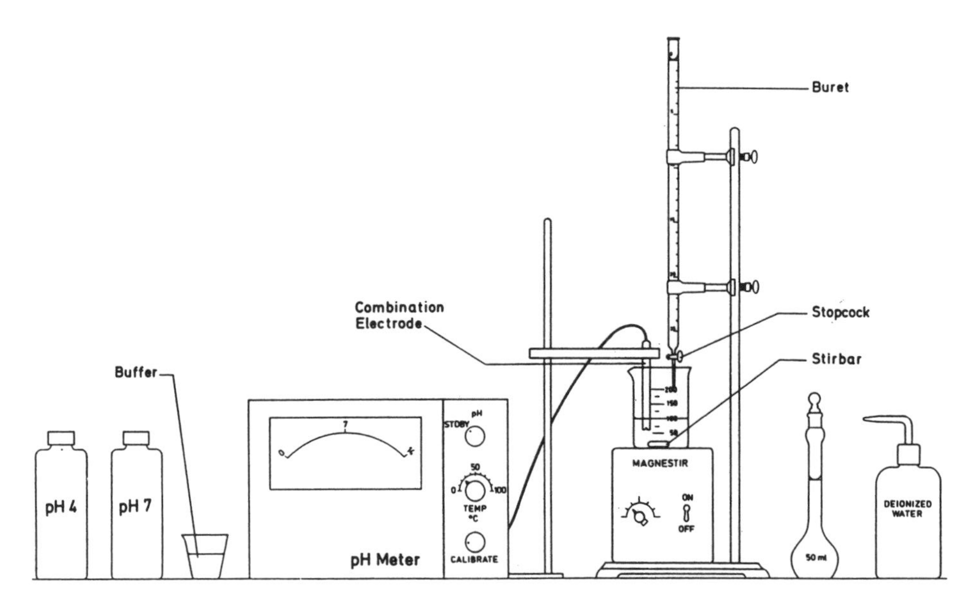

Fig. 2.3. pH meter and equipment for amino acid titration.

4. After use, rinse the electrode with deionized water. Since the glass electrode is very sensitive to dehydration, it is important that it not dry out. Keep the glass electrode immersed in deionized water when it is not in use, and leave the control switch on "standby."

Titration of an Amino Acid

REFERENCE: Wharton, D., and McCarty, R. *Experiments and Methods in Biochemistry*. Macmillan, New York, p. 185–186 (1972).

Experimental Procedure

1. Standardize the pH meter according to the instructions given above. Use standard buffers at pH 4.0 and pH 7.0; the meter should respond to either buffer to within 0.2 pH units (see Fig. 2.3 for illustration of the arrangement of the equipment). Transfer 50 mL of a 0.1 M solution of an unknown amino acid to a clean 250-mL beaker and add 50 mL of water. Place a clean stirring bar into the beaker, and place the beaker on a magnetic stirrer. Insert the rinsed combination electrode of the pH meter into the solution and record the pH. Fill a 25-mL buret with a solution of 1 N HCl and adjust the buret over the amino acid solution so that the stopcock can be

handled with ease. Proceed with the titration by adding 0.2-mL HCl increments. Stop the stirrer, and record the pH. Tabulate the data, and continue the titration until the solution reaches a pH of 1.5. Remove the electrode from the solution, wash it thoroughly, and blot it with a tissue. Pour 100 mL of deionized water into a clean 250-mL beaker, and titrate the water with the solution of 1 N HCl as before. Rinse the electrode thoroughly.

2. Repeat the titration with 50 mL of the amino acid solution and 50 mL of water using 1 N KOH. Continue the titration until the solution reaches a pH of 12.0. Repeat the titration on 100 mL of water with the KOH.
3. Plot both acid and base titration curves, and subtract the water titration curves from the amino acid titration curves to give the corrected acid and base values in milliliters. Convert the milliliters to milliequivalents of acid and base required to titrate the amino acid alone, and plot these values against pH on a separate graph.
4. From the corrected curve, establish (1) the pK_a values and (2) the isoelectric point (p*I*) of the amino acid. What amino acid is contained in the solution? Table 2.1 lists the pK_a values of some common amino acids.

TABLE 2.1 pK_a Values of Some Common Amino Acids (25°C)

Amino acid	α-COOH group	α-NH_3^+ group	R group
Alanine	2.35	9.87	–
Arginine	2.01	9.04	12.48
Aspartic acid	2.10	9.82	3.86
Glutamic acid	2.10	9.47	4.07
Glycine	2.35	9.78	–
Histidine	1.77	9.18	6.10
Leucine	2.33	9.74	–
Lysine	2.18	8.95	10.53
Phenylalanine	2.58	9.24	–
Serine	2.21	9.15	–
Tryptophan	2.38	9.39	–
Tyrosine	2.20	9.11	10.07
Valine	2.29	9.72	–

Preparation of a Buffer

Solve one of the problems from the set from Chapter 2 in Appendix II and calculate the quantities of acid and salt required to prepare the buffer.

Experimental Procedure

1. Determine the pH of H_2O, and of H_2O with the following additions using the pH meter and using pH paper.
   a. Freshly deionized water.
   b. 50 mL of deionized water + 1 drop 0.1 N HCl.
   c. 50 mL of deionized water + 1 drop 0.1 N NaOH.
2. Prepare 100 mL of 0.2 M concentration of one of the buffers for which you calculated the amounts of reagents required to obtain the pH indicated. Determine its pH with the pH meter and also with pH paper. Dilute some of the buffer 1:10, 1:100, 1:1000, and 1:10,000 and measure the pH of each solution.
3. Titrate the buffer prepared above:
   a. Put 25 mL of buffer and 25 mL of water in a 100-mL beaker. Set up the beaker at a pH meter with a magnetic stirring motor and stirring bar. Record the pH. Add 0.5 N HCl in 0.5-mL portions, recording the pH after each addition, until a total of 10 mL has been added. Titrate 50 mL of deionized H_2O with the acid. Plot titration curves for the buffer and for water. Use the water titration values to correct the buffer titration curve.
   b. Repeat the above titration of buffer and of water using 0.5 N NaOH.
4. The discussion of this experiment should consider the following questions:
   a. What is a buffer?
   b. At what pH is a buffer most effective? Within what pH range values should it be used?
   c. What determines the capacity of a buffer? How is the buffering capacity affected by dilution?

CHAPTER 3

Spectrophotometry

REFERENCES: Cooper, Ch. 2.
Robyt and White, pp. 40–49, 232–237.
Segel, pp. 1–69, 324–337.

A. Use of the Spectrophotometer to Determine the Concentration of Proteins

The determination of protein concentration is frequently required in biochemical work. Several methods are available, each having features that suit it to a particular use. Four commonly used assays will be studied using bovine serum albumin (BSA) as the reference protein (see Table 3.1).

When determining the protein concentration in an unknown sample, several dilutions should be assayed to ensure being within the range of the assay. (For this purpose 10-fold factors are generally tried first, *i.e.*, 1:10, 1:100, 1:1000). All dilutions must be taken into consideration in calculating the concentration of the original sample. The protein is added first, and then the water and all tubes are brought to the same final volume. The tubes are mixed well after each addition. The color-producing reagent is always added last, and the reaction may need to be timed accurately.

TABLE 3.1. Characteristics of the Protein Determination Methods Introduced in This Section

Method	Wavelength (nm)	Principle involved	Useful range (mg)	Reference protein	Advantages	Disadvantages
Biuret	540	Colored complex formed	0.2–2	BSA	Easy; fast	Interferences; requires a large sample
Lowry	700	Biuret reaction	0.02–0.2	BSA	Sensitive	Tedious; exact timing required
Warburg-Christian	260/280	Tyr, Trp detected	0–2	Enolase	Fast; non-destructive so that sample can be recovered	Lacks sensitivity; requires 1 mL volume
Bradford	595	Protein-dye binding	0–0.02	BSA	Fast; sensitive; easy; few interfering substances	Expensive; requires greater precision

The Beer-Lambert Law

A standard curve is generated and the results graphed. The absorbance is plotted on the ordinate and the amount of protein on the abscissa. To use a standard curve, the absorbance of the unknown is located on the ordinate, and the corresponding amount of protein is determined from the abscissa. Absorbance is a linear function of concentration which holds within the limits of the Beer-Lambert Law.

$$A = \mathrm{kcl}$$

where A = absorbance (also referred to as *optical density* [OD]); k = extinction coefficient, the absorbance of a 1 M solution of a given substance at a given wavelength in a 1-cm cuvette; l = path length (cm), usually 1.0; and c = molar concentration of sample being measured. A is a property of the sample and therefore varies with concentration and cell thickness. The property of the compound, k, is independent of concentration and cell thickness but varies with solvent and wavelength.

$$A = 2 - \log \%\mathrm{T}$$

Since the transmittance scale is linear, it is read more accurately on the Spectronic-20 than the absorbance. It is then converted to A to determine the concentration of sample in solution by using the relationship between A and T given above or by reading values off a conversion table (Table 3.2). T is the transmittance of the sample.

The range of concentrations giving a linear assay, the substances that interfere, and the sensitivity are important factors in the evaluation of a photometric or colorimetric procedure. The laboratory procedures detailed here are designed to evaluate either direct photometry of proteins in the ultraviolet region (the Christian-Warburg method) or colorimetric assays based on the biuret, Lowry, and Bradford tests. The procedures outlined for the biuret assay should familiarize the student with the operation of the spectrophotometer and provide experience in

TABLE 3.2. Conversion Table–%T *vs*. Absorbance (A)

	A					A			
%T	0.00	0.25	0.50	0.75	%T	0.00	0.25	0.50	0.75
1	2.000	1.903	1.824	1.757	51	.2924	.2903	.2882	.2861
2	1.699	1.648	1.602	1.561	52	.2840	.2819	.2798	.2777
3	1.523	1.488	1.456	1.426	53	.2756	.2736	.2716	.2696
4	1.398	1.372	1.347	1.323	54	.2676	.2656	.2636	.2616
5	1.301	1.280	1.260	1.240	55	.2596	.2577	.2557	.2537
6	1.222	1.204	1.187	1.171	56	.2518	.2499	.2480	.2460
7	1.155	1.140	1.126	1.112	57	.2441	.2422	.2403	.2384
8	1.097	1.083	1.071	1.059	58	.2366	.2347	.2328	.2310
9	1.046	1.034	1.022	1.011	59	.2291	.2273	.2255	.2236
10	1.000	.989	.979	.969	60	.2218	.2200	.2182	.2164
11	.959	.949	.939	.930	61	.2147	.2129	.2111	.2093
12	.921	.912	.903	.894	62	.2076	.2059	.2041	.2024
13	.886	.878	.870	.862	63	.2007	.1990	.1973	.1956
14	.854	.846	.838	.831	64	.1939	.1922	.1905	.1888
15	.824	.817	.810	.803	65	.1871	.1855	.1838	.1821
16	.796	.789	.782	.776	66	.1805	.1788	.1772	.1756
17	.770	.763	.757	.751	67	.1739	.1723	.1707	.1691
18	.745	.739	.733	.727	68	.1675	.1659	.1643	.1627
19	.721	.716	.710	.704	69	.1612	.1596	.1580	.1565
20	.699	.694	.688	.683	70	.1549	.1534	.1518	.1503
21	.678	.673	.668	.663	71	.1487	.1472	.1457	.1442
22	.658	.653	.648	.643	72	.1427	.1412	.1397	.1382

23	.638	.634	.629	.624	73	.1367	.1352	.1337	.1322
24	.620	.615	.611	.606	74	.1308	.1293	.1278	.1264
25	.602	.598	.594	.589	75	.1249	.1235	.1221	.1206
26	.585	.581	.577	.573	76	.1192	.1177	.1163	.1149
27	.569	.565	.561	.557	77	.1135	.1121	.1107	.1093
28	.553	.549	.545	.542	78	.1079	.1065	.1051	.1037
29	.538	.534	.530	.527	79	.1024	.1010	.0996	.0982
30	.523	.520	.516	.512	80	.0969	.0955	.0942	.0928
31	.509	.505	.502	.498	81	.0915	.0901	.0888	.0875
32	.495	.491	.488	.485	82	.0862	.0848	.0835	.0822
33	.482	.478	.475	.472	83	.0809	.0796	.0783	.0770
34	.469	.465	.462	.459	84	.0757	.0744	.0731	.0718
35	.456	.453	.450	.447	85	.0706	.0693	.0680	.0667
36	.444	.441	.438	.435	86	.0655	.0642	.0630	.0617
37	.432	.429	.426	.423	87	.0605	.0593	.0580	.0568
38	.420	.417	.414	.412	88	.0555	.0543	.0531	.0518
39	.409	.406	.403	.401	89	.0505	.0494	.0482	.0470
40	.398	.395	.392	.390	90	.0458	.0446	.0434	.0422
41	.387	.385	.382	.380	91	.0410	.0398	.0386	.0374
42	.377	.374	.372	.369	92	.0362	.0351	.0339	.0327
43	.367	.364	.362	.359	93	.0315	.0304	.0292	.0281
44	.357	.354	.352	.349	94	.0269	.0257	.0246	.0235
45	.347	.344	.342	.340	95	.0223	.0212	.0200	.0188
46	.337	.335	.332	.330	96	.0177	.0166	.0155	.0144
47	.328	.325	.323	.321	97	.0132	.0121	.0110	.0099
48	.319	.317	.314	.312	98	.0088	.0077	.0066	.0055
49	.310	.308	.305	.303	99	.0044	.0033	.0022	.0011
50	.301	.299	.297	.295	100	.0000	.0000	.0000	.0000

following the protocol for a simple colorimetric analysis for determining the concentration of protein in an unknown solution. The Lowry assay, however, requires the student to design a protocol to obtain a standard curve. It is also required that the unknown be diluted and assayed at concentrations that will result in absorbance values that will fall within the limits of the sensitivity of this assay.

Generating a Standard Curve

The "standard curve" is plotted using absorbance (*A*), rather than %T because, if the Beer-Lambert law holds, fewer experimental points are required to delineate the curve that is linear when absorbance values are used (Fig. 3.1A).

The Beer-Lambert law often does **NOT** hold at high concentrations (see Fig. 3.1B). A common reason is the depletion of one of the reagents necessary for color production. Thus readings should always be taken in the region where **ALL** reagents are in **EXCESS**. It is for this reason that assay mixtures that are too "dark" (i.e., are off-scale) must **NOT** be **DILUTED** (with a very few exceptions in which the assay is valid to *A* values greater than 1.0). The original unknown solution should be diluted *before* it is reacted and color is developed.

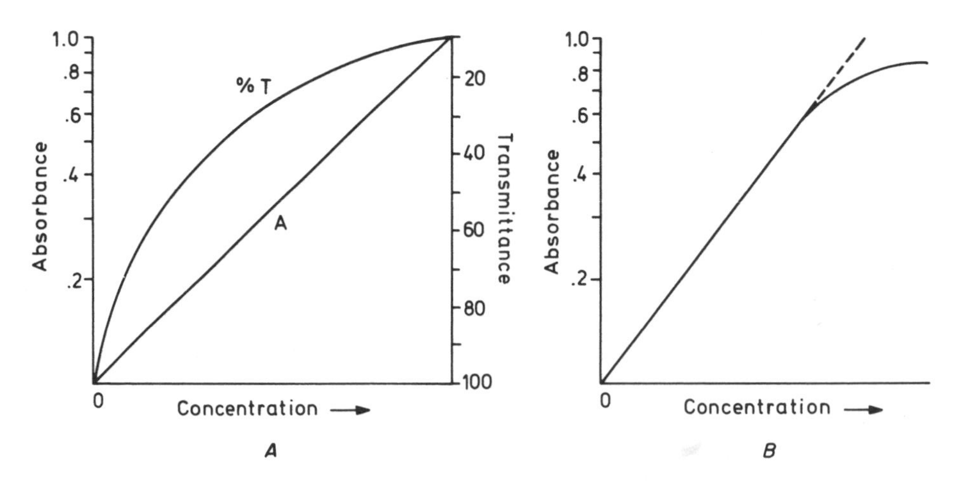

Fig. 3.1. Generating a valid standard curve. Relationship between absorbance and concentration. **A:** Plot of absorbance *vs.* concentration is linear, %*T vs.* concentration is not. **B:** Beer's law only holds within the linear portion of the curve (to A of 0.6).

Also of note is the convenience of plotting A *vs*. a given amount of material (mg, mol) present in the assay tube rather than the concentration of the original unknown solution. The range of valid A values is 0.025 to 1.0.

With every assay, a "blank" must be included. This tube contains an amount of H_2O equal to the volume of the unknown solution and the same amount of reagent as all the other tubes. The "blank" is used to set the instrument to 100% T or 0 absorbance.

B. Protein Determination Methodology

In Table 3.1 a comparison is made of some of the protein determination methods that are commonly used. All have some advantages and disadvantages, and the choice of the best method is dependent on a number of factors. Particularly important are the amount of material one has available, the time required to do the assay, and whether a UV spectrophotometer is available. If the latter is available, it is often convenient to make a rapid 260/280 reading (the Warburg-Christian method, see later in this chapter) to obtain an approximate protein concentration and then to conduct a Lowry or Bradford assay, either of which is much more sensitive. The advantage of this approach is that two to three assay tubes will be sufficient since all the tubes can be made to fall within the range of the standard curve because the approximate concentration of the unknown is already known.

The Determination of Protein by the Biuret Reaction

REFERENCE: Gornall, A.G., Bardawill, C.J., and Maxima, D. Determination of serum proteins by means of the biuret reaction. *Biol. Chem.* 177:751–766 (1949).

Reaction

The biuret reaction occurs with all compounds that contain two or more peptide bonds. The reagent consists of a solution of dilute copper sulfate in strong alkali. The purple-blue color produced is attributed to formation of a coordination complex (see diagram, page 24), between the Cu^{++} and four nitrogen atoms, two from each of two adjacent peptide chains. The name of the reaction is derived from the organic compound biuret. Dipeptides and free amino acids (except serine and threonine) do not give this reaction.

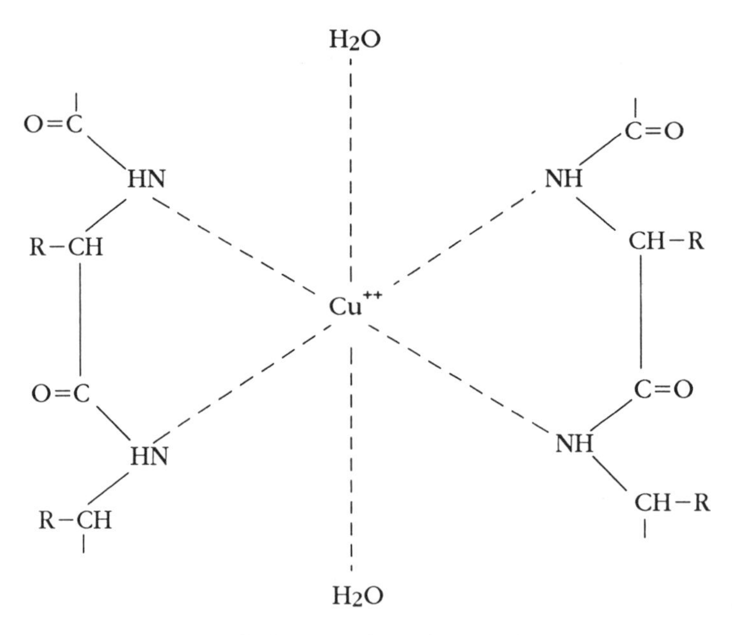

Purple-blue colored complex

Concentration Range for the Assay

0.2 to 2.0 mg.

Stability of Color

The blank is stable for several hours. The color developed on incubation is thus stable within that period but will increase slowly with longer standing. This method is both sensitive and reproducible.

Several compounds, including $(NH_4)_2SO_4$ and Tris, interfere with this assay, and care must be taken that the proper blanks have been prepared. The blank should contain an amount of interfering substance equivalent to that in the solution containing the protein.

Experimental Procedure

1. Mix sample and water to give a volume of 1.5 mL. (If preparation is insoluble, add 0.2 mL 5% sodium deoxycholate (DOC) in 0.01 N KOH. Mix protein solution with enough water to bring volume to 1.5 mL.) See Table 3.3 for the protocol to follow to generate a standard curve.

TABLE 3.3. Protein Determination by the Biuret Method—Standard Curve[a]

Additions (mL)	Tube no. 1	2	3	4	5	6	7	8
Standard BSA (2 mg/mL)	–	0.1	0.2	0.4	0.6	0.8	1.0	1.2
H_2O (deionized)	1.5	1.4	1.3	1.1	0.9	0.7	0.5	0.3
Biuret reagent	1.5 →							
	Mix the tubes on a Vortex mixer. Incubate at 37° for 15 min. Read in a Spectronic-20 at 540 nm against the blank (tube 1).							
$\%T_{540\text{ nm}}$								
$A_{540\text{ nm}}$								
mg/aliquot	0	0.2	0.4	0.8	1.2	1.6	2.0	2.4

[a]Plot $A_{540\text{ nm}}$ *vs.* mg protein. If a straight line is not obtained, the standard curve must be repeated.

2. Add 1.5 mL of biuret reagent (1.50 g $CuSO_4 \cdot 5H_2O$, 6.0 g sodium potassium tartrate, and 300 mL of 10% NaOH per L) and mix. Remember the reagent blank.
3. Incubate all tubes for 15 min at 37°C.
4. Read at $A_{540\text{ nm}}$ in a Spectronic-20 against the reagent blank. Draw a standard curve.
5. Repeat the assay including one tube (tube 2) as a check of the procedure and the reagents, since this tube should give an $A_{540\text{ nm}}$ corresponding to 1 mg of BSA on the curve generated above (see Table 3.4).
6. An unknown solution of BSA will be provided to gain experience in making dilutions and using a standard curve. A range of dilutions is made to obtain useful data from at least one of the tubes assayed.
7. Lysozyme and gelatin are included for comparative purposes since BSA was used to generate the standard curve.
8. Read the tubes against the blank and calculate the protein concentration (per mL) of the unknown sample. Be sure to take the dilution factor into account when calculating the concentration in the original sample.

TABLE 3.4 Protein Determinations on Unknowns—Biuret Method

Additions (mL)	Tube no. 1	2	3	4	5	6	7	8	9	10	11	12
Standard BSA (2 mg/mL)[a]	0	0.5	—	—	—	—	—	—	—	—	—	—
Unknown BSA	—	—	0.1	0.5	—	—	—	—	—	—	—	—
1:9	—	—	—	—	0.1	0.5	—	—	—	—	—	—
1:99	—	—	—	—	—	—	0.1	0.5	—	—	—	—
Lysozyme (2 mg/mL)	—	—	—	—	—	—	—	—	0.5	1.0	—	—
Gelatin (2 mg/mL)	—	—	—	—	—	—	—	—	—	—	0.5	1.0
Water	1.5	1.0	1.4	1.0	1.4	1.0	1.4	1.0	1.0	0.5	1.0	0.5
Biuret reagent	1.5	→										
	Mix tubes. Incubate at 37° for 15 min. Read in same instrument as standard curve at 540 nm against the blank.											
$\%T_{540\ nm}$												
$A_{540\ nm}$												
mg/aliquot	0	1.0										
mg/mL	0	2.0										

[a]At least one standard tube should be run with each series of tubes as a control on reagents and procedure.

Instructions for Use of the Spectronic-20

1. Rotate the power switch knob clockwise. Allow 5 min for instrument to warm up. See diagram of Spectronic-20 (Fig. 3.2) to locate the parts described.
2. Turn the wavelength selector knob to the desired wavelength setting.
3. With the sample holder cover closed, adjust the instrument to zero %T using the zero control knob (also power switch).
4. Standardize the instrument: (a) Insert a cuvette filled with at least 3 mL of a reference liquid (blank) or water. (b) Insert the cuvette into the sample holder, aligning the mark on the cuvette with the line on the sample holder. (c) Turn the light control knob until the meter reads 100%T.
5. Empty the cuvette, returning the liquid to a test tube for future use. Allow the cuvette to drain on a tissue, and fill it with at least 3 mL of the sample to be measured. Again align the markers on tube and sample holder, and read the percent transmittance or absorbance.

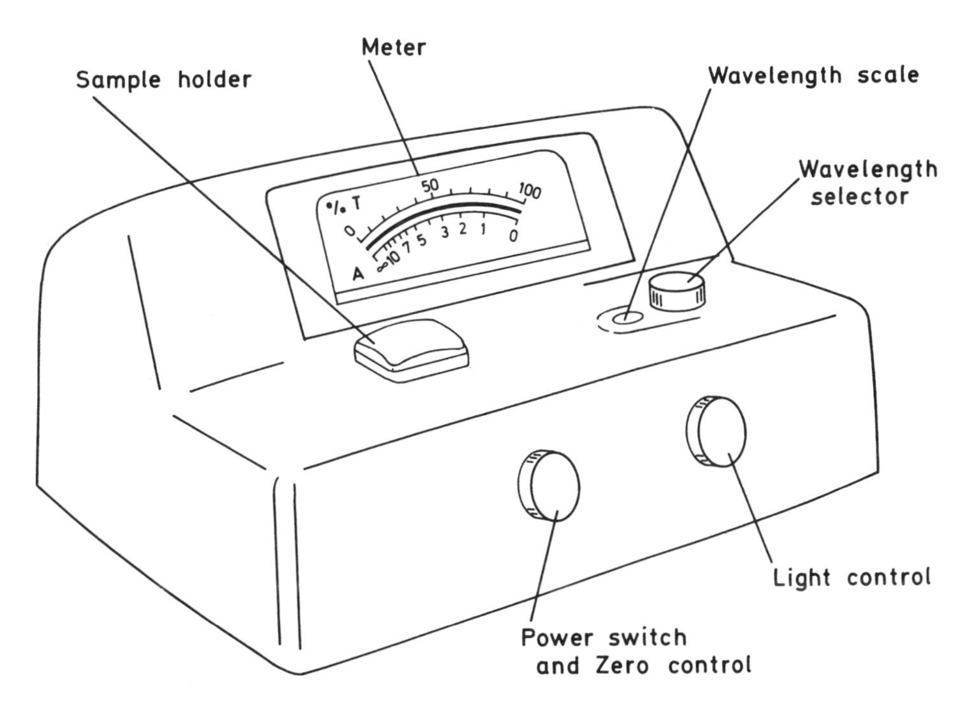

Fig. 3.2. The Spectronic-20. (Adapted from *Operating Manual,* 8th ed. © 1976. Bausch and Lomb, Rochester, NY).

NOTE: The Spectronic-20 must be standardized each time a different wavelength is used. When reading a series of tubes at a fixed wavelength, it is necessary to check periodically for meter drift from 100% T.

Lowry Method for Protein Determination

REFERENCES: Lowry, O.H., Rosebrough, N.J., Farr, A.L., and Randall, R.J. Protein measurement with the folin phenal reagent. *J. Biol. Chem.* 193:265–275 (1951).

McDonald, C.E., and Chen, L.L., The Lowry modification of the folin reagent for determination of proteinase activity. *Anal. Biochem.* 10:175–177 (1965). (Modification of the Lowry method.)

Reactions

1. Formation of the protein–copper complex as described for the biuret reaction.
2. Reduction of the phosphomolybdate-phosphotungstate reagent (Folin-Ciocalteu reagent) by tyrosine and tryptophan residues.

Reagents

1. 2% Na_2CO_3 dissolved in 0.1 N NaOH.
2. 2.7% sodium potassium tartrate.
3. 1% $CuSO_4$ in H_2O.
4. 1 N Folin-Ciocalteu "phenol reagent." (Must be diluted 1:1 with water from a 2 N commercial preparation and made fresh daily.)

Concentration Range for the Assay

20 to 200 μg.

Experimental Procedure

1. Set up a protocol following the procedure established for the determination of protein by the biuret method. Construct a standard curve using a solution containing 400 μg/mL BSA. The final volume (protein and water) must not exceed 0.5 mL, and the tubes should contain a range of protein from 20 to 200 μg.
2. Deliver an appropriate amount of the unknown protein solution to a series of tubes so that the volume of the sample plus water equals 0.5 mL. Calculate the amount of pro-

tein to use from the value obtained previously for this unknown. Use the same unknown as you did for the biuret assay.

3. Prepare per tube: 4.9 mL of 2% Na_2CO_3, 0.05 mL of sodium potassium tartrate, and 0.05 mL 1% $CuSO_4$. Prepare a quantity sufficient for all tubes. When preparing this "mix," be sure to add $CuSO_4$ last. (Mix must be made fresh daily).
4. Add 5 mL of the prepared mixture at timed intervals, and mix thoroughly on a Vortex mixer.
5. Incubate for *exactly* 10 min at room temperature. This timing is critical. Start the stop watch when adding reagent to tube 1, wait an appropriate time interval (30 sec at least) before adding reagent to tube 2. At the end of 10 min, add 0.5 mL of "phenol reagent" to the first tube (10 min 30 sec to tube 2, 11 min to tube 3, and so forth), mix rapidly on a Vortex mixer, and proceed until the reaction has been stopped in all the tubes.
6. Incubate for 30 min at room temperature. Here timing can start after addition of the reagent to the last tube.
7. Read at $A_{700\text{ nm}}$ in a spectrophotometer.
8. Plot $A_{700\text{ nm}}$ *vs.* µg protein.
9. Calculate the protein concentration of the unknown. How does this value compare with the results obtained from the biuret assay for the same unknown?

NOTE: Many compounds such as Tris buffer, sucrose, ammonium sulfate, sulfhydryl compounds, urea, etc. also interfere with this assay. Be sure to include the proper blanks when the protein is in the presence of any interfering substance.

Warburg-Christian Method for Protein Determination

The protein and nucleic acid concentrations of a solution can be estimated by making direct absorption measurements in the UV region in a spectrophotometer. The strong absorption of proteins at 280 nm is primarily due to tryptophan and tyrosine residues, and the absorption therefore will vary with the content of these amino acids. The method is not as specific as the colorimetric procedures described above, but it is nondestructive and the sample can be recovered. The most common interfering substances are nucleic acids, which absorb about 10 times as strongly (per gram) as do proteins at 280 nm. However, extinction coefficients of nucleic acids at 250 nm are about twice those at 280 nm, and since most proteins absorb weakly at 260 nm, nucleic acid contamination will usually be apparent in the protein spectrum.

Experimental Procedure

1. Prepare and plot an absorption spectrum for a solution of BSA (1 mg/mL) by measuring the absorbance at 10-nm intervals over the range 360 to 240 nm. (It is common practice to make measurements at the longer wavelengths first, since they are least destructive to the absorbing substance, and then to proceed to shorter wavelengths).
2. Prepare a similar spectrum for a solution of yeast tRNA (0.03 mg/mL).
3. Using a nomograph (see Fig. 3.3) and the absorption maximum values obtained above, determine the concentrations of protein and of nucleic acids in the two solutions. Explain why the values do not agree with the concentrations actually given. (Note that the nomograph is based on the protein enolase.)
4. For the same unknown BSA solution that was used for the biuret and Lowry methods, make a dilution to obtain a solution of *ca* 1 mg/mL. Take readings for this solution at 260 and at 280 nm in the spectrophotometer and determine the protein concentration from the nomograph. Since 260/280 readings were obtained above for a solution of known concentration of BSA, it is possible to calculate the correct concentration for the unknown.

Protein Determination by a Protein Dye-Binding Assay—Microassay Technique

REFERENCES: Bradford, M. A rapid and sensitive method for the quantitation of microgram quantities of protein utilizing the principle of protein-dye binding. *Anal. Biochem.* 72:248–254 (1976).

Bio-Rad Laboratories. *Chromatography, Electrophoresis, Immunochemistry and HPLC.* Richmond, CA, pp. 166–167 (1983).

Reaction

The color of Coomassie brilliant blue G250 in dilute acid solution changes proportionally as the dye binds to protein.

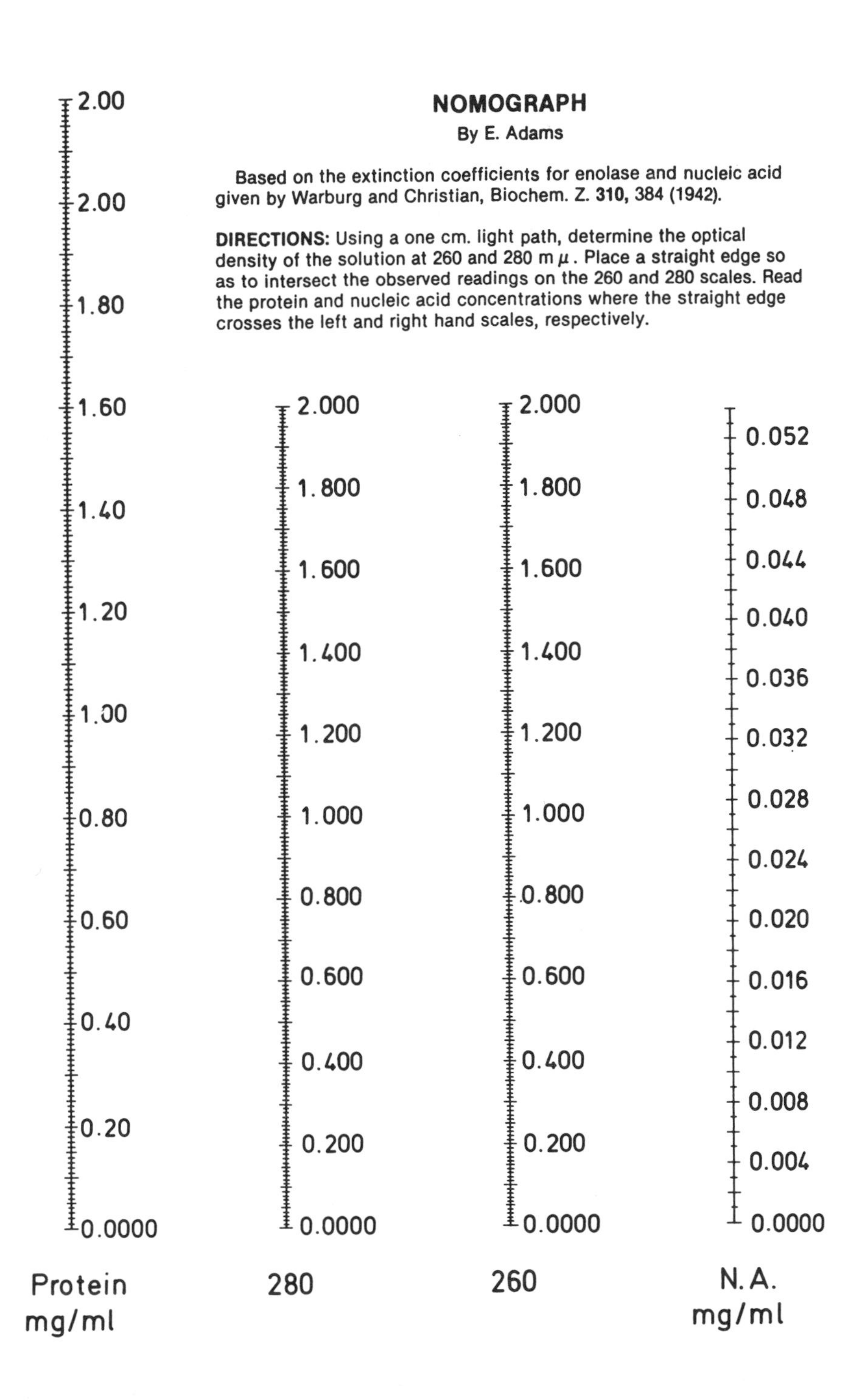

Fig. 3.3. Nomograph. (Distributed by Calbiochem., San Diego, CA.)

There are two major advantages to this method: (1) the speed and ease of performance and (2) few interfering substances.

Reagent

Coomassie blue dye dissolved in phosphoric acid and methanol (purchased from Bio-Rad Laboratories, Richmond, CA).

Concentration Range for the Assay

For microassay, 0 to 20 μg.

Experimental Procedure

1. Dilute the unknown sample to a concentration that will fall within the range of the standard curve. To do this, estimate the protein concentration by the Warburg-Christian method and make a solution containing about 25 μg/mL. Then include three assay tubes containing varying amounts of this solution as shown in Table 3.5.
2. Set up a protocol to establish a standard curve following the procedure established for previous protein assays. Use a BSA solution containing 25 μg/mL, and deliver 0, 0.1, 0.2, 0.4, 0.6, and 0.8 mL to a series of 10 × 75 mm test tubes (see Table 3.5).

TABLE 3.5. Protein Determinations by the Protein Dye-Binding Method

Additions (mL)	Tube no. 1	2	3	4	5	6	7	8	9
BSA standard (25 μg/mL)	0	0.1	0.2	0.4	0.6	0.8	—	—	—
Unknown protein[a]	—	—	—	—	—	—	0.2	0.4	0.6
H_2O	0.8	0.7	0.6	0.4	0.2	0	0.6	0.4	0.2
Dye reagent concentrate	0.2 →								
	Incubate for 5 min. at room temperature. Measure color in a spectrophotometer.								
%$T_{595\ nm}$									
$A_{595\ nm}$									
μg protein	0	2.5	5.0	10	15	20			

[a]Adjust concentration to *ca*. 25 μg/mL (take 260/280 readings to determine protein content by the Warburg-Christian method and dilute as necessary).

3. Bring the volume to 0.8 mL with deionized water. Add 0.2 mL of reagent to each tube and mix well.
4. Incubate for 5 min at room temperature. This timing is not critical, but the tubes should be read shortly after the reaction is completed.
5. Read the tubes at $A_{595\ \text{nm}}$ in a spectrophotometer in 1.0 mL cuvettes. (The smaller volumes are used to conserve sample and reagent. It is convenient to use plastic, disposable cuvettes for this purpose.)
6. Plot $A_{595\ \text{nm}}$ *vs.* μg protein to generate the standard curve, and determine the exact concentration of protein in the unknown sample.

CHAPTER 4

Separation of Molecules by Chromatography

A. Column Chromatography

REFERENCES: Cooper, Chs. 4, 5.

Heftmann, E. *Chromatography,* 3rd ed. Van Nostrand Reinhold Co., New York. Chs. 1, 4 (1975).

Finlayson, J.S. *Basic Biochemical Calculations*. Addison-Wesley, Reading, MA, pp 203–210 (1969).

Robyt and White, pp. 73–95.

Chromatography was one of the earliest tools applied to biochemical analyses. According to Heftmann (1975), it is "the [art] of separating the components of a mixture by differential migration." In the early 1900s, M.S. Tswett applied plant extracts to various adsorbents and separated the individual pigments by eluting the solid phase with various solvents. The term chromatography is derived from the Greek *chroma,* color, as separation was followed by monitoring the colors of the eluants.

Liquid–Solid Adsorption Chromatography

Liquid–solid adsorption chromatography is particularly useful for the separation of lipids. Commonly used adsorbents are MgO, Al_2O_3, silica, and alumina. A problem with the use of these substances is their tendency to pack densely; even when inert filtering aids are used, the flow rate is slow, and it is generally necessary to apply pressure to clear the column. Silica and MgO columns are used in some of the experiments in this text.

Many of the technical difficulties encountered in preparing the columns have been overcome with a liquid–solid extraction device developed by the Millipore Corporation (Waters Chromatography Division). The "Sep-Pak cartridges" are available with a choice of adsorbents and are color coded for easy identification of the packing material. The cartridge is attached to a Luer-type syringe, which then serves as the solvent reservoir. The sequence of steps in the procedure is as follows (as illustrated in Fig. 4.1):

1. The cartridge is prewet with organic solvent and rinsed with a suitable buffer.
2. The sample is applied, and unretained components are eluted.
3. A solvent appropriate to recovery of the desired component is used, and the eluate is collected.
4. If a second component that is still held in the cartridge is wanted, then another solvent can be used for its elution.

Two experiments use Sep-Pak silica cartridges:

a. In Chapter 8, a procedure for separating neutral from polar lipids is given.

b. In Chapter 9, carotenes are separated from a spinach extract for previtamin A determinations.

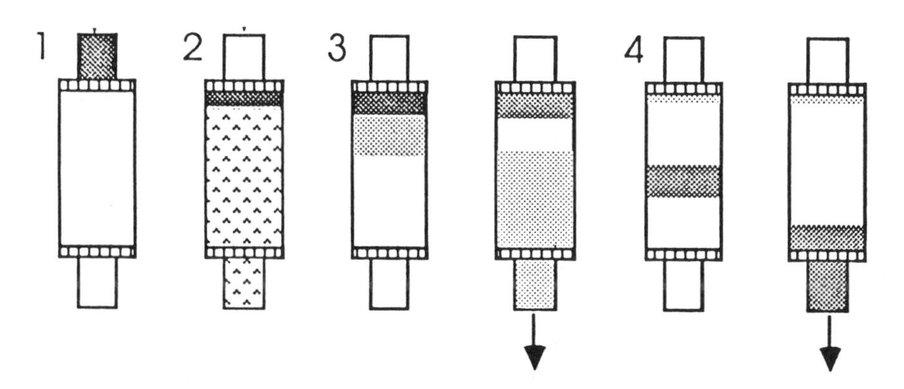

Fig. 4.1. Elution protocol for chromatography with Sep-Pak cartridges (Adapted from Millipore Corp., Waters Chromatography Division, Milford, MA).

Ion Exchange Chromatography

Ion exchange chromatography is a separation method relying on charged molecules. The procedure is dependent on the formation of electrostatic linkages between the resin and the substance being separated. Both cationic and anionic resins are available. They consist of a polystyrene or cellulose matrix with an attached functional group, and they require activation by treatment with a buffer of suitable pH. The compounds being separated likewise must be charged by reacting specific groups at some designated pH before the mixture is loaded onto the column.

The technique is particularly useful for separating compounds of similar structure, such as proteins. After the mixture has been loaded, the desired compound is recovered by washing the sample off the column. This can be accomplished by one of three techniques: (1) *Continuous gradient* elution, as described in Chapter 5 for the purification of invertase, in which increasing [NaCl] results in the recovery of proteins that adhere more tightly to the resin. A series of fractions is collected for later analysis. (2) *Stepwise elution* involves the recovery of all compounds eluting with a buffer at a given ionic strength, followed by another solution of greater strength. (3) Only one *eluant* is used if it is capable of displacing the compound selectively from the column.

To separate a given protein effectively, it is necessary to take into account the size of the sample, which will determine the dimensions of the column and the volume of eluant needed.

Gel Filtration

Gel filtration is a frequently used method. Molecules are separated from a mixture according to size by this technique. Gel filtration is useful for purification by isolating a wanted compound from a mixture as well as for the determination of molecular weight. The method is also known as *molecular sieve chromatography* because it is dependent on passing a mixture of molecules through a bed of synthetic beads with a given diameter and pores of known size. The beads are available commercially and are composed of a cross-linked dextran matrix, such as *Sephadex*. Sephadex resins are available in different exclusion sizes. The smallest particles in a mixture can enter the holes formed within the latticework of the beads, where they are retained, while the larger molecules that are unable to enter these spaces are excluded and proceed around the beads and down the column at a rapid rate. A gradient is thus established, with the largest molecules eluting first and the smallest last in the fraction tubes, which are collected in sequence. For purification purposes, it is then necessary to assay the fractions collected for a specific function in order to identify the location of the desired protein. To determine the molecular weight of an unknown protein, a calibration curve is established (see Fig. 4.2) The column is poured with the beads suspended in a suitable buffer, and the void volume, V_o, is determined by passing through a solution of blue dextran (MW 2,000,000). The elution volumes V_e, of several protein standards of known molecular weight are then measured, and the ratios of V_e/V_o are calculated. The relationship of the logarithm of molecular weight to the

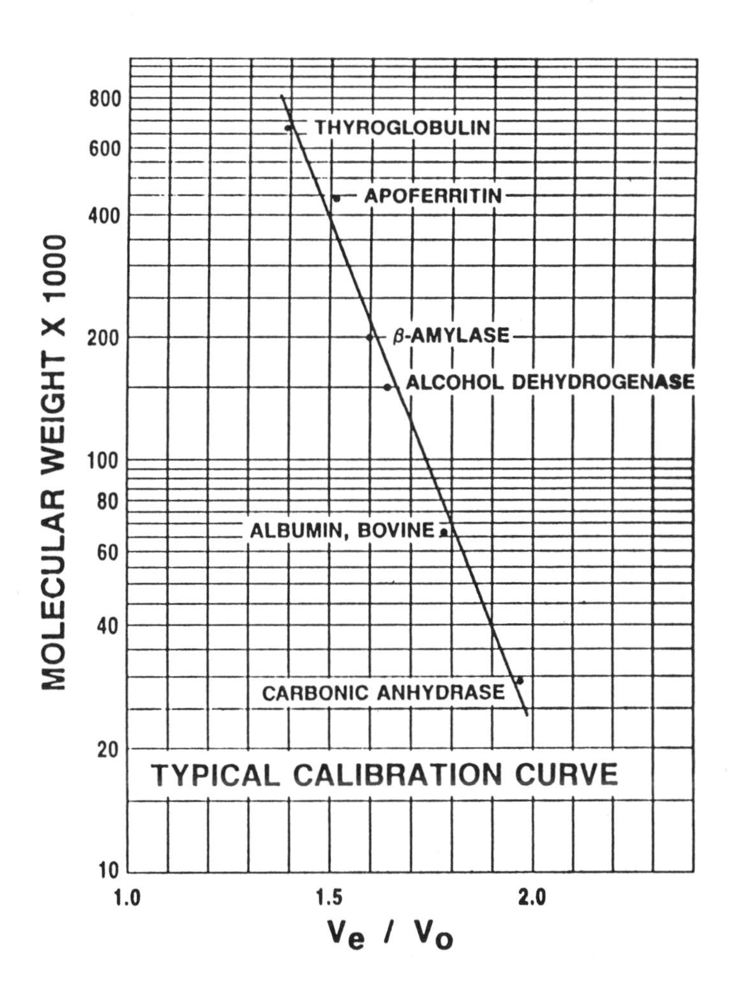

Fig. 4.2. Calibration curve typically obtained with proteins from the MW-GF-1000 kit as run on Sepharose 6B (Sigma Chemical Co., St. Louis, MO).

V_e/V_o values for the standards is linear (see Fig. 4.2). The molecular weight of an unknown is then determined by comparing its V_e/V_o ratio to that established for the standards. This method is illustrated in Chapter 5 for a determination of the molecular weight of invertase.

B. Electrophoresis

Electrophoresis is another separation method in common use. It is dependent on the charge carried by the molecules in solution to which a current is applied. This technique is described further in Chapter 5.

C. Thin-Layer Chromatography

Principles and Applications

Thin-layer chromatography (TLC) is a separation technique in which a uniform thin layer of an adsorbent material is coated onto a support such as a glass plate or a plastic sheet. The solution of compounds to be separated is spotted 2.0 to 2.5 cm from the bottom of the plate but above the level of the solvent in which development is to take place. The solvent moves through the adsorbent by capillary action, and the chromatogram is removed from the tank when the solvent is within 2.0 to 2.5 cm from the top of the plate. After separation, the solvent is removed by evaporation, and the spots are treated in various ways to make them visible (see Fig. 4.3).

Many classes of compounds can be separated by the TLC method, the successful separation being dependent on the proper selection of adsorbent and solvent system. Separation is generally dependent on the polarity of the sample, which is determined by the number and types of functional groups. Polar compounds have acid or hydroxyl groups. Polarity decreases as compounds become more hydrophobic, *i.e.*, they have fewer polar groups until they become nonpolar, as in the case of hydrocarbons and ethers.

Choosing a Solvent System

The best solvent system permits the sample spot to move halfway between the origin and the solvent front (R_f value = 0.5). If the spots do not migrate well and stay close to the origin, a more polar solvent should be used. If the spots move too far

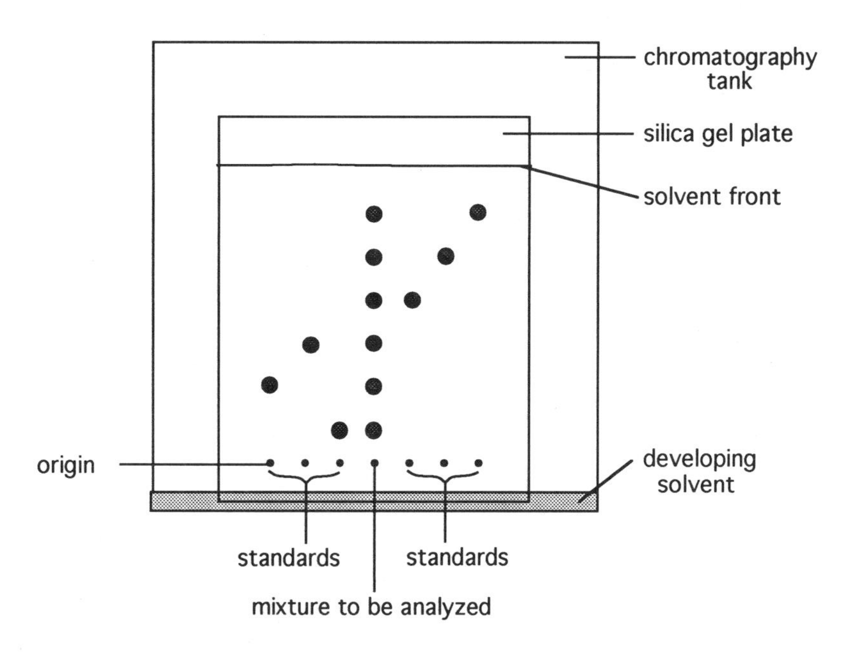

Fig. 4.3. Separation of a mixture in a suitable solvent system. (1) The TLC plate is activated by briefly heating at 110°C to remove any water that may be absorbed. (2) The "mixture" is applied toward the center of the plate, and the standards are delivered on either side as shown. (3) After development, the plate is allowed to dry, and the spots are visualized after spraying with an appropriate reagent.

and cluster near the solvent front, the polarity of the developing solvent should be decreased. In Table 4.1 common solvents used in TLC and their dielectric constants are listed in increasing order of polarity.

After spotting, the plates are developed in a tank containing a shallow layer of the developing solvent. The tank is lined with filter paper to maximize vapor saturation, which is important for even migration of the solvent across the plate. If the chamber is not adequately saturated, it is difficult to measure the distance that the solvent front has reached when the plate is removed from the tank, and the spots will not have migrated evenly.

Visualization of Chromatograms

When the plates are removed from the developing tank, the solvent front is traced with a lead pencil, and the plates are

TABLE 4.1 Dielectric Constant Values for Some Common Solvents

Solvent[a]	Polarity (dielectric constant)
n-Hexane	1.9
Petroleum ether	2.0
Cyclohexane	2.0
Carbon tetrachloride	2.2
Benzene	2.3
Toluene	2.4
Trichloroethylene	3.4
Diethyl ether	4.3
Chloroform	4.8
Ethyl acetate	6.0
Ethylene chloride	10.6
Pyridine	12.3
iso-Propanol	18.3
n-Propanol	20.1
Acetone	20.7
Ethanol	24.3
Methanol	32.6
Acetonitrile	37.0
Water	80.3

[a]From Brinkmann Instruments, Inc. "Introduction to Thin-Layer Chromatography," © 1972, Westbury, NY. p. 11.

permitted to air-dry in a hood. The spots can be detected by one or more methods of visualization. A nondestructive technique is used first. Examples are UV light for fluorescing compounds and exposure to I_2 vapor, which results in reversible staining. This first treatment can be followed by staining with a reagent that is sprayed onto the plate. A number of sprays are available for specific reactive groups, and others are of more general use. In the lipids section of this text (Chapter 8) are several experiments designed to find appropriate solvent systems for separating different classes of compounds, and some common visualizing reagents are listed.

Identification of an Amino Acid in a Mixture

In the remainder of this chapter, a mixture of amino acids will be spotted and run simultaneously with known standards. The spots will be visualized by staining with ninhydrin reagent, which reacts with primary and secondary amines as follows:

Experimental Procedure

1. *Sample applications.* Lightly draw a line in pencil 2 cm from the bottom of the silica gel-coated plate (plates can be purchased from Eastman Kodak Co., Rochester, NY). Gently place six dots along the line, 1.5 cm apart, starting 1 cm from the left edge. Record which solution will be spotted on each dot. Be sure to place your unknown sample on spot 3 or 4. Amino acid standards will be placed on the other dots (see diagram of plate below).

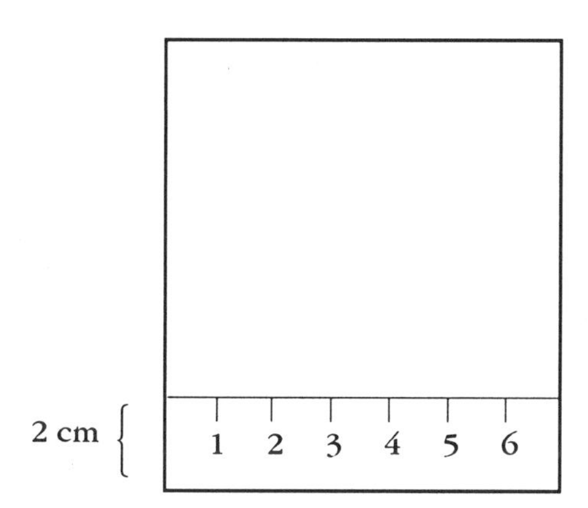

2. The calibrated capillary tubes (disposable "microcaps") are filled by capillary action and held in a vertical position during spotting. Use the 2-μL microcaps. The sample is spotted by gently touching the tip of the capillary to the silica gel. The solvent is allowed to evaporate, and the spotting is continued until the contents of the capillary have been applied. It is desirable to make the spots as small and concentrated as possible to avoid excessive diffusion as the spot is developed by the solvent system. Use a new microcap for each application (see Fig. 4.4A).
3. *Preparation of developing tanks.* The solvent system used is 95% EtOH:H_20 in a ratio of 10:1. To ensure full saturation

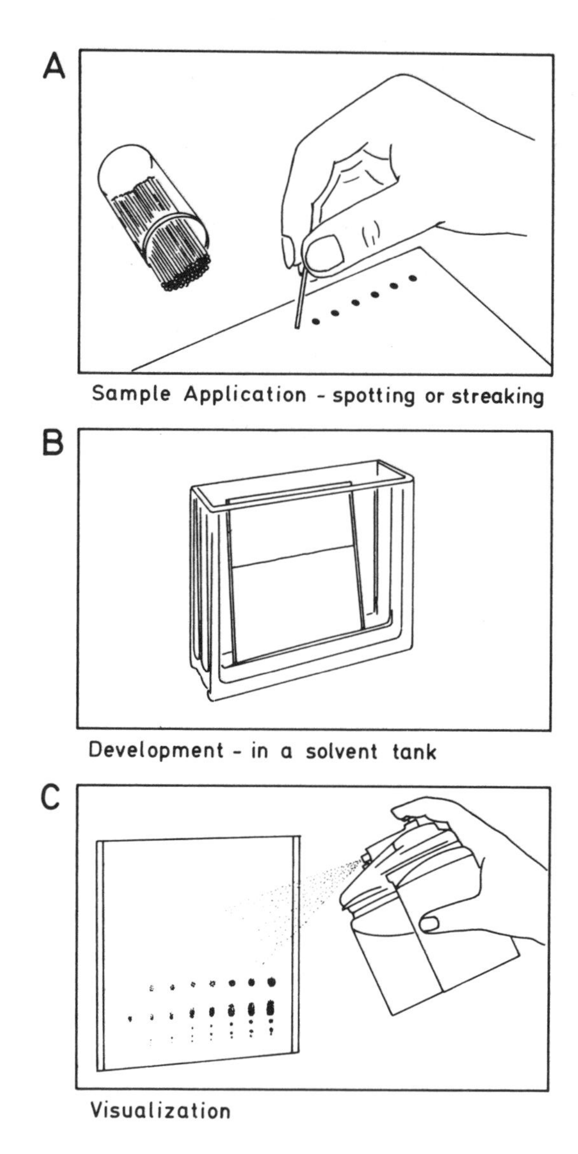

Fig. 4.4. *Separation of compounds by thin-layer chromatography.* **A:** Sample application. **B:** Development in solvent system. **C:** Visualization of the spots after staining. (Adapted from *Introduction to Thin-layer Chromatography,* © 1972, Brinkmann Instruments, Inc., Westbury, NY; Courtesy of Brinkmann Instruments, Inc.)

of the chromatography chamber with developing solvent, the chamber should be lined with Whatman chromatography paper. For this experiment, a 1-L beaker covered with aluminum foil is suitable.

4. When sample application is completed, allow several minutes for all solvent to evaporate, and then place the chromatogram in the chamber with the back against the paper-lined wall. Attempt to place the chromatogram in the solvent squarely so that the solvent front will rise evenly.

If two chromatograms are to be developed in one tank, they **MUST** be inserted at the same time. The bottom edge of the plate should be at least 1 cm below the surface of the solvent, but all sample spots must be above the solution surface (see Fig. 4.4B).

5. Allow the chromatograms to develop until the solvent front is 3 to 5 cm from the top edge (about 1.5 to 2 hr). Remove the chromatogram from the chamber and mark the solvent front quickly with a soft pencil before it disappears. Air-dry in the hood.
6. *Visualization of chromatogram.* The chromatograms will be sprayed with ninhydrin, a reagent that detects primary and secondary amines (see Fig. 4.4C) After spraying, place the chromatograms in an oven at 110°C for 5 min. After removing the plates from the oven, notice the position and color of the various spots. Lightly encircle the spots with a pencil.
7. R_f values of the spots should be calculated by using the following ratio:

$$R_f = \frac{\text{distance of migration of substance}}{\text{distance of migration of solvent front}}$$

Use the center of each spot for making measurements.

8. Identify the unknowns by matching the R_f values of the amino acid standards with those calculated for the unknowns. Report both R_f values and color development.

CHAPTER 5

Enzymology

ISOLATION, PURIFICATION, AND CHARACTERIZATION OF INVERTASE

A. Extraction and Partial Purification of Invertase

Day 1 Extraction, heat treatment, and alcohol fractionation/ store fraction as a pellet.

Day 2 Ion exchange chromatography.

B. Nelson's Assay and Activity Determinations

Day 3

1. Standard curve for glucose analysis.
2. Assay of fractions through alcohol fraction only (save remaining samples for protein determinations).

Day 4 Protein determinations by Lowry method or protein-dye binding assay of all fractions (save remaining samples for electrophoresis).

C. Electrophoresis

Day 5

1. Preparation of gels for electrophoresis.
2. Assay of column fraction and enzyme concentration curve.

3. Concentration of fraction IV—"ultraspin" or Amicon filtration method.

Day 6

1. Disc gel electrophoresis—destaining of gels as needed.
2. Product formation *vs*. time course curve.

D. Enzyme Kinetics

Day 7

1. Visualization of gels and band measurements.
2. Substrate concentration curve.
3. Inhibition by urea.

Day 8

1. Glucose oxidase method for invertase activity.
2. Inhibition by raffinose and fructose.

E. Characterization of Invertase

Day 9 Demonstration of transferase activity.

Day 10 Molecular weight determinations.

a. SDS gel electrophoresis.
b. Gel filtration method.

Day 11 Effect of temperature on enzyme activity.

Day 12 Effect of pH on enzyme activity.

A. Isolation and Purification of Enzymes

REFERENCES: Cooper, Chs. 3, 4, 5, 6, 10.

Deutscher, M.P. Guide to Protein Purification. *Methods Enzymol.* Vol. 182 (1990).

Robyt and White, pp. 79–81, 88–95, 129–142, 291–315.

Stryer, L. *Biochemistry*, 3rd ed. W.H. Freeman and Co., New York, Chs. 8, 9 (1988).

When an enzyme is to be isolated, a purification scheme must be chosen to achieve one's goal while taking into consideration a number of practical factors, such as time requirements and cost involved. Before choosing any procedure, it is necessary to find a suitable source, a sensitive activity assay, and a convenient method of protein determination.

Most purification schemes have the following procedures in common: (1) cell breakage and extraction of protein in a suitable buffer, (2) a precipitation step in some medium designed to lower the dielectric constant of the solvent, and (3) one or more chromatographic techniques that can be dependent on separation by use of ion exchange resins, molecular sieve gels, or affinity chromatography.

In each case, it is essential to test for total yield of activity and for protein at every step so that an evaluation of cost in loss of activity (recovery of total units) *vs.* gain in purity (increased specific activity) can be made. It is the goal of this module to introduce these concepts and apply representative techniques while isolating and purifying the enzyme *invertase.*

Enzymes Are Biological Catalysts

Enzymes are biological catalysts that permit reactions to occur in living cells that would otherwise proceed too slowly to permit cell growth. A reaction rate can be increased by raising the temperature or lowering the activation energy. Raising the temperature is not possible, since biological systems have a limited tolerance to heat. The enzyme invertase, isolated and studied in the following experiments, demonstrates the catalytic function particularly well. Because acid hydrolysis of sucrose occurs at relatively low temperatures, sucrose hydrolysis will be shown to occur in the presence and in the absence of enzyme, and the differences in energy requirements under the two sets of conditions will be demonstrated.

Handling of an Enzyme Preparation

1. Due to the denaturation of proteins by heat and by chemical reactions whose rates increase with temperature (*e.g.*, reactions catalyzed by proteases in the preparation), enzyme preparations should be kept near 0°C during purification. The work is generally done on ice, but it is important to protect the preparation from contact with ice that is made from tap water that contains heavy metal ions that also cause protein denaturation. If an enzyme (*e.g.*, invertase) is stable to heat, the preparation can be heated to denature other proteins, which can then be removed by centrifugation.

2. If the enzyme preparation is to be stored overnight or longer, it is usually best to freeze it. When frozen aqueous

solutions are thawed, there is usually a heterogeneity in the concentration of solute throughout the solution. *It is therefore important that frozen solutions be completely thawed and thoroughly mixed before use.* Thawing should take place at room temperature; *do not heat* the enzyme to speed the process.

3. Since proteins are subject to denaturation at surfaces because of exposure to oxygen, *strong agitation* of enzyme preparations *should be avoided*. The more dilute the protein solution, the greater is the proportion of total protein subject to surface denaturation, and consequently the greater should be the care to avoid any strong agitation during mixing. One of the best procedures to follow in mixing a protein solution is to rock the tube containing the protein solution gently back and forth, covering the open end with a piece of parafilm held in place by the thumb. *Do not use a Vortex mixer*. It is also important to keep the enzyme in the most concentrated solution possible. Since denaturation occurs in dilute solutions, it is advisable to prepare dilutions just prior to use.

Analysis of Purification Data

The values that characterize the effectiveness of a given purification procedure are (1) the specific activity, which is a measure of the purity of the enzyme; and (2) the yield or recovery of the enzyme activity. One wishes a large increase in the specific activity (more than twofold) and a reasonable yield (> 50%) for each purification step. The figures given in parentheses are close to minimal. Thus, for example, if only 1% of the protein in the crude extract is the desired enzyme, then it is theoretically possible to achieve a 100-fold increase in the specific activity.

The pertinent data necessary for the determination of specific activity and yield are obviously the activity and protein per milliliter of extract as well as the total volume of the extract. A given purification step may appear to be ineffective (*i.e.*, small increase in specific activity and/or low yield) because of (1) poor separation of the desired enzyme in relation to the other proteins; (2) inactivation of the enzyme; and (3) the separation of the necessary components of a multicomponent enzyme system into different fractions. Case (1) can be distinguished from cases (2) and (3) by measuring the total enzyme activity of each of the fractions obtained in a purification step. If the total activity of all fractions equals the total activity before fractionation, then (2) and (3) can be presumed absent in the given fraction. Such an analysis is important for the evaluation of a purification scheme and will be carried out for the yeast invertase isolated in this study.

B. Partial Purification and Characterization of Yeast Invertase

REFERENCES: Goldstein, A., and Lampen, J.O. β-D-fructofuranoside fructohydrolase from yeast. *Methods Enzymol.* 42:504–511 (1975).

Myrback, K. Invertases. *Enzymes* 4:379–396 (1960).

Robyt and White, pp. 267–271.

The enzyme invertase (β-fructofuranoside fructohydrolase; E.C. 3.2.1.26) catalyzes the hydrolysis of the 1,2-glycoside bond in the nonreducing disaccharide sucrose, giving rise to equivalent amounts of glucose and fructose:

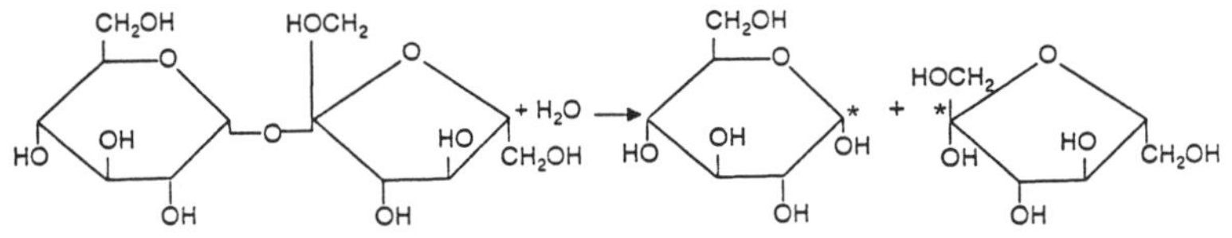

*Anomeric carbons with free -OH groups.

Thus, for each molecule of sucrose hydrolyzed, two molecules of reducing sugar are formed. The rate of sucrose hydrolysis will be determined by measuring the amount of reducing sugar formed using the Nelson procedure. *One unit* of invertase activity will be defined as the amount of substrate hydrolyzed per minute under standard assay conditions. The *specific activity* then becomes the number of units per milligram of protein.

Extraction and Partial Purification

All steps of this isolation should be carried out at 0-4°C.

Extraction

1. Weigh out 10 g of yeast and sufficient sand or fine glass beads (2–5 g) and place in a mortar.

2. Add 10 mL of toluene and grind until a smooth paste is obtained.
3. Add 16 mL of deionized water in 2-mL aliquots over the next 30 min, and continue to grind after each addition of water.
4. Transfer the contents of the mortar to 50-mL centrifuge tubes and centrifuge for 15 min at 12,000 rpm in a Sorvall refrigerated centrifuge using the SS-34 rotor.
5. Remove the middle layer (aqueous) with a Pasteur pipette. Be careful not to transfer the upper toluene phase. (This can be done by using a Pasteur pipette with the bulb squeezed shut while inserting the pipette along the wall of the centrifuge tube.)
6. Collect this protein extract in a graduated cylinder, record the volume, and save a 1.5-mL aliquot for activity and protein determinations (fraction I, *crude fraction*). Transfer the remaining solution to a 50-mL centrifuge tube.
7. Adjust the pH of the remaining solution to 5.0 by the dropwise addition of 1 N acetic acid. (This will require 2 to 4 drops only.)

Heat Treatment

1. Bring the temperature of the extract rapidly to 50°C and maintain at this temperature for 30 min. Rock the tube gently during the incubation.
2. Cool the extract rapidly in an ice bath and remove the resultant precipitate by centrifugation for 15 min at 15,000 rpm.
3. Measure the exact volume of the supernatant, and save a 2-mL aliquot for activity and protein determinations (fraction II, *heat extract fraction*).

Alcohol Fractionation

1. In a period of at least 30 min, add a volume of cold (−20°C) 95% ethanol equal to the volume of the heat extract. Keep cold and stir gently. This is accomplished by providing an "ice water jacket." Use a 100-mL beaker for the extract and a 600-mL beaker to hold the ice and water mixture. Put a stirring bar into the smaller beaker and place both on a magnetic stirrer.

2. Allow the suspension to stir for an additional 15 min, and centrifuge the resulting precipitate for 20 min at 15,000 rpm. Carefully decant the supernatant and allow the tube to drain. Cover the tube with parafilm and store the pellet in the refrigerator until the next laboratory period. Test the supernatant for Clinistix*activity before discarding. NOTE: The pellet may appear to be scant, particularly if the centrifuge tube is opaque. If a precipitate was obtained on addition of alcohol and the supernatant after centrifugation is clear, a protein pellet containing enzyme is present.
3. When ready to do the chromatography, dissolve the pellet in 5 mL of 0.05 M Tris-HC1,pH 7.3 buffer. Use a rubber policeman to stir the protein and "work" the pellet until it is all dissolved. (This process should take at least 5 min.) Centrifuge at 15,000 rpm for 20 min and discard the pellet.
4. Measure out exactly 3.5mL for the DEAE column. Save the remaining amount of this fraction for activity and protein determinations (fraction III, *alcohol fraction*).

DEAE Cellulose Column Chromatography

Ion Exchange Resins

Ion exchange chromatography is among the most widely used means of purifying proteins and separating small molecules. A chromatographic medium is made up of an inert supporting material with ionizable groups attached. The two major supporting materials are cellulose and polystyrene resins. Sephadex beads are also used. There are many ionizable groups available. Among the commonly used are diethylaminoethyl (DEAE) cellulose and carboxymethyl (CM) cellulose.

DEAE cellulose, by virtue of the positively charged amino nitrogen at pH values below 9, binds negatively charged molecules and is therefore an anion exchanger. Conversely, CM cellulose is a cation exchanger. If a solution containing negatively charged proteins is applied to a DEAE cellulose column at pH 7.5, the proteins will bind to the positively charged DEAE groups. When a salt solution is passed through the column,

---

**Clinistix* is a commercial diagnostic strip that measures glucose concentration. Use as follows: In a spot plate mix 2 drops of 0.2 M acetate buffer, pH 4.5, 2 drops of 0.5 M sucrose, and 2 drops of the supernatant. Incubate for 5 min to allow for enzyme breakdown of the sucrose, and dip the *Clinistix* into the assay mixture. Record the color development in 10 sec using a qualitative scale of − to +++.

some of the proteins will be eluted due to the increase in ionic strength. The proteins that are eluted will be a function of their charge and the salt concentration. The remaining proteins may be eluted by raising the salt concentration.

In designing an ion exchange system, there are three conditions under experimental control: the choice of resin, the pH, and the ionic strength. If the isoelectric point of the protein can be determined, then a proper choice of resin and pH can be made. The pH must be within the range of stability of both protein and ion exchange resins. Cellulose fibers are much more labile than polystyrene beads. It may be better to choose conditions so that the desired protein never adheres to the column if the protein is extremely labile. If the conditions are such that the desired protein remains on the column, then a change in the ionic strength or pH of the eluant must be made to elute the protein while leaving undesired proteins on the column. A step gradient or a continuous gradient is used. Each method has its advantages and disadvantages so that the choice of one over the other will depend on the conditions. A continuous gradient will be used in this experiment.

Preparation of a Column

Some elaborate columns are commercially available, but generally any glass tube that can be clamped off to regulate solvent flow can be used. A buret or a pipette fitted with tygon tubing and a clamp is adequate if one of suitable size is available.

The size of the column that should be used is dependent on the amount of sample that is to be applied and the resolution desired. For gel filtration, it is recommended that the sample volume be 1% to 5% of the bed volume. In ion exchange chromatography, the capacity of the resin is described by the manufacturer. The information is provided with the package at the time of purchase.

The DEAE cellulose must be carefully washed and equilibrated prior to use for best results. This procedure can be found in the Appendix. The assembly of the ion exchange column, gradient maker, and location of clamps is shown in Figure 5.1.

Elution Using a Gradient Maker

A continuous gradient is obtained when a solution of increasing salt concentration is introduced into the column with the eluant. At varying ion concentrations, different proteins are then displaced and released from the DEAE resin. A series of fractions

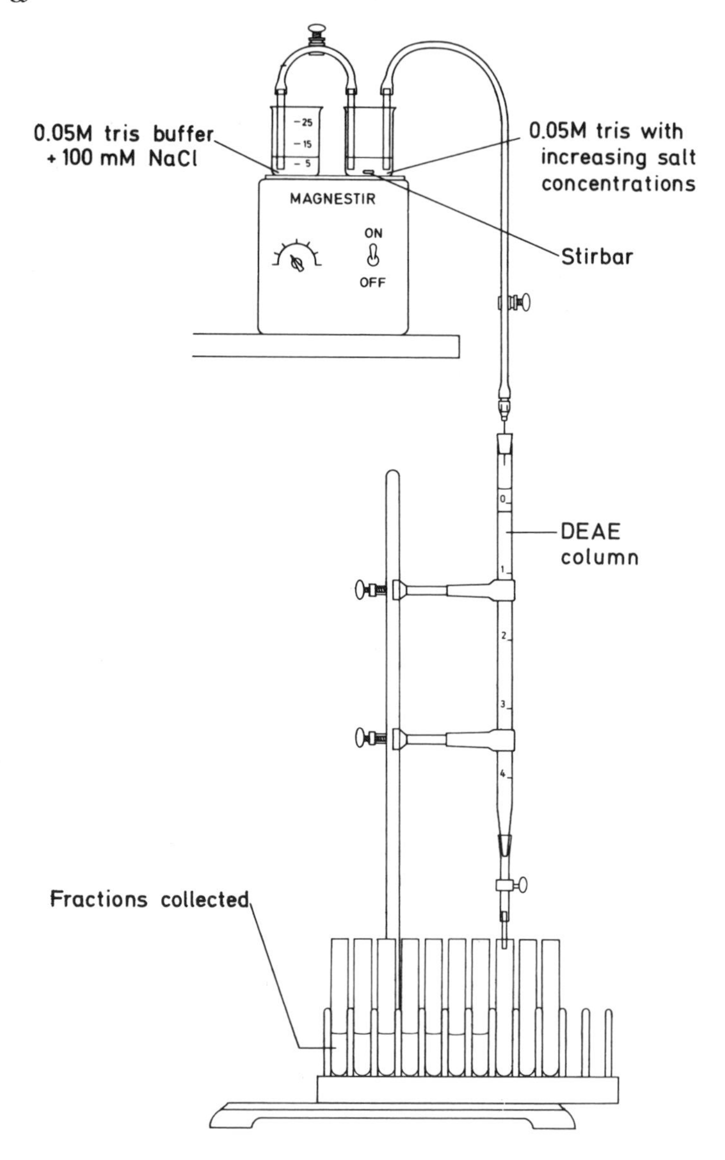

Fig. 5.1. DEAE column chromatography. The salt gradient maker is attached to the column, and 3-mL fractions are collected.

containing varying amounts of protein is collected. The amount of protein is dependent on the affinity of the individual proteins to the resin. The protein content of each fraction is then measured in a spectrophotometer at 280 nm. The concomitant salt concentration of each fraction is measured in a conductivity meter (as illustrated in Fig. 5.2) as is the presence of enzyme; the latter is determined by an activity assay. The gradient maker is mounted on a magnetic stirrer, as shown in Figure 5.1, and a stirring bar is located in the chamber that receives the NaCl solution. This beaker must therefore be centered over the mag-

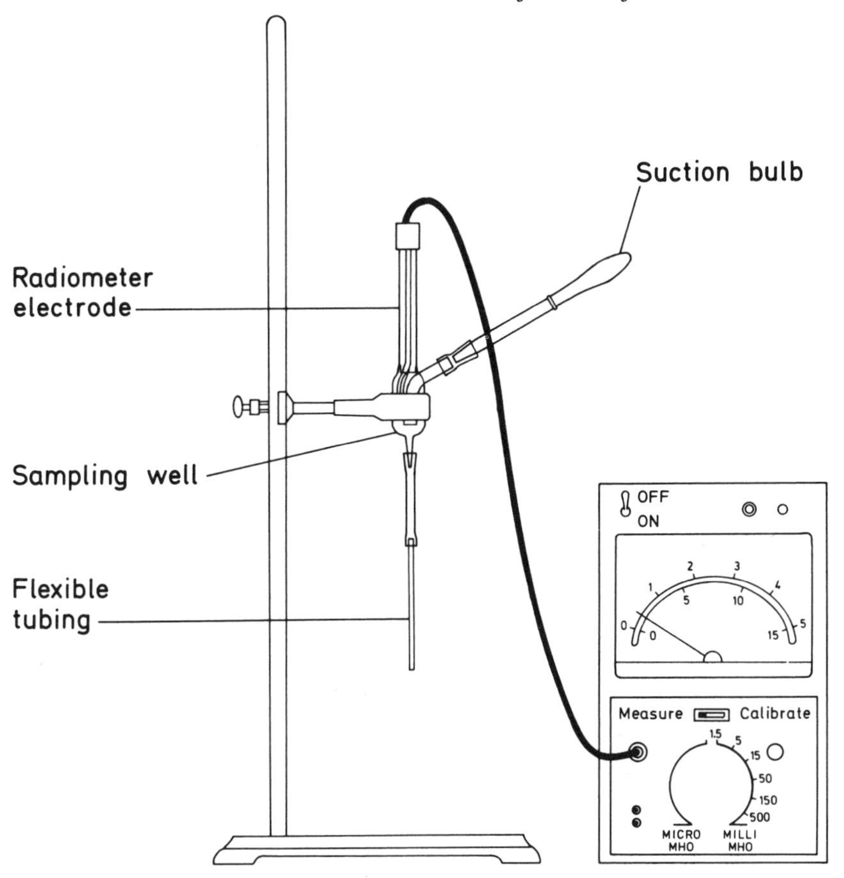

Fig. 5.2. Conductivity meter. The instrument is used to determine the salt concentration in each fraction.

netic stirring motor, and the gradient maker should be taped in place. The gradient maker is connected to the column by means of rubber tubing attached to a 20-gauge needle that is inserted through a "00" cork that fits snuggly onto the top of the column.

Experimental Procedure

1. The DEAE cellulose has been previously washed and equilibrated with 0.05 M Tris, pH 7.3, starting buffer. At room temperature, pack a column consisting of a 5-mL disposable pipette fitted with tygon tubing. This is done by placing a plug of glass wool snugly at the bottom and 0.5 cm of sand on top of the glass wool to give an even surface. Pour in approximately 3 mL of buffer, and then add the cellulose slowly until the bed height is approximately 18 cm. Wash with 5 mL of starting buffer.

2. Take a gradient maker with two 25-mL compartments; add 15 mL of 0.05 M Tris, pH 7.3, to the compartment leading directly to the column, and fill the other compartment with 15 mL of Tris buffer containing 100 mM NaCl.
3. Fill the bridge connecting the compartments with Tris buffer from the first chamber. Remove any air bubbles and clamp off the connecting tubing.
4. Apply the 3.5 mL of redissolved *alcohol fraction* (fraction III) evenly to the top of the column without disturbing the cellulose bed. Begin collecting 3-mL fractions. Allow the protein to run into the bed, and then rinse the sides of the column carefully with buffer using a Pasteur pipette. Deliver a total of 6 mL of buffer through the column. When the buffer is about 0.5 cm from the top of the cellulose, stop the column and connect the gradient maker.
5. Place a magnetic bar in the compartment without NaCl, and adjust the speed of the stirrer to achieve good mixing. Open the connection between compartments and start the column. Continue to collect 3-mL fractions until the gradient maker is empty and no more effluent flows from the column.
6. Take absorbance readings in the spectrophotometer at 280 nm, test with *Clinistix* (see footnote p. 50 for procedure) for activity in all the tubes, and measure the conductivity in each tube. The salt concentration is measured in a conductivity meter (see Fig. 5.2). A calibration curve is established by taking readings for 0.05 M Tris and 0.05 M Tris + 100 mM NaCl. Then all the tubes are read in order of increasing [NaCl]. These conductivity readings are then converted to [NaCl]. Plot absorbance, *Clinistix* activity, and salt concentration for all the tubes on one graph.
7. Pool the 2 tubes that show the highest activity, and divide this volume of enzyme among 10 tubes. One of these tubes should receive 1 mL for the protein determination. The remainder is divided equally among the other 9 tubes—cover these with parafilm and freeze (fraction IV, *column fraction*).
8. The two tubes on either side of the activity peak should be given to the instructor. The class will pool these samples, which will be concentrated later for use in the electrophoresis experiment. The concentration will be done by using

an Amicon ultrafiltration apparatus (see Fig. 5.3A) with a PM 30 membrane, which has a molecular weight cut-off level of 30,000. Fraction IV can also be concentrated by using the "ultraspin" triacetate centrifugation system (see Fig. 5.3B.)

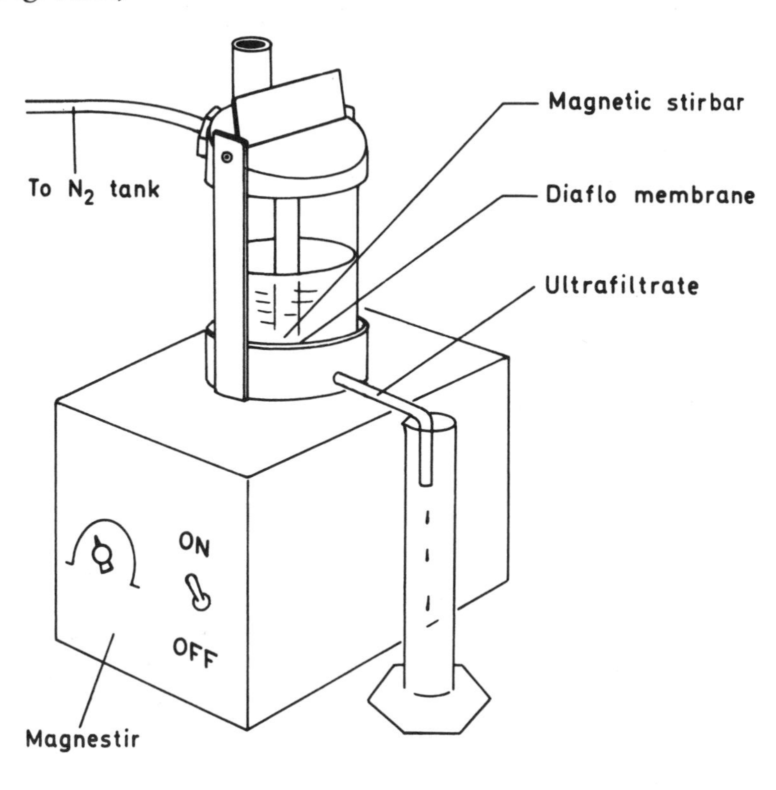

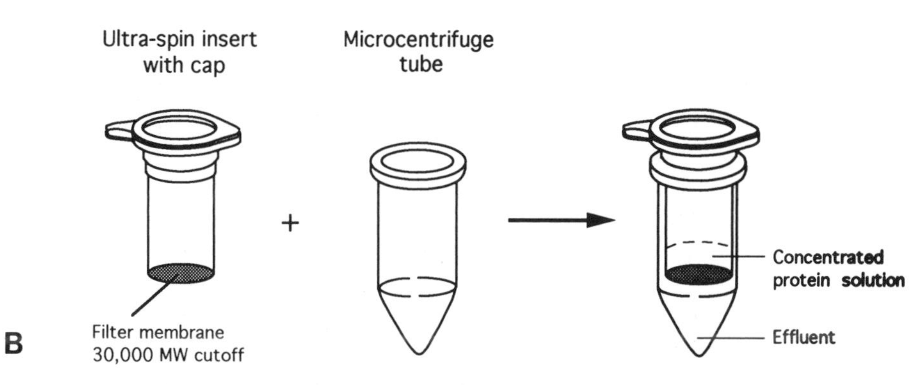

Fig. 5.3. Ultrafiltration. **A**: Ultrafiltration apparatus. (Adapted from Ultrafiltration and Microporous Filtration Systems Catalog, Publ. 550 © 1980, Amicon Corp., Lexington, MA.); **B**: "Ultraspin" Triacetate Centrifugation System (Cole-Palmer, Chicago, IL).

Ultrafiltration

Diaflo Membrane Method

Ultrafiltration refers to the efficient selective rejection of solutes by convective solvent flow through a Diaflo membrane. (Amicon Corp., Lexington, MA). In ultrafiltration, solutes, colloids, or particles of dimensions larger than the specified membrane "cut-off" are quantitatively retained in solution, whereas solutes smaller than the uniform minute pores pass unhindered with the solvent through the supportive membrane substructure.

Ultrafiltration membranes offer a selection of macrosolute retentions ranging from 500 to 300,000 MW as calibrated with globular macrosolutes. Each membrane is characterized by its nominal "cut-off," *i.e.,* its ability to retain molecules larger than those of a given size. They are made of a series of various inert, synthetic materials that provide a great selectivity and stability. They are also resistant to clogging, because retained substances are rejected at the membrane surface and kept in solution, in contrast to microporous filtration in which retained solutes are captured in the channels of the membrane filter, gradually reducing flow rates and membrane efficiency.

For effective ultrafiltration, the equipment must be optimized to promote the highest transmembrane flow and selectivity. Constant stirring is used to prevent the accumulation of a gradient of retained macrosolute above the membrane, and a pressurized inflow of nitrogen is used to speed up the process.

Ultraspin System

Another method of concentrating protein solutions is by use of a filtering system designed for microcentrifugation. The "ultraspin" tubes consist of an insert (see Fig. 5.3B) fitted with a membrane suitable to hold back molecules > 30,000 MW. This insert fits into a 1.5-mL microcentrifuge tube in which the effluent is collected upon centrifugation.

The system shown in Figure 5.3B can accommodate 0.5 mL of solution. It is manufactured by Cole-Palmer Instrument Co. (Chicago, IL).

Procedure for Using the "Ultraspin" System

1. Place insert into microcentrifuge tube and pipette 0.5 mL of fraction IV into the insert.
2. Press the cap into place and spin three to four times for periods of 10 sec each time. Check the level of solution in

the insert after each spin and repeat until 50μL remain in this upper chamber (*ca* 4–5 mm depth).

3. Remove the concentrated protein solution without disturbing the membrane by gently aspirating with an automatic pipetting device.
4. Transfer to a new microfuge tube and freeze for future use or add to the glycerol-buffer solution for transfer to an acrylamide gel for the electrophoresis procedure.

Assay of Fractions for Activity and Protein Content

Nelson's Procedure for Analysis of Reducing Sugars

1. Set up 10 test tubes and deliver the solutions according to the protocol shown in Table 5.1.
2. Add 1 mL of *Nelson's reagent*[1] to each tube.
3. Cover each tube with a marble; place all tubes in a boiling water bath for 20 min.
4. Cool tubes to room temperature, and then add 1.0 mL of the *arsenomolybdate reagent*[2] (DO NOT PIPETTE BY MOUTH). Mix by vortexing, and allow the tubes to stand for 5 min.
5. Add 7 mL of deionized water to all the tubes using an automatic dispenser, mix well on a Vortex mixer, and record the %T at 510 nm using tube 1 as blank.
6. Convert %T to $A_{510\,\mathrm{nm}}$ and prepare a graph showing absorbance *vs.* μmol reducing sugar.

Assay of Fractions for Invertase Activity

To evaluate the effectiveness of the purification procedure and to analyze the data obtained, it is necessary to determine the specific activity of the relevant enzyme. The data are tabulated as shown in Table 5.2. The extent of purification and the percent recovery are calculated from the specific activities (units of enzyme/mg protein) as follows:

---

[1]*Nelson's reagent* is prepared by mixing 50 mL of solution A with 2 mL of solution B. A contains (per liter): Na_2CO_3, 25g; NaK tartrate, 25g; $NaHCO_3$, 20 g; Na_2SO_4, 200 g. B contains (per 100 mL): $CuSO_4 \cdot 5H_2O$, 15 g; conc. H_2SO_4, two drops.

[2]Arsenomolybdate reagent (per liter): Ammonium molybdate, 50 g; conc. H_2SO_4, 42 mL; Na arsenate, 6 g.

TABLE 5.1. Standard Curve for Determination of Reducing Sugars

Additions (mL)	Tube no.									
	1 (Blank)	2	3	4	5	6	7	8	9[a]	10[b]
Glucose (4 mM)	—	0.02	0.05	0.10	0.15	0.20	0.25	0.30	—	—
Fructose (4 mM)—	—	—	—	—	—	—	—	0.20		
Sucrose (4 mM)	—	—	—	—	—	—	—	—	—	0.20
H_2O	1.0	0.98	0.95	0.90	0.85	0.80	0.75	0.70	0.80	
Nelson's reagent	1.0 →									
	Cover the tubes with marbles; place in a boiling water bath for 20 min. Cool tubes to room temperature.									
Arsenomolybdate reagent	1.0 →									
	Mix tubes on a Vortex mixer and allow the tubes to stand for 5 min.									
Water	7.0 →									
	Mix on a Vortex mixer.									
$\%T_{510\ nm}$										
$A_{510\ nm}$										
μmol reducing sugar	0	0.08	0.2	0.4	0.6	0.8	1.0	1.2	—	—

[a]Fructose as reducing sugar—similar to tube 6.
[b]Sucrose is a nonreducing sugar—no reaction.

TABLE 5.2. Purification Table for Invertase

Fraction	1 Volume (mL) Recorded	Corrected[a]	2 Protein (mg/mL)	3 Total protein (mg)	4 (Units/mL)[b]	5 Total units	6 Specific activity (Units/mg)	7 Fold purification	8 Enzyme yield
Crude								1.0	100%
Heat									
Alcohol									
Column									

[a]Each time a sample is removed for later study, the original volume is reduced. However, the yield should not be penalized for the activity thus lost; therefore, when calculating "yield" in the purification table, it is necessary to adjust the values for corrected volume for each fraction. The following is an example using an original crude volume of 15 mL and size of aliquot removed as 1.5 mL.

Fraction	Volume adjusted (mL)	Aliquot removed (mL)	Corrected volume (mL)
I	15	1.5	15.00
II	13.5 × 15/13.5	1.5	15.00
III	5 × 15/13.5 × 13.5/12	1.5	6.25
IV	6 × 15/13.5 × 13.5/12 × 5/3.5	–	10.71

[b]One unit of enzyme is the amount that catalyzes the hydrolysis of 1 μmol of sucrose per minute under standard assay conditions. Since 2 μmol of product is formed per micromole of substrate hydrolyzed, the amount of reducing sugar formed per minute is divided by 2.

1. Units per mL:

$$\frac{\mu\text{mole/min (in aliquot)}}{\text{size of aliquot (in mL)} \times 2} = \frac{\text{units / mL}}{\text{(in diluted fraction)}}$$

(Remember that one unit of enzyme is the amount that will cause the formation of 2 μmol of reducing sugar per minute).

2. Units of activity in original fraction:

Units/mL × dilution factor × corrected volume = total Units

3. Protein concentration in diluted fraction:

$$\frac{\text{mg/aliquot}}{\text{size of aliquot}} = \text{(mg/mL)}$$

4. Total protein:

(mg/mL) in diluted fraction × dilution factor × corrected volume = total protein (mg)

5. Specific activity:

$$\frac{\text{Units/mL}}{\text{Protein (mg/mL)}} = \text{(Units/mg)}$$

6. Enzyme yield:

$$\frac{\text{total Units in purified fractions}}{\text{total Units in crude extract}} \times 100\%$$

7. Fold purification:

$$\frac{\text{specific activity of purified fraction}}{\text{specific activity of crude extract}}$$

Experimental Procedure

1. *Determination of invertase activity:* To assay for enzyme activity by the Nelson reaction, follow the protocol in Table 5.3. Use the following dilutions of the four fractions (make serial dilutions with deionized water at 4°C): I

TABLE 5.3. Enzyme Activity of Fractions I, II, and III

Additions (mL)	Tube no. 1 Sucrose (blank)	2	3 Crude (1:1500)	4	5	6 Heat (1:1500)	7	8	9 Alcohol (1:3000)	10	11 Glucose standards	12
0.2 M Acetate buffer, pH 4.5	0.20	→									—	—
H_2O	0.60	0.55	0.40	0.10	0.55	0.40	0.10	0.55	0.40	0.10	1.0	0.8
Volume of diluted fraction	—	0.05	0.20	0.50	0.05	0.20	0.50	0.05	0.20	0.50	—	—
Glucose (4 mM)	—	—	—	—	—	—	—	—	—	—	—	0.2
		Start assays by the addition of substrate:										
0.5 M Sucrose	0.20	→									—	—
		Incubate all tubes for exactly 10 min at room temperature. Stop the assay by the addition of Nelson's reagent:										
Nelson's reagent	1.0	→										
		Cover tubes with marbles, place in a boiling water bath for 20 min, and continue with Nelson's procedure used to generate the standard curve.										
$\%T_{510\ nm}$[a]	100										100	
$A_{510\ nm}$	0										0	
μmol reducing sugar/10 min · aliquot												
μmol/min · mL												
Average μmol/min · mL												
Units/mL original fraction												

[a]Read assay tubes against tube 1, the sucrose blank. Read the glucose standard tube against tube number 11, the water blank.

(crude), 1:1500; II (heat), 1:1500; III (alcohol), 1:3000; IV (column), 1:600.

2. *Protein determinations by the Lowry method or the protein-dye binding assay*. Set up a protocol to determine the protein concentrations of the four fractions using the procedure described in the Protein section of this text (see Chapter 3.B).

   a. Dilute the stock solution of BSA (2 mg/mL) to 400 μg/mL and run a standard curve with five to six points. Be sure to include a water blank.

   b. Along with the standard curve, set up and assay the following series of six tubes containing 0.05, 0.2, and 0.5 mL of fractions I and II diluted 1:4. Adjust the volume to a total of 0.5 mL where necessary and proceed with the Lowry assay.

   c. The next set of tubes will determine the protein content of fractions III and IV, which were suspended in Tris buffer. Since this buffer contains substances that interfere with the Lowry reagent, it is necessary to subtract the amount of color produced by these substances from the amount due to the reaction of protein with the reagent.

      i. Set up five "Tris blank" tubes containing the following amounts of 0.05 M Tris, pH 7.3; (a) undiluted, 0.3 and 0.5 mL; and (b) buffer diluted 1:1 with water, 0.05, 0.2, and 0.5 mL.

      ii. Set up an additional five tubes containing (a) fraction III, diluted 1:1 with water, 0.05, 0.2, and 0.5 mL; and (b) fraction IV, undiluted, 0.3 and 0.5 mL.

      iii. Bring the volumes of all the tubes to 0.5 mL with water if necessary and proceed with the Lowry assay as done with the first set of tubes.

      iv. Subtract the absorbances of the "Tris blanks" from the corresponding fraction tubes to obtain corrected absorbance values. Determine the protein content for each aliquot, then calculate an average concentration per milliliter for fractions I, II, III, and IV.

   d. If the protein-dye binding assay is used, one-tenth of the above volumes are used (see Chapter 3.B).

3. *Enzyme activity of fraction IV and study of the effect of enzyme concentration on the rate of hydrolysis of sucrose*

a. Follow the protocol for an activity assay of the column fraction following the procedure described in Table 5.3 for the other fractions.

b. Set up a series of 10 test tubes. Tubes 1, 2, and 3 are glucose and sucrose blanks and the glucose standard, respectively. To tubes 4 through 9, add the following amounts of column fraction (fraction IV) diluted 1:600: 0.02, 0.05, 0.10, 0.20, 0.40, and 0.60 mL. Add 0.2 mL of 0.2M acetate buffer, pH 4.5, and bring the volumes to 0.8 mL per tube with deionized water.

c. Start the reaction by adding 0.2 mL of 0.5 M sucrose and time the reaction for 10 min. Stop the reaction by addition of 1.0 mL of Nelson's reagent, and proceed with the assay as before (see Table 5.1).

d. Tube 10 is a zero-time control tube. It is set up exactly as tube 9 but receives 1.0 mL of Nelson's reagent before the 0.2 mL of sucrose solution, thus preventing any enzymatic formation of reducing sugar.

e. Read %T for all the tubes at 510 nm, and calculate the number of Units/mL of original fraction for the purified invertase.

f. Calculate the specific activity, the fold purification, and the enzyme yield for each of the four fractions. Enter the data in a purification table as shown in Table 5.2.

g. Also, plot the number of μmol of reducing sugar formed per min (v) as a function of enzyme concentration (in mg protein as determined for fraction IV by the protein assay done previously).

C. Acrylamide Gel Electrophoresis

REFERENCES: Cooper, Ch. 6.

Maurer, H.R. *Disc Electrophoresis,* 2nd ed. De Gruyter, New York (1971).

Robyt and White, pp. 135–140.

Wharton and McCarthy, pp. 133–149.

The technique of electrophoresis is based on the movement of charged solutes in an applied electrical field. The movement of the charged particle is influenced not only by charge but also by voltage, distance between the electrodes, size and shape of the molecule, temperature, and time. The mobility of an ion is the rate at which it travels in a voltage gradient of 1 V/cm (defined as a unit of electric field). Thus, the units of mobility are $cm^2sec^{-1}\ V^{-1}$. The relationship of the mobility of a particle to the electric field is expressed as follows:

$$M = \frac{qE}{f}$$

where M = mobility of the particle; q = charge on the particle; E = electric field; and f = the frictional coefficient of the particle.

Polyacrylamide gels are used in disc gel electrophoresis. Polyacrylamide gels are polymerized products of acrylamide and bisacrylamide *(N,N'-methylene bisacrylamide)*. When free radicals are added to a solution of acrylamide and bisacrylamide, a chain reaction is initiated. Acrylamide and bisacrylamide free radicals are formed, and polymerization proceeds to completion. When ammonium persulfate is added to water, it breaks down, forming free radicals, which can then initiate the polymerization reaction. Oxygen prevents polymerization. In the absence of bisacrylamide, long polymeric chains of acrylamide are formed. This would produce a viscous solution but not the gel that is desired. When bisacrylamide is added, the long acrylamide chains are cross-linked, and a gel with an average number of cross-links per unit volume results. Thus there is a matrix riddled with pores of a certain average size. The size of the pore can be regulated by the concentration of acrylamide and bisacrylamide.

Disc stands for discontinuous and refers to both the buffer system and the gel itself. The resolving gel is made up in a buffer solution containing Tris and its hydrochloride. The lower reservoir buffer is the same as that used to make up the resolving gel. The upper reservoir buffer is prepared by mixing the amine (Tris) with a weak acid (glycine) with a pK_a at or slightly higher than the resolving gel pH. The pH of the upper reservoir buffer is adjusted to about the pK_a of the weak acid.

When the power is turned on and the ions are exposed to an electric field maintained at constant current, all the ions with net negative charge move toward the anode. The Cl ions are more negatively charged than the protein and bromphenol-blue, which are in turn more highly charged than the glycine molecules. The Cl ions then move ahead of the proteins, which move ahead of the glycine molecules. As the Cl ions move out of the gels, they leave behind them a region of low ion concentration. Lowering the ion concentration in the presence of a constant current increases the voltage gradient. In the presence of a higher voltage gradient, the protein, bromphenol-blue, and glycine ions left behind by the Cl^- move faster.

Thus the ions move through the gel in the following order: Cl^-, then protein and bromphenol-blue, then glycine. The Cl^- creates a region of low voltage gradient ahead of the protein and bromphenol-blue. Any protein moving ahead into the Cl^- zone will immediately be slowed down because of the lower voltage gradient and thus will be overtaken by the protein band. Behind the protein band is a region of higher voltage gradient created by the glycine molecules. Any bromphenol-blue or protein that moves more slowly than the protein band will find itself in a region of relatively high voltage gradient and thus will move faster and overtake the protein band. The protein and bromphenol-blue are "stacked" between the leading Cl^- and the trailing glycine ions. This creates a very fine band of protein and bromphenol-blue in the gel. Since the protein molecules are large, they will be slowed down by the gel matrix. The Cl^-, glycine, and bromphenol-blue, being small molecules, will not be retarded by the gel matrix and will thus move far ahead of the protein molecules–the bromphenol-blue remaining between the glycine and Cl^- zones marking the *front* of the glycine zone. The protein is now in a region of constant voltage gradient and is free to move at a rate determined by its size and charge.

The acrylamide gels are polymerized by addition of the ammonium persulfate, as described in Figure 5.4. The concentration of acrylamide determines the pore size that can thus be controlled. The gel concentration used in this study is 8%; when much lower concentrations are used, the pore size is increased, but the gels are quite soft and will break easily when removed from the glass forms.

The following values show the effect of polyacrylamide concentration on pore size:

% Polyacrylamide	*Pore size (nm)*
3	4.4
5	3.6
10	2.6
20	1.8
30	1.3

(1) $$S_2O_8^{=} \longrightarrow SO_4^{-}\cdot$$

Persulfate — Sulfate free radical

(2) $$SO_4^{-}\cdot + CH_2{=}CH{-}\overset{\overset{\Large O}{\|}}{C}{-}NH_2 \longrightarrow {-}CH_2{-}\dot{C}H{-}\overset{\overset{\Large O}{\|}}{C}{-}NH_2$$

Acrylamide — Acrylamide free radical

Acrylamide and bisacrylamide free radicals react with additional acrylamide and bisacrylamide molecules resulting in chain elongation. Furthermore, since bisacrylamide forms free radicals on both ends of the molecule, two growing chains are linked together forming a polyacrylamide gel:

```
                                        O
                                        ‖
              CONH2        CH2═CH—C—NH
                |                      |
(3)    —CH2—C·          +          O   CH2   ——→
                |                  ‖   |
                H          CH2═CH—C—NH
                            Bisacrylamide

                       H2N—C═O      H
                            |       |
                     —CH2—C—CH2—C·
                            |       |
                            H       C—NH
                                    ‖  |
                                    O  CH2
                                       |
                                       NH
                                       |
                             CH2═CH—C═O
```

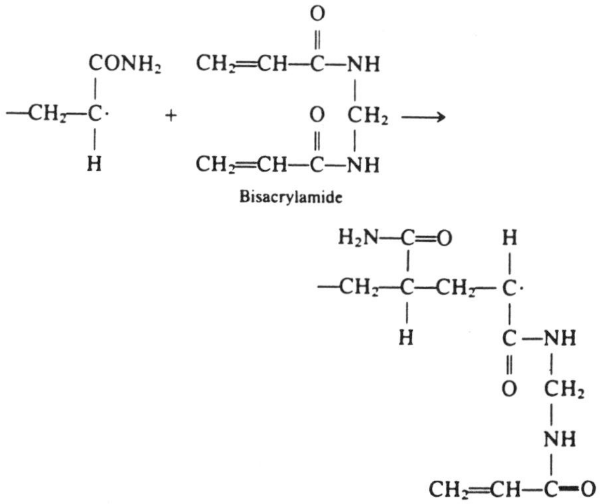

As the reaction proceeds, a lattice forms:

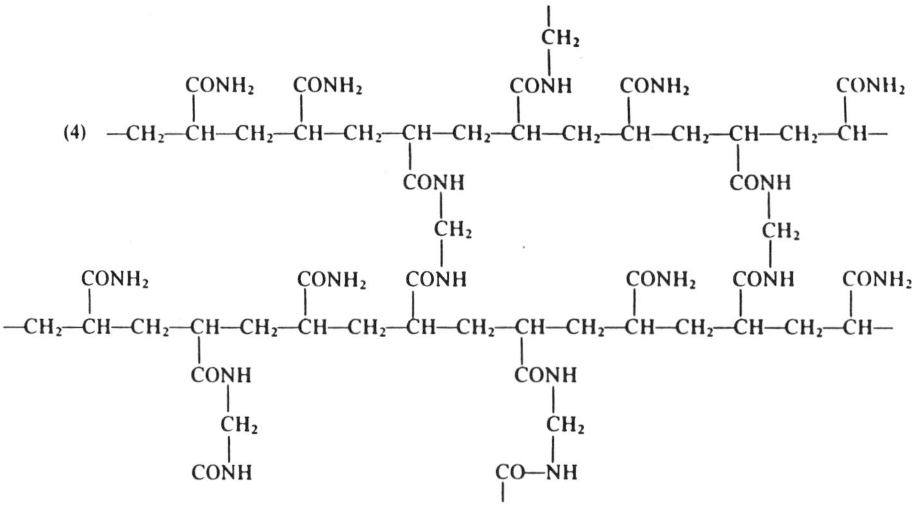

```
                                        |
                                        CH2
                                        |
          CONH2     CONH2               CONH      CONH2                CONH2
           |         |                   |         |                    |
(4) —CH2—CH—CH2—CH—CH2—CH—CH2—CH—CH2—CH—CH2—CH—CH2—CH—
                               |                          |
                              CONH                       CONH
                               |                          |
                               CH2                        CH2
                               |                          |
   CONH2              CONH2   CONH              CONH2    CONH     CONH2
    |                  |       |                 |        |        |
—CH2—CH—CH2—CH—CH2—CH—CH2—CH—CH2—CH—CH2—CH—CH2—CH—CH2—CH—
              |                      |
             CONH                   CONH
              |                      |
              CH2                    CH2
              |                      |
             CONH                  CO—NH
                                   |
```

Fig. 5.4. Polymerization of an acrylamide gel.

In the presence of water, ammonium persulfate dissociates, forming NH_4^+ and $S_2O_8^=$. The persulfate ions then break down, forming sulfate free radicals that initiate polymerization of acrylamide and bisacrylamide molecules (see Fig. 5.4).

Experimental Procedure

ACRYLAMIDE IS A NEUROTOXIN—Do not pipette by mouth or allow it to contact skin.

Preparation of the Gels for Electrophoresis

1. Cover the bottom of the gel forms with three layers of parafilm (see Fig. 5.5 for an illustration of the electrophoresis system). Set the forms in the gel form holder, making sure that they are standing perpendicular to the table (see Fig. 5.6).
2. Mix the gel ingredients in a test tube so as to obtain the desired gel concentration (see volume ratios of solutions to use in Table 5.4). Add the ammonium persulfate solution to the mixture LAST. *Rinse out the pipette with which the acrylamide solution was transferred before placing it in the pipette rinser.*
3. Fill each gel form to a depth of 6 cm with gel solution using a long Pasteur pipette. Tap the bottom of each gel form lightly to dislodge any bubbles caught at the bottom of the solution.

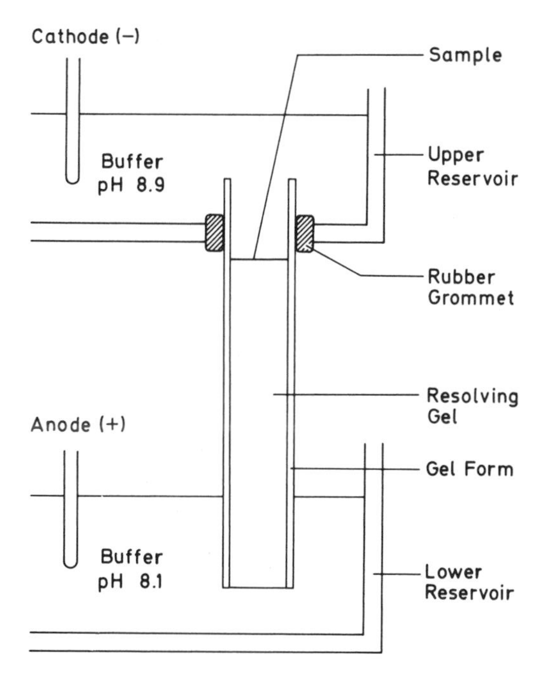

Fig. 5.5. Disc electrophoresis system.

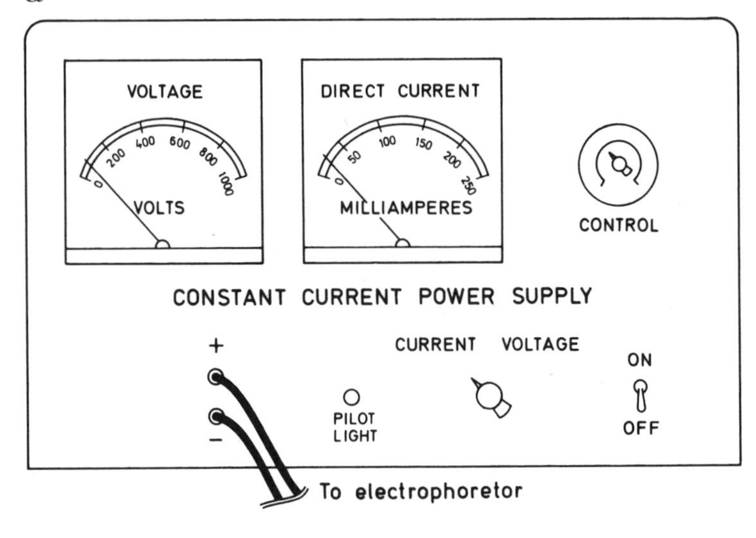

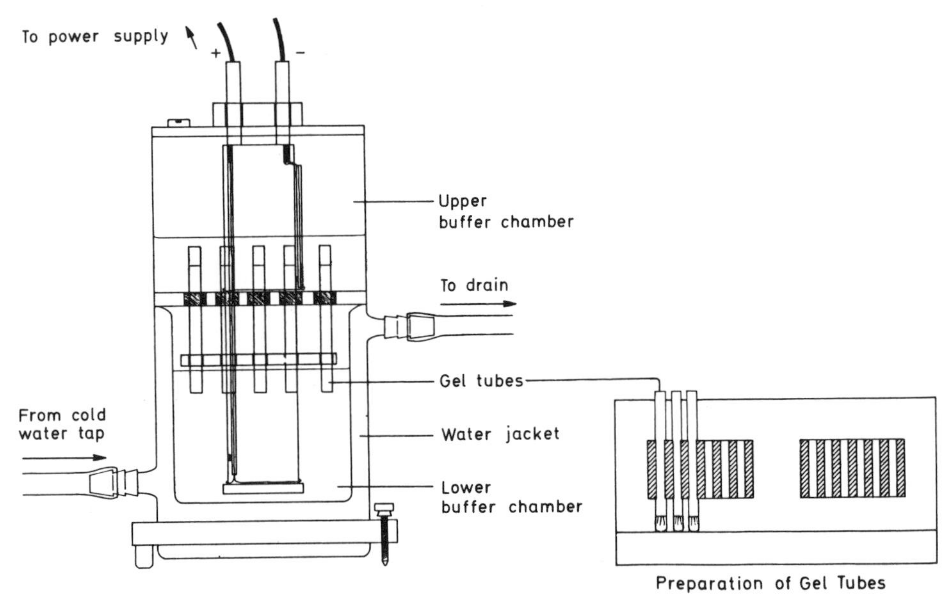

Fig. 5.6. Electrophoresis apparatus. The electrophoretor is connected to the power supply, and desired current is applied as described in the text. When the dye front bands appear to within 0.5 cm of the bottom of the gels, the current is turned off and the gels are removed.

4. Carefully layer a few drops of butanol over the top of the gel to prevent the formation of a meniscus. Allow the gels to stand for approximately 30 min so as to allow them to solidify. Now remove the butanol with a wick made from a tissue and rinse the gel surface three or four times with upper reservoir buffer. The gel tubes can now be covered with parafilm and stored at 4°C until used.

TABLE 5.4. Polyacrylamide Gel Electrophoresis[a]

Resolving gel for an 8% acrylamide concentration	Tris-glycine system			SDS system		
	Volume ratios of stock solutions	Stock solutions (components/100 mL)	pH at 25°C	Volume ratios of stock solutions	Stock solutions (components/100 mL)	pH at 25°C
A	1	32 g Acrylamide 1.0 g Bisacrylamide (32% w/v Acrylamide 1% Bisacrylamide)		1	32 g Acrylamide 1.0 g Bisacrylamide (32% w/v Acrylamide 1% Bisacrylamide)	
B	1	24 mL 1 N HCl 18.15 g Tris 0.25 mL Temed	8.9 ± 0.2	2	0.78 g $NaH_2PO_4 \cdot H_2O$ 2 g Na_2HPO_4 (anydrous) 0.2 g SDS	7.0 ± 0.2
C	2	0.2 g Ammonium persulfate		1	0.8 g Ammonium persulfate 0.13 mL Temed	
Upper buffer	10-fold dilution with water	6.32 g Tris 3.94 g Glycine	8.9 ± 0.2	2-fold dilution with water	Same as B	7.0 ± 0.2
Lower buffer	10-fold dilution with water	12.1 g Tris 50 mL 1 N HCl	8.1 ± 0.2	2-fold dilution with water	Same as B	7.0 ± 0.2

[a]The polymerization time is 30 to 45 min. Do not prepare the gel mixture until ready to pour the gels.

Preparation of the Samples to be Applied to the Gels

1. Calculate the amount of protein (as determined previously) in the crude, alcohol, and heat fractions and dilute the fractions to 1 mg/mL. The column fraction must be concentrated by ultrafiltration.
2. Prepare in separate tubes the following solutions: 50 μL diluted protein (may be increased to 75 μL and buffer omitted); 25 μL 50% glycerol; 20 μL upper reservoir buffer; and 5 μL 0.1% bromphenol-blue solution. A convenient amount of protein to layer on in this particular system and use with the stain is 20 to 100 μg.
3. Prepare five samples, one each of crude, heat, and concentrated column fractions and two of the alcohol fraction. (One will be used for an activity assay.) The volumes of all sample solutions run in a single electrophoretor should be approximately the same so that all samples have an equal opportunity of descending the same distance during exposure to the electric field.
4. Prepare the reservoir buffers.
    a. Lower reservoir buffer: Dilute to the correct concentration and permit to reach room temperature if solution is cold. Use about 400 mL (see Table 5.4).
    b. Upper reservoir buffer: Dilute to the desired concentration as done with the lower reservoir buffer. Use enough to cover at least the gel form tubes (about 250 mL).
5. Set up the electrophoretor.
    a. Remove the parafilm from the top of the gels previously prepared (see above).
    b. Remove the parafilm from the bottom of the gels, being careful not to pull at the gels with suction. Place the gel forms containing the gels in the appropriate holes in the electrophoretor. Record the position of the gels in the apparatus. Make sure that the bottom of the gels are all at approximately the same level, and release any trapped air bubbles.
    c. To make sure that the buffer goes into all the gel forms making contact with the gel, fill gel tubes using a Pasteur pipette. Gently pour the upper reservoir buffer into the upper reservoir.

d. Before adding the samples, make sure that the electrophoretor is level and that all the gels are perpendicular to the table.

e. Using a 100-μL pipetting device, *carefully* layer the sample under the upper reservoir buffer and directly over the gel. Do not pierce the gel with the tip of the pipette.

f. Place the electrophoretor cover over the upper reservoir. Connect the leads to the power supply. At this point, the power supply should be turned off and unplugged. The anode (+, *red*) should be connected to the lower reservoir electrode and the cathode (−, *black*) to the upper reservoir electrode (see Fig. 5.6).

g. With the power supply turned to the OFF position, plug in the instrument. After making sure that the dial controlling the current is turned all the way down to 0, turn on the power supply, and then slowly turn up the current until a total of 2 mAmp per gel is reached. Run at this current for 30 min. Turn the current up to 4 mAmp per gel, and run until the tracking dye reaches the bottoms of the gels.

6. Remove the gels from the forms and place on ice.

   a. Measure and record the distance from the top of the gel to the "dye front." This must be done before gels are immersed in trichloroacetic acid (TCA), as the band disappears during this treatment.

   b. Fill a syringe with an attached 20-gauge needle with cold water. Insert the needle between the gel and the gel form extruding water, thus preventing the needle from scratching the gel. Holding the gel and the needle in a horizontal position, rotate the form around the needle, thus breaking the bond between the gel and the glass form. A Pasteur pipette bulb will aid in expelling the gel from the form if placed on one end of the form and gently squeezed. Slide the gel, bottom first, into a test tube and cover with 12.5% TCA. DO NOT PLACE THE GEL THAT YOU ARE GOING TO TEST FOR ACTIVITY IN THE TCA SOLUTION. Leave the gels in TCA for 30 min to fix the proteins in the gel.

   c. With one of the alcohol fraction gels, proceed as described in the method for testing activity (see below).

d. All glassware used to handle the Coomassie blue solution should be rinsed thoroughly with either acetone or soapy water to remove all the dye before adding the equipment to the general wash.

7. Fixing and staining of gels. Remove the TCA solution from around the gels with an aspirator fitted with a Pasteur pipette. Add the Coomassie blue staining solution to the gels using a Pasteur pipette. Stain for 30 min. Then remove the stain and place the gels in destaining solution at 37°C overnight. Change the destaining solution several times over a period of 3 to 5 days. Gels can be read when the background is clear and the protein bands are clearly visible (see Appendix for composition of staining and destaining solutions).

8. Record the following values.

   a. Distance the dye moved in the resolving gel. Begin measurements from the top of the gel.

   b. Distance the protein moved in the resolving gel. Calculate the relative mobility (*Rm*) of the protein in that particular gel.

$$\text{Relative mobility} = \frac{\text{distance the protein moved}}{\text{distance the dye moved}}$$

The dye marker and protein will not migrate exactly the same distance in all tubes in the electrophoretor. The relative mobility of a protein under a given set of conditions remains constant and is the only reliable measure of how fast the protein moved through the gel.

Localization of Invertase Activity

The localization of invertase in the polyacrylamide gel is accomplished by making thin slices of the gel, incubating the slices with sucrose, and assaying with the Clinistix. Carefully slice the gel from the top every 3 mm until a distance of approximately 2.4 cm has been reached. Immerse the slices in a porcelain spot plate containing several drops of 0.2 M Na acetate buffer, pH 4.5, and several drops of 0.5 M sucrose. Chop the slices with a clean razor blade or other sharp tool to help release the protein and therefore increase the exposure of the enzyme to the substrate. After 10 to 15 min incubation, examine each well for activity using a Clinistix test.

D. Enzyme Kinetics Studies

To obtain valid Michaelis-Menten kinetic data for an enzyme, it is necessary to control all variables except the substrate that is the independent variable. The effects of enzyme concentration on velocity (see p. 76) and incubation time on product formation (see p. 76) will be studied. The curves expected as a result of conducting both experiments are of similar shape. In Figure 5.7, velocity is plotted as a function of enzyme concentration, and the characteristic curve is shown. When the linear region of the curve is passed, the enzyme is hampered for one of several reasons and conditions are unsuitable for studying enzyme activity.

In the following series of experiments, the effect of enzyme concentration and time of incubation are studied for invertase, and the appropriate conditions are used in the subsequent determination of the effect of substrate on the rate of sucrose hydrolysis. The typical Michaelis-Menten plot for an enzyme acting on a single substrate is shown in Figure 5.8. In this figure, it is evident that a determination of V_{max} is difficult as it is approached asymptotically, and the appropriate amount of sub-

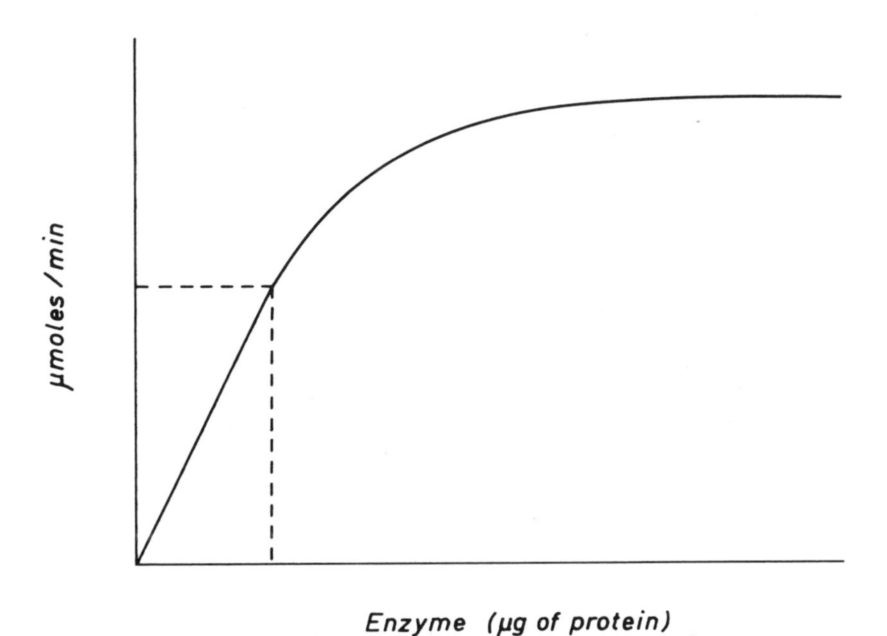

Fig. 5.7. The effect of enzyme concentration on reaction rate.

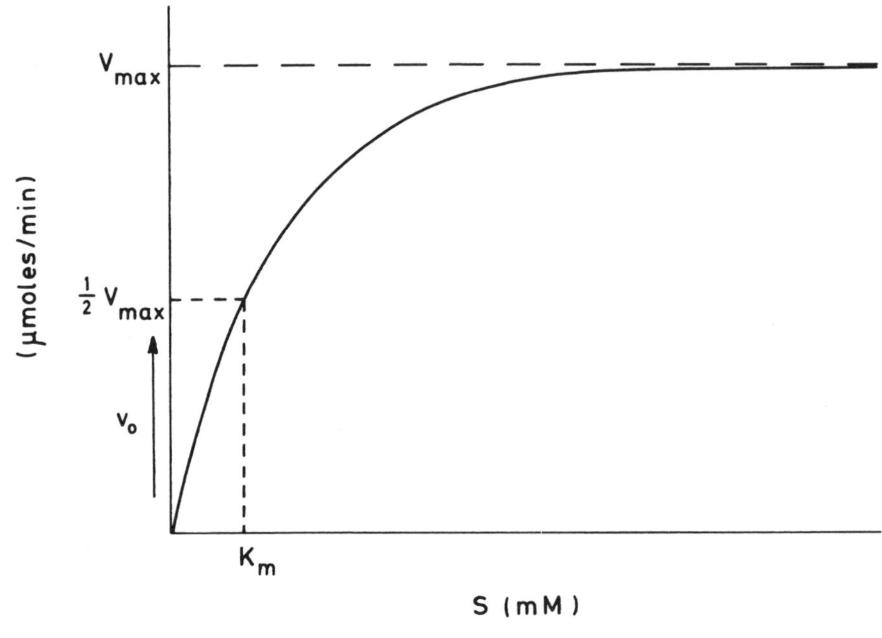

Fig. 5.8. Effect of substrate concentration on the reaction rate.

strate needed to reach V_{max} may be difficult to achieve experimentally. The reaction is described by the equation:

$$v_0 = \frac{V_{max}[S]}{K_m + [S]}$$

where v_0 = initial velocity of the reaction; V_{max} = maximum velocity of the reaction; $[S]$ = substrate concentration; and K_m = the substrate concentration at which half V_{max} is achieved.

When the reciprocal of both sides of the Michaelis-Menten equation is taken, the following equation results:

$$\frac{1}{v_0} = \frac{K_m}{V_{max}} \times \frac{1}{[S]} + \frac{1}{V_{max}}$$

This is known as the Lineweaver-Burk equation, and it defines a straight line that can be obtained experimentally by plotting $-1/v_0$ against $1/[S]$. The line intercepts the $1/v_0$ axis at $1/V_{max}$ and has an intercept of $-1/K_m$ on the $1/[S]$ axis. In addition to permitting a better determination of V_{max}, the double reciprocal plot can be used to characterize and measure the effects of inhibitors on an enzyme. The mode of inhibition can be determined by carrying out kinetic studies in the presence of the inhibitor. When the inhibition is *competitive*, the inhibition can be overcome at high concentrations of substrate so the V_{max} is unaltered, although K_m is increased. In *noncompetitive inhibition*, the active site is not affected, and the K_m remains the same but V_{max} is decreased (see Fig. 5.9).

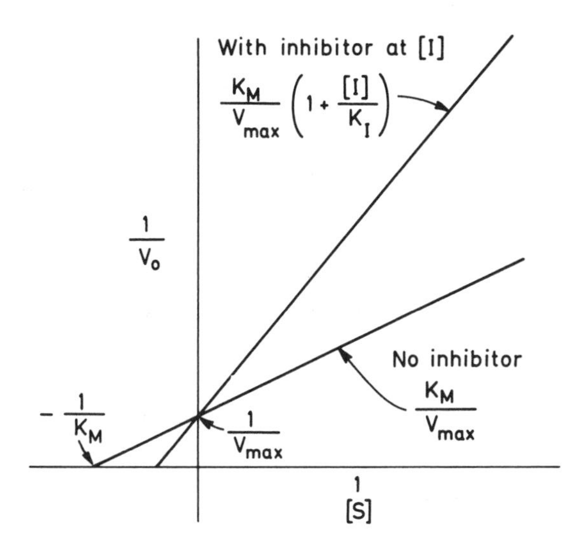

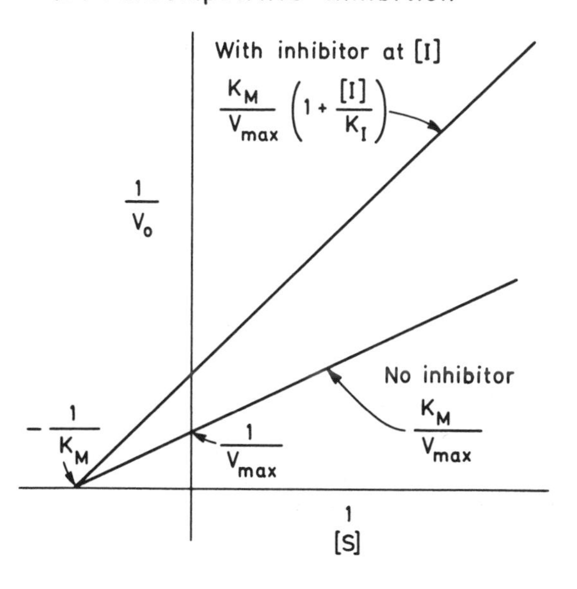

Fig. 5.9. Determination of mode of inhibition through kinetic studies.

The extent of inhibition is characterized by K_I, the inhibitor constant, which is the dissociation constant of the enzyme–inhibitor complex:

$$K_I = \frac{[E]\,[I]}{[EI]}$$

Experimentally, a fixed amount of inhibitor $[I]$ is added to the enzyme at increasing substrate concentrations. The resulting Lineweaver-Burk plots show that in the presence of the competitive inhibitor, the apparent K_m is increased over the true K_m, and the slope of the line (K_m/V_{max}) is increased by $(1 + [I]/K_I)$. In the presence of a noncompetitive inhibitor, K_m is unaltered, and the slope of the inhibited reaction is increased, as is the intercept at $1/v_0$. The effects of inhibitors on Lineweaver-Burk plots are summarized in the following table:

Mode of Inhibition	*Slope*	*Intercept on $1/v_0$ axis*
None	K_m/V_{max}	$1/V_{max}$
Competitive	$(K_m/V_{max})\,(1 + [I]/K_I)$	$1/V_{max}$
Noncompetitive	$(K_m/V_{max})\,(1 + [I]/K_I)$	$(1/V_{max})\,(1 + [I]/K_I)$

The enzyme concentration curve can be drawn from the data obtained in determining the activity of fraction IV. The enzyme kinetic studies on the purified fraction of invertase now follow.

Effect of Enzyme Concentration on Initial Velocity

See Table 5.3 for the assay of fraction IV for enzyme activity and Section B.3 (pp. 62–63) for the procedure used to generate the enzyme concentration curve.

Effect of Incubation Time on Product Formation

1. Set up a series of tubes and proceed with the assay as described in Table 5.5. Note that tube 10 represents a control tube without enzyme and that tubes 11 and 12 are a check on the standard curve for this assay. Therefore tube 1 serves as blank for tubes 2 to 10, and tube 11 is the blank for tube 12.
2. Plot the number of micromoles of reducing sugar formed as a function of time.

TABLE 5.5. The Effect of Incubation Time on Product Formation

Additions (mL)	Tube no.											
	1 Zero time[a]	2	3	4	5	6	7	8	9	10	11	12
0.2 M Acetate buffer, pH 4.5	0.2 ⟶										—	—
0.5 M Sucrose (diluted 1:25)	0.1 ⟶										—	—
H_2O	0.6 ⟶									0.7	1.0	0.8
Glucose (4 mM)	—	—	—	—	—	—	—	—	—	—	—	0.2
	Start assays by the addition of enzyme:											
Purified enzyme[b]	0.1 ⟶									—	—	—
Starting time:												
Incubation time (min)	0	1	2	4	8	10	12	15	20			
Stopping time:												
	Incubate tubes at room temperature for the designated time intervals. Stop reaction by the addition of 1.0 ml Nelson's reagent.											
Nelson's reagent	1.0 ⟶											
	Cover tubes with marbles, place in a boiling water bath for 20 min, and assay for reducing sugar as described for the Nelson's procedure.											
$\%T_{510\ nm}$												
$A_{510\ nm}$												
µmol reducing sugar formed												

[a]Nelson's reagent is added before the enzyme to this tube.

[b]Enzyme should be diluted so that 0.1 mL will yield 1 µmol/min under normal assay conditions. (i.e., 0.2 ml of 0.5 M sucrose, 0.2 ml of 0.2 M acetate buffer, pH 4.5, in a final volume of 1 mL and an incubation period of 10 min).

TABLE 5.6. Effect of Substrate on Enzyme Activity

Additions (mL)	Tube no. 1 (Blank)	2	3	4	5	6	7	8	9 (Sucrose correction)[a]	10 (Sucrose correction)[a]	11	12
0.2 M Acetate buffer, pH 4.5	0.20 ⟶										—	—
0.5 M Sucrose	—	0.02	0.04	0.06	0.08	0.10	0.20	0.40	0.20	0.40	—	—
H_2O		⟵ To bring final volume to 0.9 mL ⟶									1.0	0.8
Nelson's reagent	—	—	—	—	—	—	—	—	1.0	1.0	—	—
Enzyme[b]	0.10 ⟶										—	—
Glucose (4 mM)	—	—	—	—	—	—	—	—	—	—	—	0.2
	Start all assays by the addition of enzyme. Incubate all tubes for 10 min at room temperature. Stop reactions by the addition of Nelson's reagent.											
Nelson's reagent	1.0 ⟶								—	—	⟶	
	Cover tubes with marbles, place in a boiling water bath for 20 min and assay for reducing sugar as per Nelson's procedure.											
$\%T_{510\ nm}$												
$A_{510\ nm}$												
µmol reducing sugar/min												

[a]Tubes 9 and 10 are included to account for nonenzymatic sucrose hydrolysis.

[b]Add an amount of enzyme that is nonlimiting; choose a point about two-thirds of the way up the linear portion of the curve. See the results of the enzyme concentration experiment (Section B3) This will probably give an absorbance change of *ca* 1.0 under standard assay conditions.

Effect of Substrate Concentration on Enzyme Activity

1. A Michaelis-Menten curve will be obtained from the data generated by proceeding with the experiment as shown in Table 5.6. Note again that tube 1 is the blank against which tubes 2 to 10 should be read in the spectrophotometer, and tubes 11 and 12 again represent a standard point and a blank for the assay. From the absorbances obtained for tubes 9 and 10, corrections should be made for all the other tubes. As some spontaneous sucrose hydrolysis occurs, it is necessary to subtract the amount of absorbance due to nonenzymatic breakdown of the substrate from the tubes having the corresponding concentrations of sucrose. A convenient method of making the correction is to draw a curve by plotting $A_{510\ nm}$ *vs.* sucrose concentration for 0, 0.2 and 0.4 mL.
2. Presentation of data.
   a. Calculate v (μmol reducing sugar/min) and [S], the concentration of sucrose in each tube (mM).
   b. Make graphs of rate *vs.* substrate concentration (Michaelis-Menten curve) and $1/v$ *vs.* $1/[S]$ (Lineweaver-Burk plot).
   c. Calculate and report K_m and V_{max} values.

E. Inhibition of Invertase Activity

Noncompetitive Inhibition—Inhibition by Urea

By varying the sucrose concentration in the presence and absence of urea, the degree and mode of urea inhibition on the invertase-catalyzed hydrolysis of sucrose can be determined.

Experimental Procedure

1. For the series of tubes containing no inhibitor, see section Effect of Substrate Concentration on Enzyme Activity above.
2. For the series of tubes containing the inhibitor, set up eight test tubes with buffer and increasing amounts of sucrose as in the protocol in Table 5.6. To all tubes, add 0.25 mL of a 4 M urea solution and bring the final volume to 0.9 mL with water. Use the same amount of protein in 0.1 mL for the enzyme solution.

3. To correct for acid hydrolysis, add two tubes similar to tubes 9 and 10 of the "Effect of Substrate Concentration" study but also containing 0.25 mL of 4 M urea. Adjust the volume of water to accommodate this extra volume (to a total volume of 0.9 mL).
4. Start the reaction by adding 0.1 mL of enzyme as done for the substrate concentration series without inhibitor.
5. Incubate for exactly 10 min, timing from the addition of enzyme and proceed as in the regular Nelson's assay.
6. Plot rate of catalysis, v *vs.* substrate concentration and $1/v$ *vs.* $1/[S]$. Calculate K_m, V_{max}, V_{max} apparent, and K_I. Evaluate the effect of urea on invertase activity.

Competitive Inhibition—Inhibition by Raffinose and Fructose

Raffinose is a nonreducing trisaccharide that can be considered to be a structural analog of sucrose:

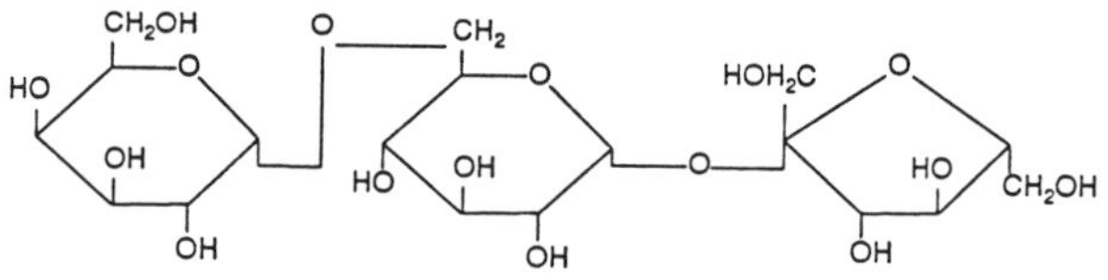

It is hydrolyzed by invertase to give the disaccharide melibiose and the monosaccharide fructose. The rate of raffinose hydrolysis may be determined by measuring the appearance of reducing sugar. Since no glucose appears as a result of raffinose hydrolysis, this compound appears inactive when the glucose oxidase procedure is used to follow the reaction.

Therefore it is possible to use this assay to measure invertase activity in the presence of raffinose. By varying the sucrose concentration in the presence and absence of raffinose and fructose, the mode of inhibition that these sugars effect can be determined.

Glucose Analysis by the Glucose Oxidase Method

The reactions involved in the assay procedure are as follows:

$$\text{Glucose} + O_2 + H_2O \xrightarrow[\text{oxidase}]{\text{glucose}} H_2O_2 + \text{gluconic acid}$$

$$H_2O_2 + \underset{\text{(reduced dye)}}{O\text{-dianiside}} + H^+ \xrightarrow{\text{peroxidase}} H_2O + \underset{\text{(oxidized dye)}}{\text{yellow pigment}}$$

The glucose oxidase reagent is made up as follows: solution 1, 4 mg peroxidase and 500 units of glucose oxidase in 100 mL of 0.1 M Na phosphate buffer, pH 7.0; solution 2, *O*–dianiside, a reduced chromogen, 6.6 mg/mL.

TO PREPARE REAGENT: Add 1 mL of solution 2 to 100 mL of solution 1 just prior to use.

Generating a Standard Curve for the Determination of Glucose

1. Set up seven tubes and follow the procedure outlined in Table 5.7.
2. Plot $A_{420\text{ nm}}$ as a function of μmol glucose.

Effect of Raffinose and Fructose on Sucrose Hydrolysis

1. Set up a series of 26 test tubes. Assay tubes 1 through 10 as indicated in Table 5.8.
2. Tubes 11 through 18 are set up with the same amounts of buffer and sucrose but receive 0.50 mL of 0.5 M raffinose in addition. The water is then adjusted to give 0.90 mL per tube. The reaction is started by addition of 0.1 mL of the same enzyme preparation. Tube 11 will serve as the blank for this set of tubes.

TABLE 5.7. Standard Curve for Glucose–Glucose Oxidase Assay

	Tube no.						
Additions (mL)	1	2	3	4	5	6	7
Glucose (2 mM)	–	0.05	0.10	0.20	0.30	0.50	0.60
H_2O	1.00	0.95	0.90	0.80	0.70	0.50	0.40
Glucose oxidase reagent	4.0 →						
	Mix; incubate for 15 min at room temperature.						
5 N HCl	0.1 →						
	Mix; incubate at least 5 min.						
$\%T_{420\text{ nm}}$							
$A_{420\text{ nm}}$							
μmol glucose							

TABLE 5.8. Effect of Inhibitors on Enzyme Activity

Additions (mL)	Tube no. 1 (Blank)	2	3	4	5	6	7	8	9 (sucrose correction)[a]	10 (sucrose correction)[a]
0.2 M Acetate buffer pH 4.5	0.2 →									
0.5 M Sucrose	—	0.02	0.04	0.06	0.10	0.13	0.15	0.20	0.10	0.20
H_2O	0.70	0.68	0.66	0.64	0.60	0.57	0.55	0.50	0.70	0.60
Enzyme[b]	0.1 →								—	—
	Incubate *exactly* 10 min at room temperature. Stop assay by placing tubes in a boiling water bath for *exactly* 2 min. *Immediately put on ice*.									
Glucose oxidase reagent	4.0 →									
	Mix; incubate for 15 min at room temperature.									
5 N HCl	0.1 →									
	Mix; incubate for 5 min.									
$\%T_{420\ nm}$										
$A_{420\ nm}$										

[a]These tubes are included to account for nonenzymatic sucrose hydrolysis.

[b]0.1 mL of purified enzyme should result in the formation of 1 μmol of reducing sugar in 10 min under standard assay conditions.

3. Tubes 19 through 26 are similarly handled except that 0.5 mL of 0.5 M fructose is used instead of the raffinose. In this series of tubes, tube 19 will serve as the blank.
4. Correct all absorbance values for nonenzymatic sucrose hydrolysis using values obtained for tubes 9 and 10 and calculate the corrected velocity at each sucrose concentration (μmol/min). Calculate the substrate concentrations, [S], for each velocity (in mM sucrose).
5. Plot v *vs.* sucrose concentration for the three curves. Calculate $1/v$ and $1/[S]$ and draw the Lineweaver-Burk plots.
6. *Evaluation of mode of inhibition*: From these curves, calculate K_m and K_m apparent values as well as V_{max} values. Also calculate K_I values for fructose and raffinose.

F. Transferase Activity of Invertase

REFERENCES: Chapman, J.M., and Ayrey, G. *The Use of Radioactive Isotopes in the Life Sciences*. Allen and Unwin Ltd., London (1981).

Cooper, pp. 65–134.

Khym, J.X., and Zill, L.P. The Separation of Sugars by Ion Exchange. *J. Am. Chem. Soc. 74*: 2090–2094, (1952).

Myrback, pp. 379–386.

Robyt and White, pp. 158–212.

Invertase catalyzes the hydrolysis of sucrose in two steps:

$$1.\ \text{Enz–Fru–Glu} \rightleftharpoons \text{Enz–Fru} + \text{Glu}$$
$$2.\ \text{Enz–Fru} + \text{HOH} \rightleftharpoons \text{Enz} + \text{Fru}$$

First the enzyme forms a covalent intermediate with fructose by eliminating glucose (step 1). Then the enzyme–fructose bond is hydrolyzed by nucleophilic attack by a water molecule (step 2). Other nucleophiles can replace the water in the second step. For example, the covalently bound fructose can be solvolyzed off the enzyme by an alcohol to give a fructoside instead of free fructose:

$$\text{Enz–Fru} + \text{ROH} \rightleftharpoons \text{RO–Fru} + \text{Enz}$$

In such a combination of reactions, the enzyme transfers the fructose in sucrose from the glucose to the alcohol. For this reason, invertase is classified as a transferase.

In this experiment, the transfer of fructose to methanol will be demonstrated. The formation of the methyl fructoside will be shown by separating it from the substrate (sucrose) and other products (glucose and fructose). This is done using an anion exchange column equilibrated with borate. Borate anions can react with sugars having vicinal diols–that is, two adjacent hydroxyl groups on the same side of the sugar ring as shown below:

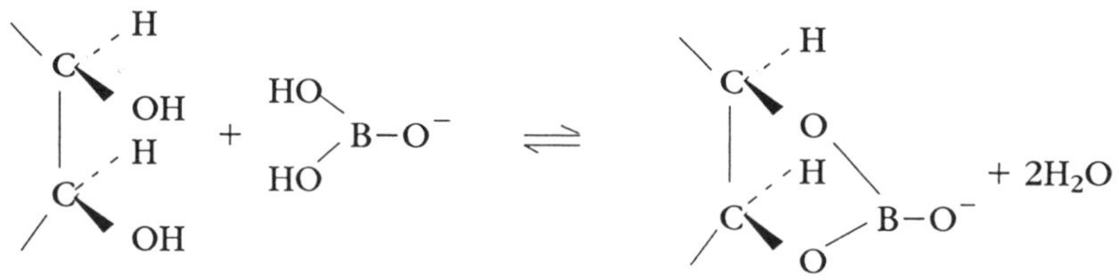

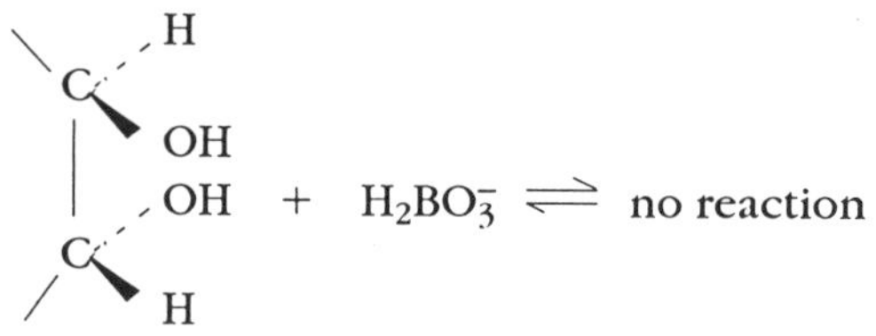

C(H)(OH)–C(H)(OH) + (HO)$_2$B–O$^-$ $\rightleftharpoons$ C(H)(O)–C(H)(O) >B–O$^-$ + $2H_2O$

C(H)(OH)–C(OH)(H) + $H_2BO_3^-$ $\rightleftharpoons$ no reaction

When invertase acts on sucrose containing tritiated fructose in the presence of methanol, the enzyme transfers the [^{3}H]fructose to methanol. The methylfructoside formed should not stick to the resin since one of the hydroxyl groups is involved in the -OCH_3 bond, and the methylfructoside will pass through the anion exchange column. The column effluent containing the radioactive methylfructoside can then be collected in scintillation vials and detected by measuring ^{3}H decay in a scintillation counter. The following experiment is designed to demonstrate the formation of methylfructoside in the presence of invertase and to introduce the student to the use of the scintillation counter.

Liquid Scintillation Counting

Liquid scintillation counting is a technique for assaying radioactivity by placing the sample in a solvent containing complex aromatic compounds called *primary* and *secondary fluors*. The particles given off by the radioactive sample transfer some of their energy to the solvent molecules, which in turn excite the primary fluor. When this fluor returns to its ground state, a photon is emitted that excites the secondary fluor. The light impulse from the secondary fluor is of a longer wavelength and is detectable by the phototube in the scintillation counter. This fluorescence is converted to electrical current and recorded as counts per minute.

When working with radioactive substances, a number of safety measures must be observed:

1. Cover the work area with absorbent paper with plastic backing.
2. Keep all materials that are to be used in the experiment on this paper.
3. Use rubber gloves throughout the experiment.
4. Use automatic pipettors with disposable tips to pipette the radioactive material.
5. Carry out the reactions in disposable test tubes.
6. When finished, all empty test tubes, gloves, and contaminated glassware should be wrapped in the absorbent paper, the package taped tightly, and disposed of in a container provided for radioactive waste.
7. Liquid radioactive waste should be disposed of in a bottle set aside for that purpose.

Transferase Activity of Invertase

The formation of methylfructoside will be demonstrated by comparing the reaction products resulting from incubation of labeled sucrose in the presence of invertase with an incubation mixture having no enzyme. As the borate column does not retain methylfructoside, the appearance of this product will be expected first, followed by the gradual appearance of excess substrate. In the control column, all the radioactivity will be from the unhydrolyzed [^{3}H]sucrose alone.

Experimental Procedure

1. The transferase activity will be demonstrated by showing the synthesis of radioactive methylfructoside.

   a. Set up two 100 × 13-mm tubes as follows:

	(1)	*(2)*
Sodium acetate buffer, 0.2 M, pH 4.5	0.10 mL	0.10 mL
0.5 M [^{3}H]sucrose (1 μCi/mL, 0.02 μCi/mmol)	0.25	0.25
H_2O	–	0.05
Methanol, 15 M	0.15	0.15
Enzyme	0.05	–

   b. The reaction is started by adding 0.05 mL of enzyme (1:1 dilution of fraction IV) to tube 1.

   c. Incubate both tubes for 15 min at room temperature.

   d. STOP the reaction by boiling the tubes for exactly 2 min and then immediately transfer the tubes to an ice bucket to cool.

   e. Add one drop of bromthymol blue to each tube and titrate with 0.1 N NaOH to a blue end point. (This will take 3 to 4 drops of NaOH using a Pasteur pipette.)

2. Prepare anion exchange columns as follows:

   a. Place a small amount of glass wool into the tips of two Pasteur pipettes (14 cm in length), and fill them with anion exchange resin to a height of 7 cm. Attach a piece of tygon tubing about 5 cm in length, and then attach a small plastic clamp to the tubing.

   (The resin, AG-1-X2, a Biorad product, has been equilibrated with a saturated borate solution and the pH adjusted to 7.0.)

   b. Apply each of the two reaction mixtures to the top of one of the two columns while they are clamped shut. Open the clamp and collect the first 10 drops into one scintillation vial and discard this effluent appropriately. Then start collecting 5 drops per scintillation vial. Add small amounts of deionized H_2O to the top of the column as needed and continue collecting until 10 vials have been collected from each column.

3. Add 9 mL of Opti-Fluor (a commercially obtained solution designed to detect radioactivity in water-based samples) scintillation fluid to each vial, and count each vial for 10 min in the scintillation counter. Subtract the counts per min (cpm) obtained for the control vials (having no enzyme) from the corresponding reaction vials.
4. Calculate μmol of methylfructoside produced from the corrected cpm:
   a. Count a standard to obtain the counter efficiency, and divide this value in cpm by the dpm value provided by the manufacturer for this standard.
   b. Calculate product formation:

$$\mu\text{mol methylfructoside} = \frac{\text{cpm}}{(\text{counter efficiency})(2.2 \times 10^6)(0.02)}$$

   where: 2.2×10^6 = dpm and 0.02 = the specific activity of the substrate in μCi/μmol sucrose.

5. Plot the velocity (μmol methylfructoside/min) *vs.* volume eluted (in mL). Assume that 1 drop = 0.03mL.

G. Determination of the Molecular Weight of Invertase

Molecular Sieve Chromatography

By molecular sieve chromatography, it is possible to separate molecules on the basis of molecular weight (MW). Agarose Biogel A-1.5M, a Biorad product, is made up of small porous beads of cross-linked dextran that swell in water. Very small molecules can freely enter all pores of the beads and are accessible to the entire volume. Very large molecules are not able to penetrate any of the pores and are eluted in the void volume, V_o. The elution volume V_e is inversely proportional to the molecular weight and can be determined for proteins of known molecular weights to establish a reference curve from which the molecular weight of an unknown can be estimated.

An Agarose Biogel column will be prepared, and its void volume will be determined using blue dextran. This resin can resolve molecules of molecular weights up to 1.5×10^6. A standard curve will be constructed by using the elution data obtained for the proteins listed below. Two proteins can be loaded on the column at the same time if they are of sufficiently different molecular weights so that one will have cleared the column before the second appears in the fraction tubes. From the V_e to V_0 ratio obtained for invertase, it will then be possible to estimate its molecular weight.

Protein or marker	*MW*	*Method of detection*
Glucose	180	Clinistix test
Cytochrome *c*	14,000	Color
Hemoglobin	64,000	Color
Alcohol dehydrogenase	150,000	A_{280}
Catalase	230,000	O_2 generated on addition of 3% H_2O_2
Ferritin	480,000	Color
Invertase[a]	?	Clinistix activity on addition of sucrose and acetate buffer
Blue dextran	*ca* 2,000,000	Color

[a]Use 0.5 mL of undiluted fraction IV.

Experimental Procedure

1. Allow the Biogel to swell for several days before pouring the column. (The beads used in this experiment are suspended in buffer when purchased and need no presoaking). Degas the suspension in a suction flask attached to an aspirator. Set up a column having a volume of about 25 mL.

2. Arrange for the buffer (0.05 M Tris, pH 7.5) to flow into the column from a reservoir (see Fig. 5.10) so as to keep the pressure constant.

3. *Calibration of the column*. Determine the elution volumes for blue dextran (MW = 2,000,000) and the internal volume of the columns by running together 0.5 mL of a 2 mg/mL solution of blue dextran and 0.5 mL of 40 mg/mL glucose (MW = 180). The appearance of the blue dextran is visualized, whereas glucose is detected by use of the Clinistix test. The suggested order of addition of the samples to the column is:

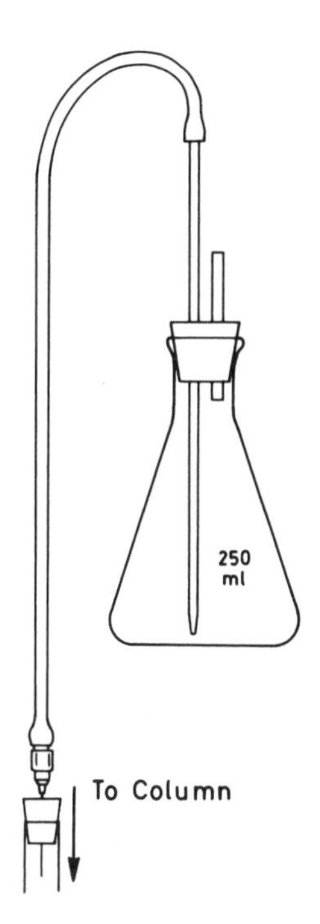

Fig. 5.10. Reservoir for column elution.

a. *Calibration:* blue dextran + glucose.

b. Cytochrome *c* + ferritin.

c. Hemoglobin + catalase.

d. Alcohol dehydrogenase + invertase.

4. The combined samples should be added in a total of 1.0 mL and have a concentration sufficient to visualize color or have an $A_{280\ nm}$ of 0.5 to 1.0 in the peak tube. Fractions of 1 mL will be collected using a Gilson fraction collector. Two proteins may be combined and run at the same time if they are of sufficiently different size so that the larger has cleared the column before the smaller begins to appear, as indicated above. Invertase appearance will be measured with the Clinistix test, and the appearance of the standards can be determined by absorbance at 280 nm, appearance of color, or O_2 evolution in the case of catalase.
5. Plot the relative elution volume, V_e/V_o *vs*. log MW for the five proteins used, and estimate the molecular weight of invertase from this curve (see Fig. 5.11 for a typical calibration curve).

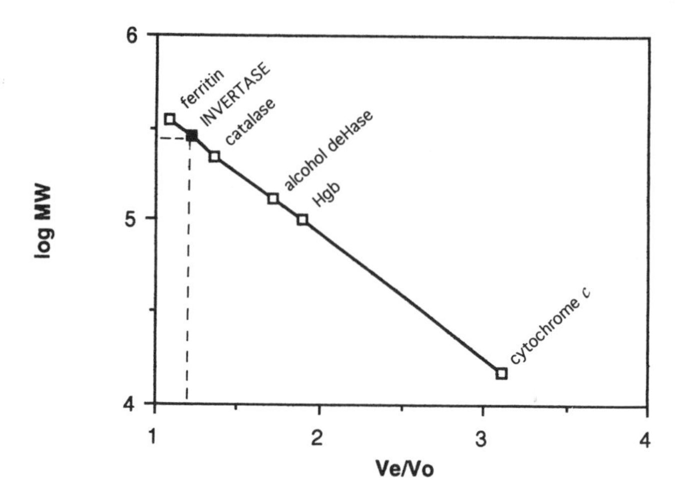

Fig. 5.11. Molecular weight calibration curve obtained with a 1.5 M gel filtration column. The relative elution volume V_e/V_o is determined for standards and unknown, and the molecular weight of the unknown is determined from the curve established with the standards.

Sodium Dodecyl Sulfate Polyacrylamide Gel Electrophoresis

The standard disc gel electrophoresis method has two disadvantages: (1) the procedure often gives rather diffuse bands, making resolution only fair; and (2) there is no simple way to correlate the relative mobility with physical properties such as molecular weight and charge on the molecule. These difficulties can be overcome by first denaturing the protein with sodium dodecyl sulfate (SDS), a detergent, and then performing the electrophoresis. This method will give sharp bands. The detergent forms a complex with the denatured protein, masking its intrinsic charge and giving it a charge roughly proportional to the number of detergent ions bound to the protein, which in turn is directly proportional to the size of the protein. Under these conditions, the relative mobility of a protein is proportional to the log of its molecular weight.

The approximate molecular weight of invertase measured by SDS electrophoresis will be determined by comparing its mobility with those of the following six proteins, which are purchased as a kit from Sigma Chemical Co. (St. Louis, MO).

Molecular Weight Standard Mixture[a]

Protein	*MW*
Carbonic anhydrase	29,000
Albumin, egg	45,000
Albumin, bovine plasma	66,000

Phosphorylase b	97,000
β-galactosidase	116,000
Myosin	205,000

[a]This solution already contains SDS, mercaptoethanol, bromophenol-blue, and glycerol in a sodium phosphate buffer and can be loaded directly on the column.

Experimental Procedure

1. Prepare four gel tubes per group as for disc gel electrophoresis according to the protocol shown on Table 5.4 for the SDS system.
2. After filling the tubes, layer two to three drops of butanol on top, and rinse the tops of the gels with SDS buffer (1:1) after they have polymerized. Store the gels until the next laboratory period at 4°C.
3. Transfer 65 μL of suspension buffer to each of two microfuge tubes and add 35μL of concentrated (10×) fraction IV (invertase). The suspension buffer contains the following:

0.1 M Na phosphate buffer, pH 7	10 μL
5% SDS	20 μL
β-Mercaptoethanol	5 μL
50% Glycerol	20 μL
0.1% Bromphenol-blue	10 μL

4. Set up two gel tubes, each with MW standard mix, and two with invertase.
5. After electrophoresis is complete and the gels have been stained, generate a standard curve by plotting the log MW against the relative mobility (R_m) values calculated for the proteins listed above.
6. Use SDS buffer (1:1) for both upper and lower reservoirs and run initially at 3 mAmp/tube for 30 min to permit proteins to enter gels. Then turn voltage up to achieve 7 mAmp/tube and stop the electrophoresis when the dye front bands appear to be about 1 cm from the bottom of the gel tubes.

7. Remove the gels, record the distances of the dye fronts and transfer to test tubes. Fix the protein bands by incubating the gels in 12.5% TCA for approximately 30 min, remove the TCA, and stain in 0.25% Coomassie blue stain for 30 min. Destain gels as needed to obtain a clear background with destaining solution (see Appendix I for composition of destaining solution).
8. Measure the bands and calculate the R_m values for the standards and for invertase.
9. Plot a standard curve as described in step 5 above.

H. Effect of pH on Invertase Activity

Some of the amino acid residues of proteins are ionizable, and their state of ionization is important for maintaining protein conformation, substrate binding, and/or catalysis. The rate of an enzyme-catalyzed reaction is therefore dependent on the pH of the solution. One example of this phenomenon is the case in which only the partially protonated form of an enzyme is active, represented by

$$EH_2^+ \underset{pK_{a1}}{\overset{H^+}{\rightleftharpoons}} \underset{\text{(active)}}{EH} \underset{pK_{a2}}{\overset{H^+}{\rightleftharpoons}} E^-$$

When the substrate can also be ionized and is primarily bound by the enzyme when in a particular ionized state, or when the enzyme is irreversibly inactivated at high or low pH, the system becomes more complex. Since the substrate and the products are not ionized in this study, a rate *vs.* pH profile will show how ionization of the protein influences enzymatic activity.

Invertase appears to have two groups of ionizable residues that influence the ability of the enzyme to hydrolyze sucrose. One, with a pK_a of about 7 is thought to represent the imidazole group of histidine; this branch of the pH activity curve appears not to be dependent on substrate concentration. The acid branch of the curve has a pK_a of about 3 and describes a group that is involved in the binding of substrate and demonstrates a "competitive-type" of inhibition (by H^+) when the enzyme activity is determined at the lower pH values at varying substrate concentrations. This effect will be demonstrated in the second part of the following study.

Experimental Procedure to Determine Effect of pH on Invertase Activity

1. Set up two series of 12 test tubes and deliver 0.2 mL of the appropriate buffers listed below to each. One set of tubes will receive invertase (see item 3 below), the other an equivalent amount of water. The tubes receiving no enzyme will be the blanks corresponding to each pH within the series.

Buffers pH	
2.5	0.2 M Succinate
3.0	0.2 M Succinate
*3.5	0.2 M Succinate
*3.5	0.2 M Acetate
4.0	0.2 M Acetate
4.5	0.2 M Acetate
5.0	0.2 M Acetate
5.5	0.2 M Acetate
*6.0	0.2 M Acetate
*6.0	0.2 M Phosphate
6.5	0.2 M Phosphate
7.0	0.2 M Phosphate

*Note the overlapping pH values. Plot data for both buffers and observe the effect exerted by different ions on enzyme activity.

2. Again include a water blank and a standard glucose tube in the experiment.
3. Add an appropriate amount of invertase to one half of the tubes so as to yield an absorbance of 0.6 under standard assay conditions.
4. Bring the volume in all the tubes to 0.8 mL with deionized water.
5. Initiate the reaction by adding 0.2 mL of substrate to all of the tubes.
6. Incubate for exactly 10 min, and proceed with the Nelson's assay. Read $A_{510 \text{ nm}}$ for each tube incubated with invertase after adjusting the spectrophotometer to 100% transmittance with the tube incubated without enzyme at the same pH.
7. Plot activity (μmol/min) as a function of pH. Include both points where ions differ but pH values overlap.

Effect of Substrate Concentration at Different pH Values on Enzyme Activity

This experiment will be carried out by the group. Each student will study one of the buffers, and all students will use the same enzyme preparation.

1. Using varying sucrose concentrations, determine the rate of catalysis at these pH values: 2.5, 3.5, 4.5, 5.5, and 6.5. Use 0.2 M succinate buffer only in this study to eliminate the effect of varying ions.
2. Use a protocol similar to the one on Table 5.6 and substitute the appropriate buffers for the 0.2 M acetate at pH 4.5 used in the standard assay.
3. Plot $1/v$ *vs.* $1/[S]$ for each pH studied. Make one graph for pHs 4.5, 5.5, and 6.5 and on a second graph plot pHs 2.5, 3.5, and 4.5. Analyze and interpret the results obtained.

I. Effect of Temperature on Enzyme Activity and Determination of the Activation Energy of a Reaction

Any chemical reaction of reactants going to products will eventually come to equilibrium. At equilibrium, there is no net conversion of reactants to products or products back to reactants; the proportion of products to reactants remains fixed. If the energy of the products is less than the energy of the reactants, products will predominate; if higher, there will be more reactants at equilibrium.

Enzymes act as catalysts. They greatly increase the rate of reaction, but they never shift the chemical equilibrium, they just bring the system to equilibrium more quickly.

The rate at which chemical reactions come to equilibrium varies widely, from microseconds to millions of years. There is always an energy barrier between reactants and products that must be crossed for the reaction to take place. This barrier is the activation energy, E_a. The lower the activation energy, the faster the rate of reaction. Catalysts lower the activation energy and increase the rate, without changing the relative energies of reactants and products. Figure 5.12 shows the change in energy between reactants and products.

In the invertase reaction, sucrose is converted to glucose and fructose. The products glucose and fructose have lower energy than sucrose, but ordinarily sucrose does not hydrolyze sponta-

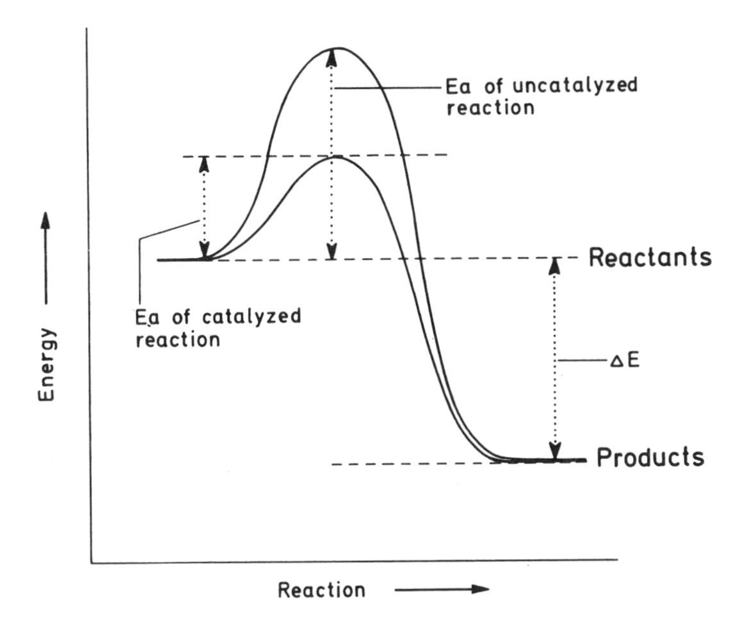

Fig. 5.12. Change in energy for enzyme catalyzed reaction.

neously, because the activation energy barrier is too high. Invertase vastly accelerates the reaction. Acid also accelerates the reaction, but not as extensively as invertase because it does not lower the activation energy as much. In this experiment, the activation energies of the invertase and acid-catalyzed reactions will be measured.

The activation energy (in calories/mol) can be calculated from the Arrhenius equation:

$$k = Ae^{-E_a/RT}$$

where: E_a is the activation energy, in cal/mol; k is the rate of the reaction, in μmol/min; R is the gas constant, 2.0 cal/deg mol; T is the absolute temperature, °C + 273°; and A is a constant.

The activation energy, E_a, can be determined by measuring the rate of the reaction at different temperatures and plotting log k *vs*. 1/T. Then the slope of this Arrhenius plot becomes

$$\log k \times \mathrm{T} = -\frac{\mathrm{E_a}}{2.3\mathrm{R}}$$

Another characteristic used in describing the effect of temperature on the rate of a reaction is Q_{10}, which is the factor by which the rate of reaction is increased for every 10°C increment in temperature. It can be calculated as the ratio of V_{max2}/V_{max1}

where V_{max2} is the velocity of the reaction at T_2 and V_{max1} is the rate at T_1 (which is 10°C lower).

The ratio Q_{10} can also be calculated by using the following equation:

$$E_a = \frac{2.3R\, T_2T_1 \log Q_{10}}{10}$$

Experimental Procedure

The rate of the enzyme and acid-catalyzed reactions at 10 different temperatures from 0° to 100°C will be determined. Use an ice-water bath for 0°, room temperature, a boiling-water bath for 100°, and water baths for 15°, 37°, 50°, 60°, 70°, 80°, and 90°C. For each temperature prepare two test tubes, one with enzyme and one without. The acetate buffer will be the source of acid for the "acid-catalyzed" tubes.

1. Set up the test tubes at room temperature as for standard assay conditions, with buffer and enzyme.
2. Dilute the enzyme with water and use enough to give an absorbance of 0.2 to 0.3 under standard assay conditions. Prepare a water blank, and read all the tubes against this tube.
3. Bring the volume to 0.8 mL with water. Start the reaction by adding 0.2 mL of 0.5 M sucrose at timed intervals. Allow 30 sec for equilibration of reactants to the bath temperature before starting the reaction.
4. Incubate the tubes for exactly 10 min at the appropriate temperatures.
5. Stop the reaction by adding 1 mL of Nelson's reagent, and proceed as in the regular invertase assay. Record the exact temperature of each water bath. A thermometer has been placed in each bath.
6. Plot a graph of the rate *vs.* temperature.
   a. Two curves will be obtained. They will show the value for v, (μmol/min) observed at the various temperatures for tubes (i) containing enzyme or (ii) undergoing the acid-catalyzed reaction only.
   b. When the acid catalysis product in each tube is subtracted from the value obtained for the reaction in the

presence of enzyme, the effect of temperature on enzyme activity is observed. The curve shows a typical increase in μmol/min formed as the temperature is raised until an optimum is obtained. This optimum is *ca* 50°C for invertase. This is followed by a fairly steep drop in activity at temperatures > 50°C due to protein denaturation.

7. Plot another graph showing the log of the rate (k) *vs*. 1/T (where T = the absolute temperature, 273° + °C) for both the acid- and enzyme-catalyzed reactions.
   a. Calculate the activation energy of the enzyme and of the acid-catalyzed reactions from the slope of the log rate *vs*. 1/T plots. Determine the slope only from the linear portion of each curve.
   b. Calculate Q_{10}, the effect of increasing the temperature by 10°C, on the rate of the reaction as described above.

CHAPTER 6

Nucleic Acids

Day 1

1. Isolation of leu-tRNA synthetase.
2. Role of leu-tRNA synthetase in protein synthesis.

Day 2

1. Charging of leu-tRNA and leu-tRNA synthetase kinetics.
2. Introduction to the use of radioisotopes.

Day 3

Isolation of DNA from calf thymus.

Day 4

1. Determination of T_m.
2. Acid hydrolysis and determination of % G + C.
3. Determination of reducing sugars in DNA preparation.
4. Diphenylamine and orcinol assays of DNA preparation.

Day 5

Isolation of plasmids from *E. coli* cells and digestion with restriction enzymes.

Day 6

Cast, load, and run agarose electrophoresis on digested plasmid DNA.

REFERENCES: Alexander, R.R., Calvo, J.M., and Freundlich, M. Mutants of *Salmonella typhimurium* with an altered leucyl-transfer ribonucleic acid synthetase. *J. Bacteriol.* 106:213–220 (1971).

Chapman, J.M. and Nyrey, G. *The Use of Radioactive Isotopes in the Life Sciences*. Allen and Unwin, London (1981).

Cooper, Ch. 3.

Lehninger, A.L. *Biochemistry,* 2nd ed. Worth Publishing Co., New York pp. 873-876 (1975).

Lehninger, A.L. *Principles of Biochemistry.* Worth Publishing Co., New York (1982).

Rawn, J.D. *Biochemistry.* Neil Patterson, Burlington, VT, Ch. 26 (1989).

Robyt and White, pp. 353–372.

Segel, pp. 354–366.

Soll, D. and Schimmel, P.R. Aminoacyl-tRNA Synthetases. *Methods Enzymol.* 15:489–536 (1976).

Stryer, pp. 649–685.

Watson, J.D., et al. *Molecular Biology of the Gene,* 4th ed. Benjamin Cummings Publishing Co., Menlo Park, CA, Ch. 9 (1987).

A. Introduction

Use of Radioisotopes in Biological Research

Rapid advances have occurred in biochemistry since compounds containing radioactive isotopes became available as metabolites that could be used to study biological processes. Because a labeled compound is not distinguishable from its normal form, the tracer can be followed *in vivo* as well as *in vitro,* and only a very small fraction of the total amount need be the radioactive form. The use of radioisotopes has facilitated the elucidation of biosynthetic as well as degradative pathways and often made possible the discovery of the steps involved in intermediary metabolism. The most commonly used radioisotopes in the biochemistry laboratory are the following β-particle-emitting radionuclides:

Isotope	*Half-life*
^{14}C	5760 years
^{3}H	12.3 years
^{32}P	14 days

Methods using tracer compounds are sensitive and do not alter the properties of the test compound. These are distinct advantages for conducting experiments in the biological sciences.

Safety Measures and Proper Handling of Radioisotopes

The hazards involved in using weak β-emitting radioisotopes are small, because very small quantities are needed to follow most biological reactions. Nevertheless, precautions when using these materials should be taken, and the work and storage areas as well as the equipment used should be monitored for possible spills. Contamination of ^{32}P or ^{35}S can be measured by using a portable Geiger-Müller counter with a thin mica film attached to a wand, which can be used to scan the surface where possible spillage may have occurred. For ^{14}C and ^{3}H monitoring wipe tests are performed. The test surface is wiped with a swatch of moistened filter paper, which is placed in a scintillation vial to which an appropriate solution is added. The vial is placed in a scintillation counter and counted.

In liquid scintillation counting for assaying radioactivity, the sample is placed in a solvent containing complex aromatic compounds called *primary* and *secondary fluors.* The particles given off by the radioactive sample transfer some of their energy to the solvent molecules, which in turn excite the primary fluor. When this fluor returns to its ground state, a photon is emitted that excites the secondary fluor. The light impulse from the secondary fluor is of a longer wavelength and is detectable by the phototube in the scintillation counter. This fluorescence is converted to electrical current and recorded as counts per minute. The instrument used to detect the radioactive decay in the samples is a scintillation counter, of which there are a number of models. The counter consists of an automatic sample changer that holds 100 or more vials. The activity in the vials will be counted for some preselected time period. Several channels are usually available, but the instrument commonly comes with plug-in modules that are preset for a particular radionuclide.

The data from a Beckman-type scintillation counter are generally recorded on a tape that typically records the following information:

1. *Channel no. 1, 2,* or *3*—selected for radioisotope to be counted, *i.e.*, ^{3}H, ^{14}C, ^{32}P.

2. *Vial no.*—position of sample on the chain (from 1 to 100 or more, depending on sample capacity of the instrument).
3. *Counting time*—period during which the sample was counted. This can be preset (*i.e.*, for 10 min) so that counts are variable, or the counts may be preset (*e.g.*, 10,000), in which case the time it took for that number of counts to accumulate will appear as the variable.
4. *Counts per minute*—calculated on the basis of total counts recorded during the counting period.

It is usual practice to include a blank vial to record background counts, and a standard. The standard is a commercial preparation having a known amount of radioactive material.

The units of measurement of radioactivity recorded by the scintillation counter are counts per minute (cpm), which represent the number of disintegrations that are detected by the instrument but not the actual number of β-particles emitted or disintegrations per minute (dpm). This value is dependent on the efficiency of the counter, which is determined as follows: efficiency = cpm/dpm.

A typical commercial ^{14}C standard is purchased with radioactive carbon emitting 41,500 dpm. If the counter records 36,000 cpm, then the efficiency would be *ca* 87% for this instrument.

Another unit of measurement associated with the use of radionuclides is the *specific activity* of a compound, which is the amount of radioactivity per amount of compound. In biological experiments, specific activity is generally given in μCi/μmol (1 μCi = 2.2×10^6 dpm). One μmol of radioactive compound can go a long way since the counter is quite sensitive, and, furthermore, it is possible to obtain a larger number of counts by counting for a longer period of time. It is usual to make a solution containing mostly unlabeled (or "cold") compound, and only some small fraction of the final amount is represented by the radioactive form.

When working with radioactive substances, a number of **safety measures** must be observed in the laboratory.

1. Cover the work area with absorbent paper with plastic backing.
2. Keep all materials that are to be used in the experiments on this paper.

3. Use latex gloves throughout the experiment.
4. Use automatic pipettors with disposable tips to pipette the radioactive material.
5. Carry out the reactions in disposable test tubes.
6. When finished, empty test tubes, gloves, and contaminated glassware should be wrapped in the absorbent paper, taped into a tight package, and disposed of in a container provided for radioactive waste.
7. Liquid radioactive waste should be discarded in a bottle set aside for that purpose.

B. Role of an Amino Acyl-Transfer RNA Synthetase in Protein Synthesis

An amino acyl-transfer RNA (AA-tRNA) synthetase attaches its cognate amino acid to the 3′ terminal end of its specific RNA. There is at least one specific enzyme and tRNA for each amino acid. The esterification of a given amino acid to its corresponding tRNA is thought to proceed in two steps with the involvement of three substrates. The overall reaction is described by the following equations:

$$\text{AA} + \text{ATP} + \text{E} \rightleftharpoons \text{E·AA-AMP} + \text{PP}_i \qquad (1)$$

$$\text{E·AA–AMP} + \text{tRNA} \rightleftharpoons \text{AA-tRNA} + \text{AMP} + \text{E} \qquad (2)$$

Equation 1 describes an "activation" step that results in the formation of an enzyme-bound aminoacyl–adenylate complex and pyrophosphate. In the second stage of the reaction, the amino acid is transferred to tRNA with the release of AMP and free enzyme. The formation of AA-tRNA is determined by assaying for the esterification of a radioactive amino acid to tRNA. In this series of experiments, partially purified leu-tRNA synthetase will be isolated, and the formation of leu-tRNA^{leu} will be assayed by measuring the amount of [^{14}C]leucine complexed with the tRNA.

The assay used to follow the leu-tRNA^{leu} synthesis depends on the formation of a precipitate in the presence of trichloroacetic acid (TCA). The precipitate is retained by Millipore filter pads when the reaction mixture is filtered. Thus the amount of ^{14}C deposited on the Millipore filters is indicative of the amount of leu-tRNA^{leu} formed. Neither the leucine nor the tRNA is retained on the pad during filtration. The Millipore filters are then allowed to dry, scintillation fluid is added, and the vials are counted in a scintillation counter.

Isolation of Leu-tRNA Synthetase

Experimental Procedure

1. Growth and harvest of *E. coli* K-12 cells.
   a. (i) Prepare 1 L of minimal salts medium as follows and autoclave in a 2-L Erlenmeyer flask; K_2HPO_4, 10.5 g; KH_2PO_4, 4.5 g; $(NH_4)_2SO_4$, 1.0 g; sodium citrate (dihydrate), 0.97 g; $MgSO_4$ (anhydrous), 0.05 g; and deionized H_2O, 950 mL.
      (ii) Prepare the following solution in a 125-mL Erlenmeyer flask and autoclave separately; glucose, 2 g; and deionized H_2O, 50 mL.
   b. After autoclaving, add the glucose solution to the salts medium and inoculate with 20 mL of an overnight culture that will have been grown in nutrient broth.
   c. Incubate at 37°C with shaking, and allow culture to grow until an $A_{540\,nm}$ value of 0.7 to 0.8 is reached. This should yield 3 to 4 g of cells nearing the end of their logarithmic growth phase.
   d. Transfer the culture into wide-mouthed 250 mL plastic bottles and centrifuge at 8000 rpm for 20 min in a GSA rotor in the refrigerated Sorvall centrifuge.
   e. Resuspend the pellets in 20 mL of 0.05 M Tris buffer at pH 7.5 containing 0.1 mM dithiothreitol (DTT), transfer the suspension to a 50-mL plastic centrifuge tube, and centrifuge at 12,000 rpm for 15 min in a SS-34 rotor. Discard the supernatant.
   f. Obtain the wet weight of the cells. Cover the tubes with parafilm and freeze. Proceed with step 2 or step 3.
2. Preparation of "activating enzyme" and $(NH_4)_2SO_4$ fractionation.

Cell Breakage—Grinding With Alumina

a. Transfer the weighed cell paste into a chilled mortar, and grind with twice its weight of levigated alumina. Add 0.5 to 1.0 mL of 0.05 M Tris buffer, pH 7.5, containing DTT as needed to obtain a smooth paste.

b. When the mixture has become pasty and viscous and a "popping" sound is heard, take up the extract in about 5 mL of the same buffer per gram of cells. Use another 5 mL of buffer to rinse the mortar and pestle.

c. Centrifuge the suspension at 15,000 rpm for 30 min in 15-mL plastic centrifuge tubes, and save the supernatant. (This is a crude enzyme extract.)

d. Measure the volume of the crude enzyme preparation, and transfer it to a small beaker. Place the beaker in an ice bath on a magnetic stirrer, and slowly add enough solid $(NH_4)_2SO_4$ to achieve a concentration of 75%. (Determine the appropriate amount of $(NH_4)_2SO_4$ to use from Table 6.1, which shows the weight of salt required; *e.g.*, for the 75% concentration, 47.6 g should be added to 100 mL of solution).

e. Centrifuge the ammonium sulfate precipitate at 15,000 rpm for 20 min, discard the supernatant, and resuspend the pellet in 3 mL of 0.05 M Tris buffer, pH 7.5, containing DTT.

3. If a sonifier is available the following method can be used.

Cell Breakage—Sonication

a. Transfer the weighed cell paste to a 30-mL plastic beaker, and add 10 mL of 0.05 M Tris buffer, pH 7.5, containing DTT. Stir gently with a rubber policeman until the cells are evenly suspended. Keep the suspension cold in an ice bucket.

b. Adjust the Branson sonifier to a setting of 3 or 4, and turn the tuning dial until a steady high-pitch sound is obtained (generally occurs at *ca* 3 amp).

   Avoid bubbling air through the solution, and avoid contact between the probe of the sonifier and the walls of the tube.

c. Use 10 mL of water in a small beaker to set the power.

d. Sonicate the cell suspension for a total of 2 min, giving 15 sec bursts. Allow the cell suspension to cool in an ice bucket between treatments. The fine-tuning may need adjustment as cell breakage proceeds.

e. When finished, rinse the probe by allowing the instrument to run for 30 sec while the probe is immersed in a beaker of water.

TABLE 6.1. Fractionation With Solid Ammonium Sulfate[a]

Initial concentration of ammonium sulfate (% saturation at 0°C)	Solid ammonium sulfate to add to 100 mL of solution																
	20	25	30	35	40	45	50	55	60	65	70	75	80	85	90	95	100
0	10.6[b]	13.4	16.4	19.4	22.6	25.8	29.1	32.6	36.1	39.8	43.6	47.6	51.6	55.9	60.3	65.0	69.7
5	7.9	10.8	13.7	16.6	19.7	22.9	26.2	29.6	33.1	36.8	40.5	44.4	48.4	52.6	57.0	61.5	66.2
10	5.3	8.1	10.9	13.9	16.9	20.0	23.3	26.6	30.1	33.7	37.4	41.2	45.2	49.3	53.6	58.1	62.7
15	2.6	5.4	8.2	11.1	14.1	17.2	20.4	23.7	27.1	30.6	34.3	38.1	42.0	46.0	50.3	54.7	59.2
20	0	2.7	5.5	8.3	11.3	14.3	17.5	20.7	24.1	27.6	31.2	34.9	38.7	42.7	46.9	51.2	55.7
25		0	2.7	5.6	8.4	11.5	14.6	17.9	21.1	24.5	28.0	31.7	35.5	39.5	43.6	47.8	52.2
30			0	2.8	5.6	8.6	11.7	14.8	18.1	21.4	24.9	28.5	32.3	36.2	40.2	44.5	48.8
35				0	2.8	5.7	8.7	11.8	15.1	18.4	21.8	25.4	29.1	32.9	36.9	41.0	45.3
40					0	2.9	5.8	8.9	12.0	15.3	18.7	22.2	25.8	29.6	33.5	37.6	41.8
45						0	2.9	5.9	9.0	12.3	15.6	19.0	22.6	26.3	30.2	34.2	38.3
50							0	3.0	6.0	9.2	12.5	15.9	19.4	23.0	26.8	30.8	34.8
55								0	3.0	6.1	9.3	12.7	16.1	19.7	23.5	27.3	31.3
60									0	3.1	6.2	9.5	12.9	16.4	20.1	23.9	27.9
65										0	3.1	6.3	9.7	13.2	16.8	20.5	24.4
70											0	3.2	6.5	9.9	13.4	17.1	20.9
75												0	3.2	6.6	10.1	13.7	17.4
80													0	3.3	6.7	10.3	13.9
85														0	3.4	6.8	10.5
90															0	3.4	7.0
95																0	3.5
100																	0

[a]Reprinted from Segel, p. 400.

[b]Values are final concentrations of ammonium sulfate (% saturation at 0°C).

f. Centrifuge at 15,000 rpm for 30 min in 15-mL plastic centrifuge tubes, and save the supernatant. (This is a crude enzyme extract.)

g. Measure the volume of the crude enzyme preparation, and transfer it to a small beaker. Place the beaker in an ice bath on a magnetic stirrer and slowly add enough solid $(NH_4)_2SO_4$ to achieve a concentration of 75% (see Table 6.1). Add the salt slowly, allowing about 30 min for this procedure, and make sure that all the salt is dissolved.

h. Centrifuge at 15,000 rpm for 20 min, discard the supernatant, and resuspend the pellet in 3 mL of 0.05 M Tris buffer, pH 7.5, containing DTT.

4. Separation on a Sephadex column.

a. Pour a G-25 (coarse) Sephadex column and calibrate it using blue dextran to determine the void volume, V_o. Place glass wool in the bottom of a glass column, and add some Tris buffer. Using a glass rod, gently squeeze out any air bubbles and allow some of the buffer to run out. Add a small amount of sand to achieve a flat surface above the glass wool. With the column about one-third full of buffer, add a 10-mL volume of Sephadex. Be certain that everything is at the same temperature! Once the gel has settled and equilibrated with buffer (about 30 mL), the V_o can be determined by applying 1.5 mL of blue dextran. Collect the eluate in a graduated cylinder, and note when the first blue color appears; this is V_o. Continue to collect eluate until all the blue dextran has eluted, and record the elution volume (V_e).

b. Apply 1.5 mL of the resuspended $(NH_4)_2SO_4$ fraction, and begin collecting 1.0-mL fractions.

c. Take $A_{280\,nm}$ readings, and pool the tubes containing the protein. The volume of the pool should be equal to the volume of blue dextran collected during the calibration. This procedure separates the proteins from the $(NH_4)_2SO_4$ and any endogenous leucine that will be retained on the column. The leucine and the salt must be removed from the enzyme preparation to perform the activity assays and the protein determinations.

d. Freeze this crude "activating enzyme" preparation in 1.0-mL aliquots.

Charging of Leu-tRNA by Leu-tRNA Synthetase

The following experiments are designed to demonstrate one of the important steps in protein synthesis. The charging of an amino acid by its AA-tRNA synthetase will be accomplished in the following two experiments by studying the kinetic properties of the synthetase.

The conditions required to determine the K_m and V_{max} values for an enzyme are studied in some detail in the Enzymology module. Here we have set up the protocols that will enable you to carry out the experiments to obtain the data from which to calculate these parameters for leu-tRNA synthetase.

Experimental Procedure

Enzyme Concentration Curve for Leu-tRNA Synthetase

1. Determination of protein content of the "activating enzyme" preparation.
   a. Using the dye-binding assay, set up a series of six tubes containing from 0 to 20 μg of BSA to establish a standard curve (see Chapter 3 for procedure).
   b. Assay the enzyme as follows: extract diluted 1:9, 0.02, 0.05, and 0.10 mL; and extract diluted 1:99, 0.02, 0.05, and 0.10 mL.
   c. Calculate the average mg/mL for the extract.
2. Carry out the first experiment according to the protocol given in Table 6.2.
   a. Calculate the correct amount of protein to use as "activating enzyme." Make the appropriate dilution in Tris buffer containing DTT.
   b. Start the reaction by addition of the tRNA, and time for *exactly* 10 min. The reaction is stopped by adding 2.0 mL of 8% TCA. Incubate for 20 min in an ice bath.
   c. As is evident in the protocols in Tables 6.2 and 6.3, the precipitate is allowed to form for a period of 20 min following addition of TCA. The leu-tRNA^{leu} is then collected by passing the solution through a Gelman glass fiber filtering apparatus containing a filter 0.45 μm in size. The equipment used is shown in Figure 6.1, in which the proper position of the filter pad is illustrated. The pads to be used are presoaked in 5% TCA and inserted between the funnel and the glass stem as

TABLE 6.2. Enzyme Concentration Curve for Leu-tRNA Synthetase

Additions (mL)	1 (Blank)	2	3	4	5	6	7	8	9	10 (Controls)
"Activating enzyme" (μg protein)	–	5	10	20	40	60	80	100	–	100
Charging "mix"[a]	–	0.2 ⟶								
[^{14}C]leucine (10 μCi/μmol, 0.5 μCi/mL)	–	0.2 ⟶								
H_2O	1.0 ⟵ Bring to a final volume of 0.8 mL. ⟶									
Bulk tRNA (1.2 mg/mL)	–	0.2 ⟶								–[b]
	Incubate for exactly 10 min at 37°C.									
8% TCA	2.0 ⟶									
	Incubate in an ice bath for 20 min. Pass through a Millipore filter, and wash pads with 5% TCA.									
cpm/10 min incubation										
Corrected cpm/min incubation										
μmol Leu-tRNA/min										

[a]Charging "mix" contains in 40 mL: 10 mL of 1.0 M Tris, pH 7.4; 10 mL of 0.2 M $MgCl_2$ + 12 mg DTT; 10 mL 0.01 M EDTA; and 10 mL 0.05 M ATP, pH 7.0.
[b]Add an additional 0.2 mL of water to this tube for a final volume of 1.0 mL.

shown; the suction flask is connected to an aspirator to speed up the filtration. After the solution is passed through, the funnel is rinsed three times with 3-mL aliquots of 5% TCA. The funnel is removed with the suction left on, and the pad still resting on the scintered-glass stem is rinsed thoroughly using a wash bottle containing 5% TCA.

Then the filter pad is transferred to a scintillation vial using tweezers and allowed to dry at 110°C for 10 min or until the pads are dry. The vials are cooled, and 8 mL of scintillation fluid is added. Then the caps are fastened on tightly, and the vials are placed on the chain in the

TABLE 6.3. The Effect of Substrate Concentration on Enzyme Velocity

Additions (mL)	Tube no. 1	2	3	4	5	6	7	8	9	10 (Controls)
tRNA-bulk (1.2 mg/mL)	0.01	0.02	0.04	0.08	0.10	0.15	0.20	0.25	—	0.25
H_2O	0.49	0.48	0.46	0.42	0.40	0.35	0.30	0.25	0.50	0.35
[^{14}C]leucine (10 μCi/μmol, 0.5 μCi/mL)	0.2	→	→	→	→	→	→	→	→	→
Charging mix	0.2	→	→	→	→	→	→	→	→	→
Enzyme[a]	0.1	→	→	→	→	→	→	→	→	—
	Incubate for exactly 10 min at 37°C.									
8% TCA	2.0	→	→	→	→	→	→	→	→	→
	Incubate for 20 min in an ice bath. Pass through a Millipore filter, and wash pads with 5% TCA.									
Bulk-tRNA (mg/aliquot)										
[S][b]										
1/[S]										
cpm/10 min incubation										
Corrected cpm/min incubation										
μmol leu-tRNA/min (v)										
1/v										

[a]Add an amount of enzyme that is nonlimiting (about 2/3 of the way up the linear portion of the concentration curve).

[b]Assume an average MW of 30,000 for Leu-tRNA. Convert mg tRNA/aliquot to μmol/mL (remember that the assay volume is 1 mL). Calculate the concentration (in mM) of leu-tRNA, [S], in each tube assuming that 10% of the bulk tRNA represents leu-tRNA.

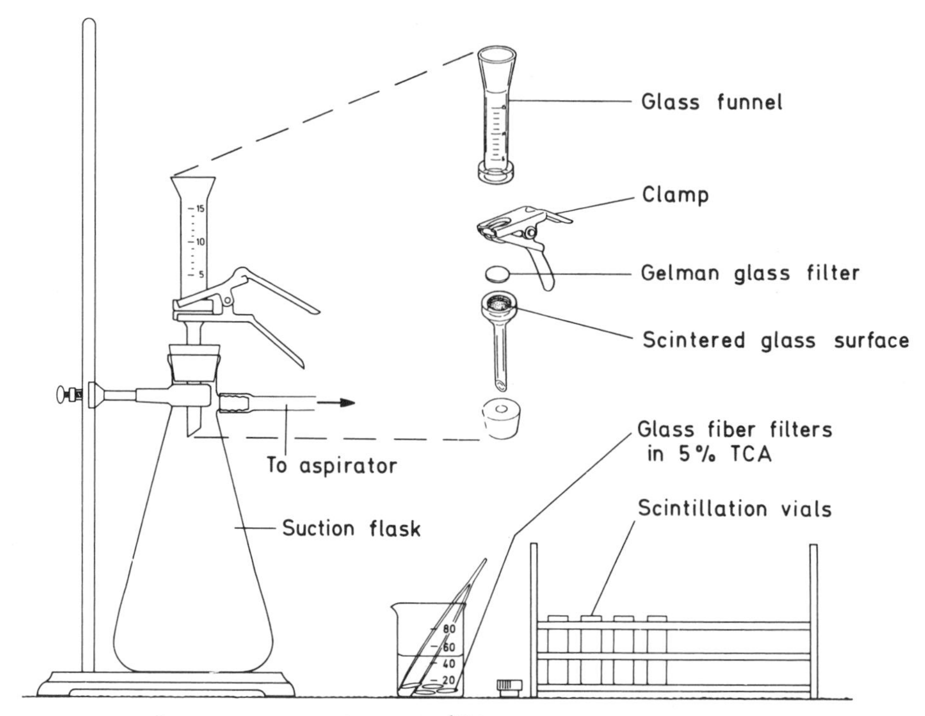

Fig. 6.1. Collection of leu-tRNA^{leu}following incubation with activating enzyme. Filtration apparatus for collecting labeled leu-tRNA^{leu}. A plastic-backed protective pad is used to cover the bench top to collect any radioactive spills, and the filter pads are transferred to scintillation vials arranged in order in the rack; the caps are numbered but are only placed onto their respective vials after drying as they do not tolerate the heat to which the vials are exposed during the drying period.

counter. Each vial is counted for 10 min. The counter records the data on a tape, as described earlier in this chapter.

3. Plot an enzyme-concentration curve.
   a. Correct the cpm/min (counts per min/min of enzyme incubation) obtained by subtracting whichever control tube gives a background value that is higher (tRNA [tube 9] or protein [tube 10]).
   b. Calculate the velocity in "µmol leu-tRNA/min" from the cpm/min incubation data as follows:

$$\underset{(\mu\text{mol leu tRNA per min})}{\text{velocity}} = \frac{\text{cpm/min incubation}}{(\text{counter efficiency})\underset{\text{dpm}}{(2.2 \times 10^6)}\underset{\text{sp. act. of leucine}}{(10\mu\text{Ci}/\mu\text{mol})}}$$

   c. Graph "µmol leu-tRNA/min" *vs.* µg protein.

Michaelis-Menten Kinetics for Leu-tRNA Synthetase

1. Determination of K_m and V_{max} values for leu-tRNA.
   a. Carry out the experiment according to the procedure outlined in the protocol given in Table 6.3 (note that 25 to 50 μg of protein will generally fall about two-thirds of the way up the linear portion of the curve).
   b. The reaction is started by the addition of enzyme and stopped by the addition of 8% TCA.
2. Graph a Michaelis-Menten curve showing velocity as a function of tRNA concentration and calculate values for V_{max} and K_m from a Lineweaver-Burk plot.
3. Use the values of V_{max} and protein concentration to calculate the specific activity of the enzyme.

C. Isolation and Physical Properties of DNA

REFERENCES: Lehninger, Ch. 31.

Marmur, J., Schildkraut, C.L., and Doty, P. Biological and physical aspects of reversible denaturation of deoxyribonucleic acids, in *The Molecular Basis of Neoplasia.* University of Texas Press, Austin, pp. 11–43 (1961).

Kirk, J.T.O. Determination of the base composition of deoxyribonucleic acid by measurement of the adenine/guanine ratio. *Biochem. J.* 105:673–677 (1967).

Bruening, G., Criddle, R., Preiss, J., and Rudert, F. *Biochemical Experiments.* Wiley-Interscience, New York, pp. 245–261. (1970).

Kornberg, A. *DNA Replication.* W.H. Freeman, New York (1980).

Stryer, Ch. 27.

Zubay, G. *Biochemistry,* 2nd ed. Macmillan, New York, pp. 236–242 (1988).

The molecular weight of DNA molecules of some cells probably exceeds 10^9g/mol, making them the largest linear covalent structures in nature. Seldom are they isolated intact, however, because of the action of the deoxyribonucleases present during the isolation procedures and because such large molecules are highly sensitive to shear forces. Merely pipetting DNA may result in breakage. Care should be taken to keep both of these factors to a minimum during isolation of the DNA. Molecules approaching 10^7g/mole can be obtained by conventional purification techniques.

A distinctive property of DNA is its behavior upon "denaturation." The native form of cellular DNA is a helical, double-stranded structure. When the native DNA is disrupted, the molecule loses its highly ordered structure, and single strands as well as random coils result. The disruption is accompanied by significant changes in optical properties and in molecular dimensions. These changes can be followed by absorbance measurements at 260 nm or by viscometric procedures.

The absorbing chromophores in the UV region are the purine and pyrimidine bases. Hypochromicity results from the stacking interactions between these bases, and when DNA is denatured the bases become "unstacked" and the UV absorbance increases, approaching but never equaling the absorbance of a mixture of deoxynucleotides of the same composition as that found in the native DNA. This increase in absorbance is known as a *hyperchromic shift.*

The T_m is the melting temperature of double stranded DNA, that is, the point at which one-half of the DNA molecule is double stranded while the other half of the unwinding molecule is single stranded. T_m is at the midpoint of the hyperchromic shift. It is dependent on the G-C content of the DNA and on the ionic strength of the solution. The G-C content is characteristic of a given species and a plot of T_m *vs.* mol % G+C shows a linear relationship (Fig. 6.2A, B). In the following experiments, DNA will be isolated from calf thymus tissue and T_m determined by following $A_{260\ nm}$ at increasing temperatures.

Isolation of DNA From Calf Thymus

The DNA will be extracted from thymus gland tissue. Caution must be exercised to prevent denaturation during the isolation. For this reason, all glassware that will come in contact with the DNA should be rinsed with 1 mM EDTA, and the preparation should be kept cold (use an ice bucket unless otherwise indicated). The use of a chelating agent, such as EDTA, serves to remove Mg^{++}, which is required for the action of DNase. In the absence of the required ion and in the cold, this degradative enzyme becomes nonfunctional. It is also important to handle the DNA gently to avoid mechanical shearing.

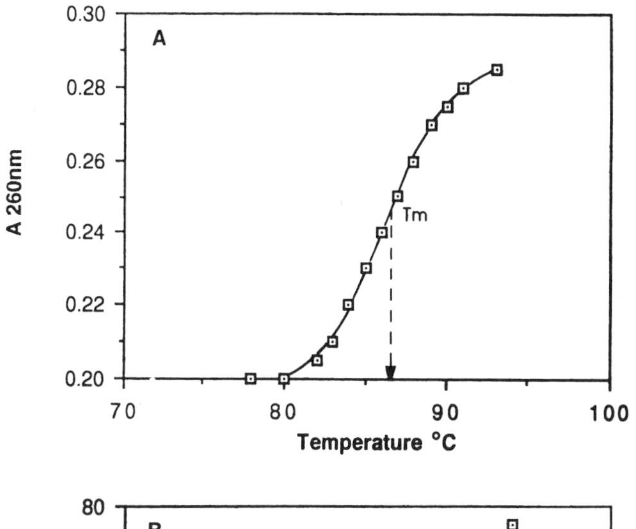

Fig. 6.2A. Typical melting curve for calf thymus DNA. The T_m will vary with the species of DNA examined.

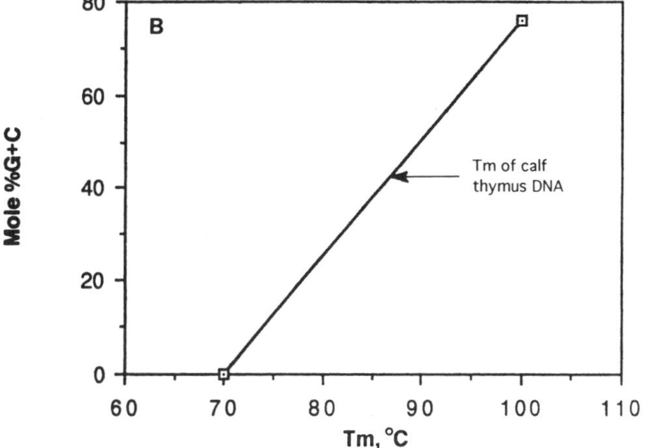

Fig. 6.2B. Relationship between % G+C content of DNA and T_m based on T_m determinations for DNA samples from different sources. (Adapted from Lehninger, p. 815 [1975].)

Experimental Procedure

1. Weigh 15 g of calf thymus (avoid the white fatty tissue), and chop it into small pieces.
2. Place the sample in a chilled Waring blender. Add 50 mL of citrate saline buffer (*CS buffer*: 9 mg/mL NaCl in 0.01 M trisodium citrate), and homogenize for 5 min at high speed at 4°C.
3. Pour the homogenate into 50-mL plastic centrifuge tubes, and centrifuge at 6500 rpm for 15 min.
4. Remove and discard the top lipid layer with a plastic spatula. Pour off and discard the supernatant, remove

pellets, and return to the blender. Add 50 mL of CS buffer, and blend at high speed for 3 min.

5. Resuspend the combined pellets in 90 mL of 0.15 M Na citrate buffer, pH 7.0.
6. Place the suspension in a beaker on a magnetic stirring apparatus and **SLOWLY** add 8 mL of 20% SDS (sodium dodecyl sulfate) solution dropwise with a Pasteur pipette. (The addition should take about 15 min.) Stir the suspension for an additional 5 min. The suspension will become increasingly viscous.
7. Place the suspension in a 55°C water bath and incubate for 15 min, stirring occasionally with a glass rod.
8. Add 8 g of NaCl, keeping the solution at 55°C. Continue to stir for 10 min or until the salt is dissolved.
9. Cool to room temperature in an ice bath, and pour the sample into a 250-mL separatory funnel.
10. Add 100 mL of a chloroform:isoamyl alcohol mixture (24:1), and shake vigorously for 10 min.
11. Pour both layers into a plastic centrifuge bottle, and centrifuge at 10,000 rpm for 15 min. Use a GSA rotor.
12. Carefully remove the top layer to a graduated cylinder, measure the volume of the extract, and transfer it to a 400-mL beaker. Discard the white interface and bottom layers.
13. Slowly add 2 volumes of 95% EtOH while gently stirring with a glass rod.
14. Wind out the gelatinous precipitate onto a glass rod.
15. Add another 20 mL of EtOH and collect any additional precipitate that may wind onto the rod.
16. Gently press out the solvent by turning the rod against the side of the beaker.
17. Rinse the wound DNA with 95% EtOH and then with acetone (AR grade) until the washings are no longer turbid.
18. Remove the DNA from the rod, and dry it in a small beaker in a desiccator overnight.* On the following day, weigh the

*Since DNA is a very large molecule, it goes into solution with difficulty. It is therefore advisable to allow 24 hr for a sample to go into solution. To make a *DNA stock* solution, weigh out 300 mg of DNA, crush the sample gently, and transfer the powder to a 50-mL beaker. Add 20 mL of 0.1 M Na phosphate + 1 mM EDTA, pH 7.0, buffer and insert a magnetic stirring bar. Place the beaker on a stirrer in a refrigerator or cold room.

sample and record the dry weight. Calculate the percent crude recovery on a weight basis. After the diphenylamine assay has been performed, calculate the net yield and purity of the product.

Determination of T_m for DNA

The T_m is determined from the midpoint of a melting curve, which is obtained by heating the DNA at various temperatures and observing the increase in absorbance at $A_{260\ nm}$ that results.

Experimental Procedure

1. Centrifuge the stock solution of DNA at 12,000 rpm for 10 min, and discard the pellet. Using the Warburg-Christian method for protein determination, take 260/280 absorbance readings on the supernatant (a dilution of about 1:50 should be within range to read the absorbance) to determine the approximate concentration of DNA in the stock solution. In making any dilution of the stock solution, be sure to mix gently but thoroughly. From these data, calculate also the amount of protein present in the DNA isolated.
2. Dilute an aliquot of the DNA stock solution to obtain a concentration having an $A_{260\ nm}$ of between 0.1 and 0.3 in 0.1 M Na phosphate buffer, pH 7.0. Dilute a solution of standard DNA to the same concentration.
3. Transfer both DNA solutions to quartz cuvettes. Use the buffer as a blank.
4. Place the cuvettes in the carrier of a spectrophotometer, which has a water-jacketed cuvette holder.
5. Set the instrument to zero at a wavelength of 260 nm, and read the absorbance of the samples at room temperature. As the temperature is raised, the cuvettes must be checked for the presence of air bubbles; these must be eliminated prior to making measurements.
6. Set the thermostat of the water bath to 75°C, and allow 5 min for the temperature to equilibrate after the pilot light (which indicates that the bath is heating) shuts itself off. Set the instrument to zero against the buffer, and take readings on the samples.
7. Increase the thermostat setting to 80°C, and proceed as above. Repeat this procedure for the 83°, 86°, 89°, 92°,

95°, and 100°C settings on the thermostat. Reset the instrument to zero against the buffer each time, and determine the *actual* water temperature by reading the thermometer inserted in the water bath alongside the thermostat.

8. Since there is some heat loss during the transfer of the water from the bath to the jacket surrounding the cuvettes, it is necessary to correct the water bath temperature measurements to the cuvette housing temperatures. For an *LKB ultraspec II* spectrophotometer, corrections for these temperature differences can be made by using a temperature conversion graph (Fig. 6.3) generated from the following data:

Water bath temperature (°C)	*Cuvette temperature (°C)*
76.5	70.5
81.0	75.0
86.0	78.5
91.0	82.5
96.0	89.5
100.0	94.0

9. Plot temperature *vs.* absorbance to determine the T_m of the DNA and of the standard. Use the following equation to determine the mole percentage G+C (Marmur, J., and

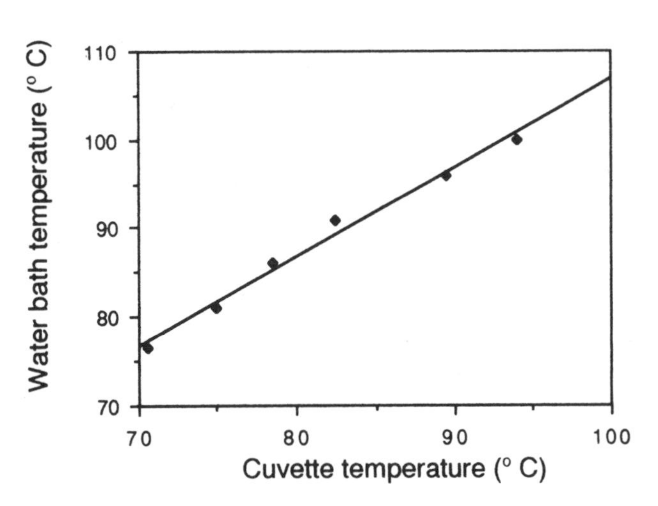

Fig. 6.3. Temperature correction curve for solutions in a spectrophotometer with circulating water-jacket adapters for the cuvettes.

Doty, P. Determination of the base composition of DNA from its thermal denaturation temperature. *J. Mol. Biol* 4:109–118 [1962]):

$$T_m = 69.3 + 0.41\,(\%\,G{+}C)$$

Determination of Base Composition and Polysaccharide Content of the Calf Thymus DNA

When DNA is subjected to mild acid hydrolysis, the sugar–purine bonds are cleaved, leading to the formation of apurinic acids and the release of adenine and guanine. This procedure will be employed to analyze the DNA for contamination with polysaccharides and to determine the G+C ratio of calf thymus DNA. The same sample of hydrolyzed DNA can be used in both studies.

Experimental Procedure

Acid Hydrolysis and Determination of Polysaccharide Content

1. Dilute an aliquot of stock DNA (15 mg/mL) with water to a concentration of 3 mg/mL. Prepare 10 mL of this solution.
2. To 3 mL (*measure carefully*) of the DNA solution, add 0.33 N HCl dropwise until the pH is between 2 and 3 (pH paper may be used for this test).
3. Place a marble on top of the tube to avoid evaporation, and heat the acidified DNA for 40 min in a boiling water bath.
4. Add 3.6 M KOH until the pH is about 7.0. *Measure and record the new volume.* Note the resultant change in the concentration of the DNA.
5. Follow the protocol on Table 6.4 to assay for reducing sugars, including 0.20-, 0.50-, and 1.0-mL samples of hydrolyzed DNA and untreated DNA. See Table 5.1 for the Nelson's reducing sugar assay.
6. Calculate the percent reducing sugar in the hydrolyzed DNA.
7. To the remaining DNA solution, continue adding 3.6 M KOH until the pH reaches 11 to 12 (test again with pH paper).

TABLE 6.4. Polysaccharide Content of DNA

Additions (mL)	Tube no. 1	2	3	4	5	6	7	8 (Blank)
Hydrolyzed DNA[a]	0.2	0.5	1.0	–	–	–	–	–
Untreated DNA (3mg/mL)	–	–	–	0.2	0.5	1.0	–	–
Glucose (4 mM)	–	–	–	–	–	–	0.2	–
H_2O	0.8	0.5	–	0.8	0.5	–	0.8	1.0
Nelson's reagent	1.0 →							
	Mix. Boil all the tubes for 20 min. Cool to room temperature.							
Arsenomolybdate reagent	1.0 →							
	Mix. Incubate for 5 min at room temperature.							
H_2O	7.0 →							
$\%T_{510\ nm}$								
$A_{510\ nm}$								
μmol RS[b]/aliquot of DNA:								
μmol RS/mL of DNA:								
μmol RS/mg of DNA:								
mg RS/mg of DNA[c]:								

[a]See text for concentration of this solution.
[b]RS-reducing sugar.
[c]Assume that the MW of the reducing sugar is 180g/mol.

This sample is now ready for base-ratio analysis by paper chromatography. An untreated sample of DNA will be analyzed for comparison.

Chromatography of DNA Bases

1. Cut a 20 × 30 cm sheet of chromatography paper. Begin to apply spots 1.5 cm from the left edge of the paper and 2.5 cm from the bottom. Allow about 1.5 cm between spots, and be sure to spot the hydrolyzed sample near the center of the paper. (See TLC section in Chapter 4 for spotting technique.)
2. Spot 0.10 mL of the hydrolyzed DNA sample in a band about 1.5 cm wide. (Use a 10-μL microcap, and deposit a series of droplets across the designated area).

3. Spot 0.05 mL of unhydrolyzed DNA (3mg/mL) adjusted to pH 11.
4. Apply 2 μL each of adenine, cytosine, and thymine (5 mg/mL in 0.02 M KOH) and 6 μL of guanine as standards. Include a spot that contains a mixture containing 2 μL of each of the four bases.
5. Develop the chromatogram overnight in a tank containing 145 mL of water-saturated butanol. Allow the chromatogram to dry in a fume hood, and view the spots under a UV lamp in the dark. Circle with a soft lead pencil the UV-absorbing spots, which will be deep purple in color.
6. Cut out these spots, and elute them from the paper with 2 mL of 0.05 M HCI. Also cut out and elute a clear region of paper of about the same size to provide a spectrophotometric blank. (Cut the designated areas into fine strips, and transfer the strips to test tubes with tweezers.)
7. Warm the tubes in a hot water bath, but avoid boiling. Periodically mix the suspensions on a Vortex mixer and return the tubes to the water bath. Continue this procedure for *ca* 15 min; then filter out the paper particles by passing the suspensions through Pasteur pipettes containing some glass wool.
8. Presentation of data and calculation of percent G+C.
   a. Construct absorption spectra of the solutions obtained above for the four standards and for the spots obtained from the hydrolyzed DNA. These spectra are obtained by reading the eluates at 5-nm intervals from a wavelength of 220 to 300 nm. These data should confirm the identity of the bases released from the DNA sample. Verify this by comparing the peaks with the absorption maxima listed below.

Base	*Absorption maximum (nm)*	*Extinction coefficient*
Adenine	263	13,100
Cytosine	274	10,200
Guanine	275	7,350
Thymine	264	7,890

   b. Using the absorbance values for the hydrolyzed DNA at 263 and 275 nm for A and G, respectively, calculate the percent G+C as follows:

i. Determine the concentrations of A and G extracted from the spots according to the Beer-Lambert law. (Refer to the list above for the extinction coefficients of the bases.)

ii. % G+C = (G/G + A) × 100 (since the amount of G is equal to that of C and the amounts of A and T are likewise equal.)

Determination of DNA and RNA Contents of the Isolated DNA Product

Diphenylamine Assay for DNA: Experimental Procedure

1. Construct a standard curve by preparing a series of tubes containing from 0.015 to 0.30 mg of DNA per tube using a 0.15 mg/mL standard DNA solution. Bring the final volumes in the tubes to 2 mL with H_2O. Include two tubes respectively containing 50 and 100 μg of RNA as controls.
2. Add 2.0 mL of diphenylamine reagent to each tube and mix. Heat the marble-topped tubes in a boiling water bath for 15 min. The diphenylamine reagent must be prepared shortly before use. (Dissolve 0.75 g of diphenylamine in 50 mL of glacial acetic acid. Add 0.75 mL of conc. H_2SO_4. Just prior to use, add 0.25 mL of cold 1.6% acetaldehyde. [Prepare in a fume hood.])
3. Allow the tubes to come to room temperature and read % transmittance at 600 nm. If cloudiness develops, warm the tubes for a few minutes at 50°C.
4. Plot $A_{600\text{ nm}}$ *vs.* mg DNA.
5. To determine the DNA concentration in the preparation, estimate the proper amount of sample that would fall on the linear assay range. (Use the 260/280 measurements made earlier to approximate this value.) Assay three tubes at different concentrations of unknown, assuming that the nucleic acid content is about 75% DNA. Average the values that fall within the limits of the standard curve.

Orcinol Assay for RNA: Experimental Procedure

1. Construct a standard curve by preparing a series of tubes containing from 10 to 100 μg of RNA per tube using a 50 μg/mL standard RNA solution. Bring the final volumes in the tubes to 2.0 mL with 5% TCA. Include two tubes respectively containing 150 and 300 μg of DNA as controls.

2. Add 2.0 mL of orcinol reagent and mix. Heat the marble-topped tubes in a boiling water bath for 15 min. (The orcinol reagent deteriorates after 30 min and should be prepared just prior to use. Dissolve 0.5 g of orcinol in 50 mL of 0.1% $FeCl_3$ in conc. HCl. [Prepare in a fume hood.])
3. Remove the tubes and allow them to cool. Read % transmittance at 640 nm.
4. Plot $A_{640\ nm}$ *vs.* μg RNA.
5. To estimate the unknown amounts of RNA in the preparation, calculate the proper amount of sample to assay again by making use of the Warburg-Christian nomograph and the 260/280 measurement. Assume that about 25% of the DNA preparation is RNA. Assay the sample again at three different concentrations, and average the values that fit on the linear portion of the standard curve. Note that this will give a high estimate because of the DNA interference in this assay.

Calculation of Yield and Purity of Calf Thymus DNA

1. From the diphenylamine, orcinol, and Nelson's assays and from the 260/280 absorbance readings (Warburg-Christian method), the percentage DNA, RNA, polysaccharide, and protein, respectively, can be calculated for the DNA isolated.
   a. Calculate the amount of nucleic acids in the thymus sample. Express the values in mg/g wet weight of tissue and in percent yield of crude DNA.
   b. Set up a purification table.

Component	*Method*	*Concentration (mg/ml)*	%[a]
DNA	1. 260/280 2. Diphenylamine assay		
Protein	260/280		
RNA	Orcinol assay		
Glycogen	Nelson's assay		
"Other"			

[a]Based on a solution prepared from crude DNA (15 mg/mL).

2. Evaluate the effectiveness of the purification procedure. Discuss possible further steps that could be taken to achieve greater purity.

D. A Brief Introduction to Recombinant DNA Methodology

A discussion of recombinant DNA methodology and some applications can be found in Chapter 7, in which is a series of experiments on this topic. In this chapter, a brief two-laboratory sequence of experiments is presented to learn to isolate a plasmid (pBR322) from *Escherichia coli* strain HB 101 and to use restriction enzymes and agarose gel electrophoresis to characterize the resulting DNA fragments.

Small-Scale Isolation of Plasmid pBR322 DNA

Plasmid pBR322 carries two antibiotic-resistance genes (see restriction map, Fig. 6.4) for ampicillin and tetracycline. When *E. coli* HB 101 is grown in a medium containing one of these antibiotics, the only cells that will be able to grow are those with plasmids that confer resistance to the bacteria.

Experimental Procedure—Boiling "Miniprep" Method

1. Inoculate 3 mL of sterile LB medium* containing 100 μg/mL of ampicillin. Incubate at 37°C overnight in a shaking water bath. Three milliliters of such a plasmid-carrying culture grown overnight typically yields 4 to 6 μg of plasmid DNA.
2. Pour the grown culture into two plastic microcentrifuge tubes (do not use glass, as DNA will adhere to the walls of the tube). Spin for 2 min in a microcentrifuge.
3. Decant the supernatant completely, and resuspend the cell pellets in 0.5 mL of sucrose solution containing 8% sucrose, 5% Triton X-100 (a detergent), 50 mM EDTA, and 50 mM Tris, pH 8.0.
4. Add 25 μL of a freshly prepared solution of lysozyme (10 mg/mL in 10 mM Tris, pH 8.0). Mix by shaking gently.
5. Place the tube in a boiling water bath for 60 sec. Cool on ice.
6. Centrifuge for 15 min in a microcentrifuge

---

*LB medium contains: bacto tryptone, 10 g; bacto yeast extract, 5 g; and NaCl, 10 g; pH 7.5.

Restriction Map of pBR322 DNA

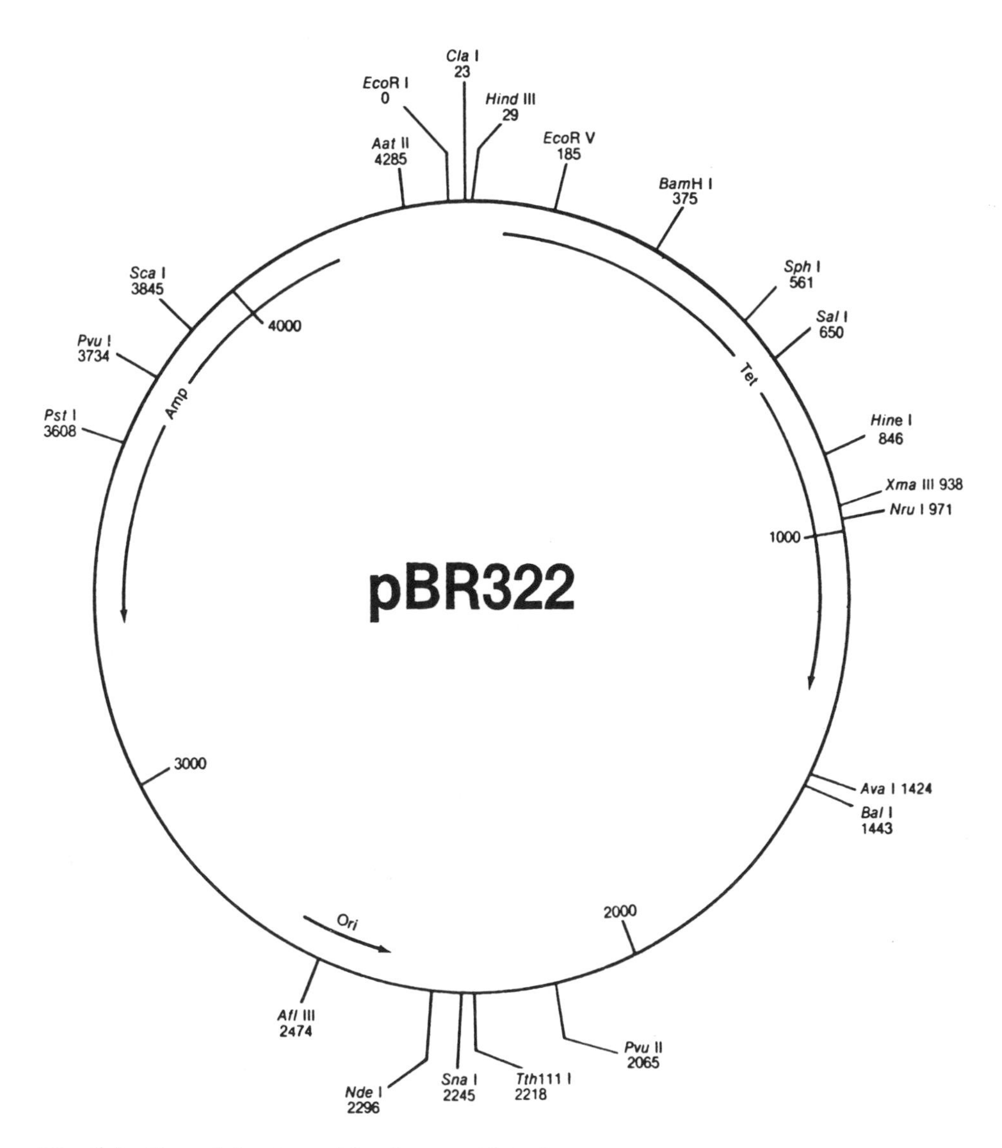

Fig. 6.4. Map of the sites of first bases within the recognition sequences are expressed in nucleotide numbers counting from the origin, which is the center of *Eco*R1. This is also the terminal restriction enzyme recognition sequence at 4360. (From *BRL Catalogue and Reference Guide,* p. 86 [1983], Bethesda Research Laboratories, Inc., Gaithersburg, MD; reprinted with permission of Bethesda Research Laboratories.)

7. Remove the pellets by gently teasing them out of the tube with a toothpick, leaving the clear supernatant behind.
8. Add 25μL of 2.5 M NaCl and 0.6 mL of cold isopropanol to the supernatants. Mix thoroughly and incubate for 15 min in a dry-ice/EtOH bath (This should permit the plasmid DNA to precipitate out of solution.)
9. Centrifuge again in the microcentrifuge for 15 min.
10. Decant and discard the supernatants. Resuspend the plasmid pellets in 50 μL of sterile TE buffer (10 mM Tris, 1 mM EDTA, pH 8.0). Add 6 μL of 2.5 M NaCl and 125μL of cold EtOH. Incubate in a dry-ice/EtOH bath for 10 min, and centrifuge for 10 min. Decant and discard the supernatants.
11. Cover the tubes with parafilm. Punch several holes through the coverings. Place the tubes in a small beaker in a desiccator containing silica gel. (This desiccant has a bright blue color when dry; as it absorbs water, it becomes a faint pink. It can be reactivated by heating at 110°C until it returns to its original color.) Pull a vacuum by attaching the desiccator to an aspirator, and allow the pellet to dry in this fashion for 10 to 15 min.

Digestion of Plasmids With Restriction Enzymes

The enzymes used in molecular cloning are known as *restriction enzymes*. They are endonucleases that have been isolated from prokaryotes, and they recognize specific sequences of nucleotides within double-stranded DNA.

There are three types of restriction enzymes, which are classified according to their mode of action. Among the commonly used enzymes are those that recognize sequences that are four to six nucleotides in length with a twofold axis of symmetry. *Hind* III (purchased from Bethesda Research Laboratories, Gaithersburg, MD), which will be used here, is one such enzyme. It recognizes the hexanucleotide sequence shown below and cuts the plasmid at positions that are four nucleotides apart in the two strands, as indicated by the arrows:

5′-A ↓ AGCT T-3′

3′-T TCGA ↑ A-5′

The resulting DNA fragments thus have protruding "sticky" five-prime (5′) ends, which means that they can form base pairs with other single-stranded DNA termini. Thus new recombinant DNA molecules can be formed from fragments that have been cut by a given enzyme and have been rejoined by base-pairing at another site.

The other enzyme chosen for this experiment is *Pvu* II (also purchased from Bethesda Research Laboratories) which does not generate DNA fragments with protruding 5′ tails but cuts at the axis of symmetry and yields "blunt-end" fragments:

5′-CAG ↓ CTG-3′

3′-GTC ↑ GAC-5′

Information on the optimum conditions required for each enzyme is supplied by the manufacturer. In general, the conditions that vary are the composition of the buffer and the temperature of incubation. For convenience in conducting the experiment done here, two enzymes with similar requirements for temperature and buffer concentration were chosen.

Experimental Procedure

1. Resuspend the dry pellet in 50 μL of sterile TE buffer, pH 8.0. *Use sterile pipettor tips* to prevent contamination of restriction enzymes. Proceed with one of the two DNA plasmid fractions.
2. Set up a series of four sterile microcentrifuge tubes and follow the protocol given in Table 6.5.
3. Incubate the four tubes in a 37°C H_2O bath for 30 min.
4. Add 12 μL of 5× tracking dye to stop the reaction (50% glycerol + 0.25% bromphenol blue).
5. Store these tubes at 4°C until the next laboratory period, when agarose gel electrophoresis will be performed.

Agarose Gel Electrophoresis of Plasmid DNA Fragments

Agarose gel electrophoresis is used to separate DNA fragments, which can then be identified by comparing their electrophoretic patterns with the patterns obtained with λ *Hind* III–cut DNA fragments of known composition. The DNA fragments band within the gel and are stained with ethidium bromide,

TABLE 6.5. Protocol for Digestion of Plasmid pBR322 by Restriction Enzymes

Additions (μL)	Tube no. 1	2	3	4
DNA plasmid solution	5	5	5	5
10× Enzyme buffer[a]	5 ⟶			
Sterile deionized H_2O	40	36	36	34
RNase (1 mg/mL)	—	2	2	2
Hind III (*ca.* 10 to 20 Units)[b]	—	2	—	2
Pvu II (*ca.* 10 to 20 Units)[b]	—	—	2	2

[a]10× Buffer contains 0.6 M NaCl; 60 mM Tris, pH 7.5; 60 mM $MgCl_2$; and 10 mM DTT.

[b]One unit of enzyme is that amount required to digest 1 μg of λ DNA in 1 hr under the appropriate conditions in a volume of 50 μL. Note that enzyme is added in excess to achieve complete digestion during a shorter incubation period.

which is incorporated into the gel. Low concentrations of this fluorescent dye permit detection of small amounts of DNA (down to 1 ng), which can be visualized directly under UV light.

The migration rate of the fragments depends on their size, the concentration of agarose, the conformation of the molecules undergoing electrophoresis, and the current applied. As in disc gel electrophoresis, the distances migrated, as measured in R_m values, are inversely proportional to the log of their molecular weights. Thus, it will be possible to construct a standard curve using λ DNA fragments of known size and to determine the sizes of the pBR322 digests by comparison.

Experimental Procedure

1. Prepare the 0.8% agarose solution by adding 2.4 g of powder to 300 mL of TBE buffer. (See Appendix I for composition.) Dissolve the agarose by gently heating either over a Bunsen burner or in a microwave oven; cool to 50°C.
2. Add 15 μL of ethidium bromide from a stock solution (10 mg/mL) to obtain a concentration of 0.5 μg/mL in the gel. **(WEAR GLOVES WHEN HANDLING; IT IS A POTENT MUTAGEN).**
3. A horizontal slab gel will be used. The plastic form, which is open at both ends, must be taped securely to provide the mold into which the agarose is poured. Insert the comb,

which will form the sample wells. (Be sure that the teeth do not cut through the bottom of the gel. There must be at least a 1-mm layer of agarose between the bottom of the teeth and the base of the gel.) Pour the cool agarose solution carefully to avoid air bubbles.

4. Allow the gel to set completely, about 30 min; *carefully* remove the comb and the two strips of tape.
5. Transfer the slab gel to the electrophoresis tank, and add enough TBE buffer to cover the gel to a depth of at least 1 cm.
6. Load the entire samples from the four microfuge tubes into each of the wells formed by the comb, using an automatic micropipettor and a disposable tip for each sample. The glycerol in the solutions will prevent the liquid from rising, but be careful to inject the sample slowly and avoid contaminating neighboring wells.
7. To a fifth well add 10 μL of commercial λ DNA marker in a tracking dye solution.
8. When all the samples have been loaded, connect the electrophoretor to a power supply (be sure to connect the + [red] lead *to the bottom* of the apparatus) and apply 150 V for about 3 hr.
9. Turn off the power supply, and disconnect the apparatus carefully. Then, wearing gloves to avoid contact with ethidium bromide, move the gel to a UV viewing box to note the fluorescing bands.
10. Photograph the gels under UV light, and make measurements of the bands from the photographs.

Visualization and Calculation of Fragment Sizes

The λ DNA *Hind* III fragments should show six bands clearly:

DNA fragment	*Kilobase pairs (kb)*
1	23.13
2	9.42
3	6.56
4	4.36
5	2.32
6	2.03

1. Measure the bands, and calculate R_m values for the known λ DNA and for the digests of pBR322.
2. Construct a standard curve by plotting R_m *vs*. log kb, and determine the exact sizes of the fragments obtained. Compare these values with those expected on the basis of the restriction map of pBR322 and the known loci for *Hind* III and *Pvu* II.

CHAPTER 7

Recombinant DNA Methodology

Day 1 Alkaline "miniprep" procedure on an overnight culture of *E.coli* JM 101/pUC9.

Day 2

1. Digestion of pUC 9 and λ DNA by treatment with *Eco*R1 endonuclease.
2. Small-scale electrophoresis.
3. Calculation of plasmid DNA yield.

Day 3

1. Ligation of fragments.
2. Preparation of competent cells.

Day 4

1. Transformation into competent cells.
2. After 18 to 20 hr incubation:
   a. Record number of white and blue colonies growing on X-gal plates.
   b. Transfer of white colonies into ampicillin containing growth medium.
   c. Overnight incubation at 37°C.

Day 5

1. Boiling "miniprep" procedure on overnight culture to isolate recombinant plasmids.
2. Preparation of agarose gel for electrophoresis.

Day 6

1. Agarose gel electrophoresis.
2. Calculation of fragment sizes and determination of the orientation of the inserts.

A. Introduction

The new methods that have been developed since the discovery of restriction endonucleases have opened further opportunities for the study of gene function and for the development of products of practical value in agriculture, medicine, and industry. Certain strains of *Escherichia coli* can carry extra chromosomal pieces of DNA that have sufficient homogeneity to enable replication within the cell and thus make copies of genes carried on these plasmids. Under suitable conditions, the plasmids may thus serve as vehicles for transferring genetic material from one source to another, with *E. coli* serving as a reservoir.

The availability of a series of restriction enzymes, enzymes that recognize and cut specific nucleotide sequences, makes it possible to do fine genetic mapping. When the proper conditions are provided, it is also possible to construct plasmids containing a desired gene, which one can then transfer to another strain or amplify by permitting the formation of multiple copies. If gene expression in *E. coli* is possible, cells carrying the plasmid can be grown in large quantity to produce the gene product. It can then be recovered by harvesting the cells, or, if it is excreted, the gene product can be isolated from the culture medium.

Recombinant DNA technology depends on the isolation of a piece of DNA of interest by digestion with an enzyme that specifically recognizes a nucleotide sequence. Each DNA strand cut at the same site leaves a short piece of single-stranded DNA. When a plasmid (cloning vector) is then treated with the same restriction enzyme, the "sticky ends" that result will have complementary nucleotide sequences, and base pairing between the insert DNA and the vector DNA can occur. These are covalently joined in a reaction with DNA ligase, and a recombinant DNA molecule is formed.

The recombinant DNA molecule is introduced into a host cell that has been treated to make it receptive to the newly formed plasmid. The cells containing the plasmid will be selected on the basis of some property, such as resistance to an antibiotic, which the plasmid confers on the host. The experiments that follow represent typical procedures used to transfer genetic information from a donor source to a new host.

REFERENCES: Anderson, W.F., and Diacumakos, E.G. Genetic engineering in mammalian cells. *Sci. Am.* 245(1): 106-121 (1981).

Del Sal, G., Manfioletti, G., and Schneider, C. The CTAB-DNA precipitation method: A common miniscale precipitation of template DNA from phagemids, phages or plasmids suitable for sequencing. *Biotechniques* 7:514-519 (1989).

Fritsch, E.F., Sambrook, J., and Maniatis, T. *Molecular Cloning—A Laboratory Manual*, 2nd ed. Cold Spring Harbor Laboratory Press, Cold Spring Harbor, NY (1989).

Rawn, Ch. 30.

Stryer, pp. 117-139.

Watson et al., pp. 595-618.

Watson, J.D., Tooze, J., and Kurtz, D.T. *Recombinant DNA—A Short Course*. W.H. Freeman, New York (1983).

The purpose of the experiments in this module is to introduce recombinant DNA techniques. Genetic material from a plasmid vector (pUC9) will be isolated from *E. coli* JM 101, which carries this plasmid. The plasmid and a sample of λ DNA obtained commercially will be digested with the restriction endonuclease *Eco*R1, which cuts both strands of the DNA molecule as follows:

5′ G ↓ AATT   C 3′

3′ C   TTAA ↑ G 5′

The fragments resulting from the treatment will have homologous "sticky" ends (i.e., several bases that protrude and can bind to their complements in other fragments) which will make ligation of λ DNA and pUC9 DNA possible. Following treatment with ligase, the DNA fragments will be introduced into "competent" *E. coli* cells (*i.e.*, bacterial cells that have

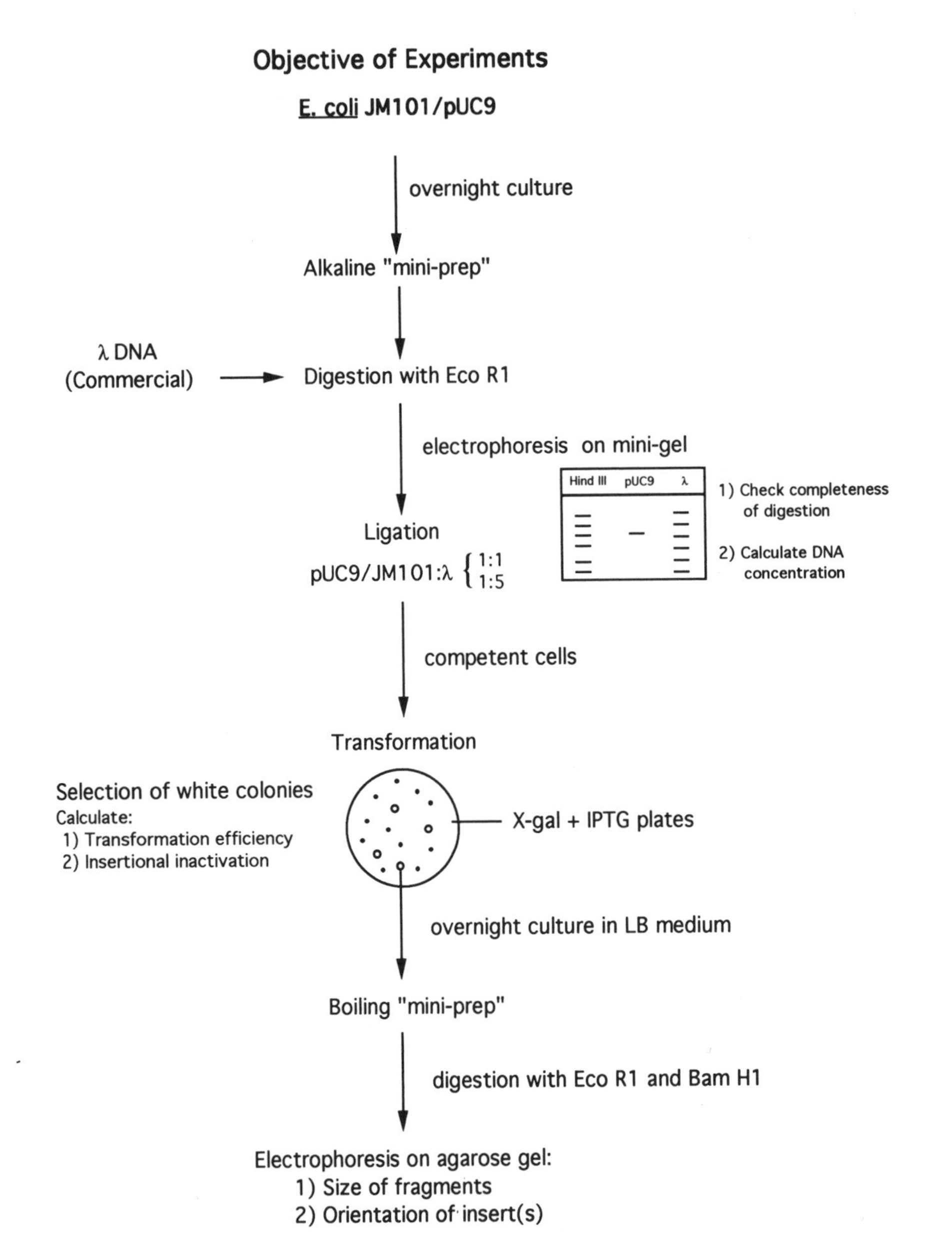

been treated to make them take up foreign DNA)—a procedure called *transformation*. The diagram above illustrates the sequence of experiments that will be performed.

Since the *Eco*R1 site in pUC9 is located in the *lac*Z gene (Fig. 7.1), a successful ligation will incorporate a fragment of λ DNA that will disrupt normal translation of the *lac*Z gene and prevent the formation of the enzyme β-galactosidase.

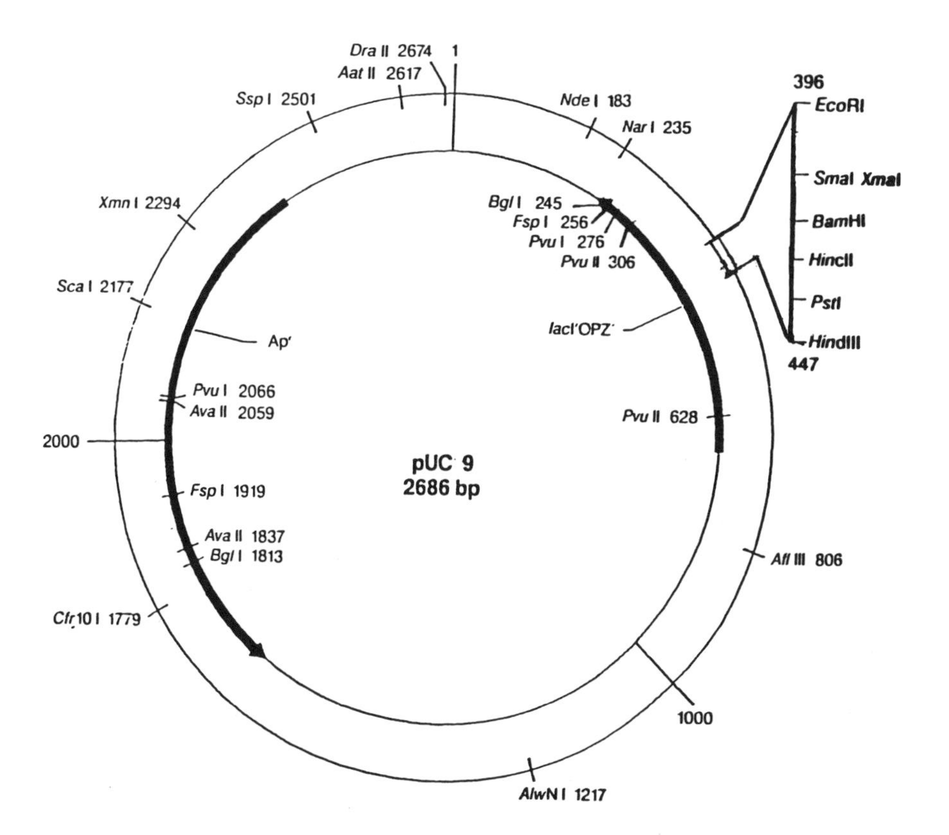

Fig. 7.1. pUC9 restriction map (modified from BRL catalog, Bethesda Research Laboratories, Bethesda, MD, 1988).

Bacterial cells containing recombinant plasmid DNA will be ampicillin resistant and β-gal(−). When spread onto agar plates containing X-gal, and IPTG as an inducer, cells able to hydrolyze the substrate (5-bromo-4-chloro-3-indolyl-β-galactoside) will form blue colonies, whereas cells that contain the inserted DNA fragments will grow but will not act on this substrate and therefore will form white colonies. In addition, the agar plates contain ampicillin, which permits growth of only those cells having the plasmid and therefore the *Amp*^{r} gene.

Some white colonies, potentially carrying plasmids of interest (*Amp*^{r} white colonies, disrupted *lacZ* gene), will be isolated and cultivated in an appropriate growth medium, the cells will be harvested and the plasmid DNA will be isolated. This DNA will be digested with restriction enzymes and the fragments separated by electrophoresis on agarose gels. The resulting bands can be visualized by ethidium bromide staining and the fragments identified by comparison with bands obtained on the same gel with fragments of known size. Also, the orientation of the inserts will be determined. Because the electrophoretic mobility of a DNA fragment is inversely proportional to the logarithm of the number of base pairs, the sizes of the unknown fragments can be calculated relative to the known fragments, and an estimate of the yield of DNA can be made.

B. Commonly Used Terms and Materials

Ampicillin: An antibiotic used to inhibit growth of *E. coli* cells that do not carry a plasmid containing the *Amp*^{r} gene.

Cloning: The insertion of an exact copy of a piece of DNA into a plasmid.

Competent cells: Bacterial cells artificially made susceptible to transformation, usually by treatment with Ca^{++}.

Expression vector: A plasmid that allows transcription and translation of a gene cloned into it.

***Hind* III "ladder"** : Commercial λ DNA digested with *Hind* III restriction enzyme and used in agarose gel electrophoresis to standardize fragment sizes.

IPTG: Compound that induces transcription of the *lac*Z gene.

Ligation: Procedure to join complementary ends by means of the specific enzyme, DNA ligase.

Plasmid: A small piece of circular DNA used for cloning. A plasmid usually has a "selectable" marker, such as a drug-resistance gene, that allows for selection of cells containing the plasmid and for selection against those cells lacking the plasmid.

Restriction enzyme: An enzyme isolated from bacterial cells that cuts DNA at a specific, palindromic, four to eight base pair site (called the *recognition sequence*), creating sticky or blunt ends.

Sticky ends: The overhangs of DNA (one to four bases) left on each strand after the action of some restriction enzymes (*e.g.*, *Eco*R1, *Bam*H1).

Subcloning: The cloning of a smaller piece of DNA derived from an already cloned piece of DNA.

Transformation: A procedure by which DNA (usually plasmid DNA) is introduced into competent cells.

X-gal plates: A solid culture medium containing a synthetic chromogenic β-galactosidase substrate on which transformed cells can grow and from which colonies are selected for further study.

C. Preparation of Plasmid DNA

Harvesting of E. coli MJ 101/pUC 9 Cells

1. Use an overnight culture grown in LB medium containing ampicillin (60 μg/ml).
2. Pour 1.5 mL of culture into each of two microfuge tubes. Centrifuge in a microfuge for 1 min at 4°C (12,000 *g* at full speed). *Always* orient the tube with the cap joint facing outward to identify the location of the pellet if it is too small to see readily.
3. Decant the medium and proceed with one of the small-scale DNA preparations outlined below.

Alkaline Lysis "Miniprep" Method for the Preparation of Plasmid DNA

The bacterial cells that grow in the LB medium with ampicillin carry the pUC9 plasmids that have the ampicillin resistance gene (see Fig. 7.1), because *E. coli*, which is sensitive to this antibiotic, would not be able to grow by itself. In the preparation of plasmid DNA by an "alkaline-lysis" method, SDS is added to denature proteins, thus rupturing the bacterial cell walls, and NaOH is added to denature both chromosomal and plasmid DNAs. Potassium acetate neutralizes the solution, thereby creating an environment conducive to plasmid DNA reannealing. Bacterial proteins and DNA that have now precipitated out of solution are removed by centrifugation. The plasmid DNA that remains in solution is then purified and concentrated by addition of EtOH, and the resulting precipitate is resuspended in an appropriate volume of buffer.

Method I (Fritsch et al., 1989)

1. Resuspend the bacterial pellet (obtained in step 3 above) in 100 μL of ice-cold Solution I (50 mM glucose; 25 mM

Tris·HCl, pH 8.0; 10 mM EDTA, pH 8.0 [GET]) by vigorous vortexing.

2. Add 200 μL of freshly prepared Solution II (0.2 N NaOH, 1% SDS [SDS]). Make this solution by mixing 500 μL of 0.4 N NaOH with 500 μL of 2% SDS for a final concentration of 1% SDS in 0.2 N NaOH. Close the tube tightly, and mix the contents by inverting the tube rapidly several times. Make sure that the entire surface of the tube comes in contact with Solution II. Do not vortex. Store the tube on ice for 5 min.

3. Add 150 μL of ice-cold Solution III (5 M potassium acetate, 60 ml; glacial acetic acid, 11.5 ml; H_2O, 28.5 ml [K acetate working solution]). The resulting solution is 3 M with respect to potassium and 5 M with respect to acetate. Close the tube and vortex it gently in an inverted position for 10 sec to disperse Solution III through the viscous bacterial lysate. Incubate the tube on ice for 3 to 5 min.

4. Centrifuge for 5 min at 4°C in the microfuge. Transfer the supernatant to a fresh tube. Add 1000 units RNAse (1 μL) and incubate 5 min at 37°C. Measure the volume.

5. Add an equal volume of the yellow phenol:chloroform solution **(WEAR GLOVES TO DO THIS PROCEDURE, AND WORK IN THE HOOD)**.

   a. First check the pH of the TE buffer layer on top of the reagent (see Appendix I for composition). The phenol reagent can be used as long as the pH is > 7.6.

   b. Insert the pipettor tip into the phenol layer (yellow). While holding down the plunger, eject the TE buffer that collects in the tip, and then draw the phenol reagent into the tip.

   c. Vortex for 15 sec and centrifuge for 2 min.

   d. Transfer the clear top layer to a fresh tube using a P200 automatic pipettor.

6. Precipitate the double-stranded DNA with two volumes of 100% EtOH at room temperature. Mix by vortexing. Allow the mixture to stand for 2 min at room temperature.

7. Centrifuge for 15 min in the microfuge.

8. Decant the supernatant and invert the tube on a Kimwipe. If any droplets of EtOH remain inside of the rims, they can be gently removed by blotting with a fresh Kimwipe.
9. Gently rinse the pellet of double-stranded DNA with 1 mL of cold 70% EtOH. Again centrifuge for 5 min. Remove the supernatant as described above, and allow the pellet of nucleic acid to dry by vacuum desiccation for 10 min. To desiccate, lift cover of tube and cover the top tightly with parafilm. Punch a few holes in the cover to allow air to escape without disturbing the pellet.
10. Redissolve the nucleic acids in 20 μL of TE buffer pH 8.0. Vortex briefly; incubate for 5 min at 37°C. Store the DNA at −20 °C.

NOTE: The typical yield of high-copy number plasmids, such as pUC, prepared by this method is about 3 to 5 μg of DNA/mL of original bacterial culture.

Method II: Precipitation With CTAB (Del Sal et al., 1989)

1. Resuspend the cells in 100 μL of cold Solution I (for composition, see Method I). Vortex or pipette up and down several times until suspension is uniform.
2. Add 200 μL of Solution II (for composition, see Method I) to each tube, and close tube tightly. Mix by inverting the capped tubes several times. **DO NOT VORTEX!** Place the tubes on ice for 5 min.
3. Add 150 μL of Solution III (for composition, see Method I). Close tubes tightly and mix thoroughly. Again place the tubes on ice for 5 min.
4. Spin for 5 min in a microfuge at 4°C. Pour the supernatants carefully into fresh microfuge tubes. Pour away from the pellets to avoid suspending the debris.
5. Add a 0.6 volume (*ca* 300 μl) of isopropanol to precipitate the plasmid DNA. Mix.
6. Centrifuge for 15 min in the microfuge at 4°C. Pour off the supernatants. Drain the last few drops by inverting the tubes on a Kimwipe.
7. Resuspend the DNA in 100 μL sterile TE buffer, pH 8.0. Add 1 μL (1000 units) RNase T1 to each tube, and incubate at 65°C for 30 min.

8. Add 10 μL of 5% CTAB (hexadecyltrimethylammonium bromide) and mix. The solution should become cloudy.
9. Centrifuge for 5 min in the microfuge at 4°C. Decant the supernatants carefully.
10. Resuspend the DNA pellets in a total of 300 μL of 1.2 M NaCl.
11. Add 2 volumes (*ca* 600 μL) of cold 100% EtOH to each tube to precipitate the DNA. Vortex. Incubate 2 min at room temperature.
12. Centrifuge for 15 min in the microfuge at 4°C.
13. Carefully decant the supernatants. Invert the tubes on a Kimwipe to allow the remaining liquid to drain. Handle tubes gently to avoid dislodging the pellets.
14. Add 1 mL of 70% EtOH to each tube to rinse the pellets, but do not resuspend them. Decant and drain off the supernatants as before.
15. Allow the pellets to dry for 10 to 15 min in a vacuum desiccator. Cover each tube with parafilm, and punch the cover with a few holes to permit air to escape.
16. Resuspend the DNA in 20 μL of TE buffer, pH 8.0. Store at −20°C until the next laboratory period.

The yield of plasmid DNA should be similar to that obtained with the previous method.

D. Digestion of DNA With Restriction Enzymes and Agarose Gel Electrophoresis

The plasmid DNA and a sample of λ DNA are digested with the endonuclease *Eco*R1 to obtain complementary cohesive ends for vector and insert(s) that will subsequently be joined by ligation.

Agarose Gel Electrophoresis

Agarose gel electrophoresis is used to separate DNA fragments, which can then be identified by comparing their electrophoretic patterns with the patterns obtained with λ *Hind* III-cut DNA fragments of known composition. The DNA fragments band within the gel and are stained with ethidium bromide, which is incorporated into the gel. Low concentrations of this fluorescent dye permit detection of small amounts of DNA (down to 1 ng), which can be visualized directly under ultraviolet light.

The migration rate of the fragments depends on the size, the concentration of agarose, the conformation of the molecules undergoing electrophoresis, and the current applied. As in disc gel electrophoresis, the distances migrated, as measured in R_m values, are inversely proportional to the log of their molecular weights. Thus it will be possible to construct a standard curve using λ DNA fragments and determine the sizes of the fragments by comparison.

The molecular weight standards are from a commercial preparation of linear fragments of known sizes. As the migration of these fragments is mainly dependent on size, the smaller fragments move faster, and the larger ones form bands nearer to the point of application. The standard curve is established by plotting the log of the molecular weight (in kilobase pairs [kb]) against the R_m values (distance of the bands from the origin see Fig. 7.2).

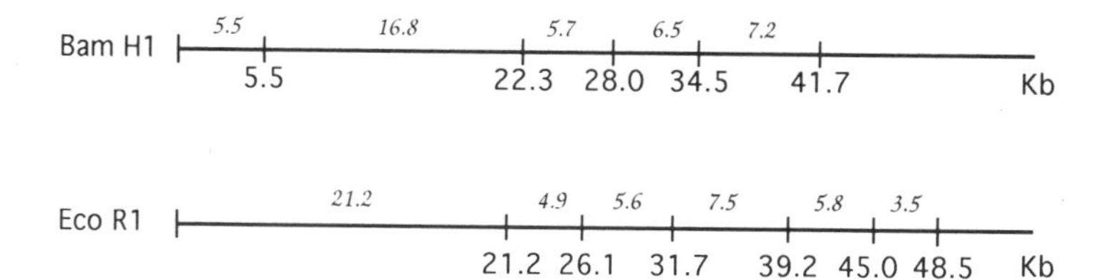

Lambda DNA-
Hind III Ladder

Fragment #	Number of Base Pairs (in Kb)
1	23.130
2	9.416
3	6.557
4	4.361
5	2.322
6	2.027

Fig. 7.2. Restriction map and *Hind* III digests of λ DNA.

The sizes for the vector pUC9 and the λ fragments that will become the DNA inserts in this procedure are determined from such a curve. Since pUC9 has one *Eco*R1 recognition site, a single band of size 2.7 kb is expected when complete digestion is achieved. The *Hind* III "ladder" is also used to estimate the amount of DNA in the plasmid preparation by comparing the intensity of the pUC9 band to the intensities of the bands obtained with a known amount of the standard. For calculation of the amount of DNA in an unknown band:

1. Compare the fluorescence of the plasmid band to the λ fragment band of similar intensity.
2. Calculate the amount of DNA represented by this band as follows:

$$\frac{(\text{Size of comparable } \lambda \textit{ Hind} \text{ III band}) \times (\text{amount of } \lambda \textit{ Hind} \text{ III in the well})}{(\lambda \text{ genome size} = 49 \text{ Kb})}$$

For example, if 4 μL of plasmid preparation was added to the well, the fluorescence of the plasmid band matches the 4.3-kb λ *Hind* III band in intensity, and 500 ng of λ *Hind* III standard was applied, then the preparation contains

$$\frac{4.3 \text{ kb} \times 500 \text{ ng}}{49 \text{ kb}} = 43.9 \text{ ng}/4\ \mu\text{L} = 11 \text{ ng}/\mu\text{L}$$

Procedure for Digestion of Plasmid DNA

1. Set up three sterile 0.5-mL tubes as follows. Be sure to deliver all solutions carefully to the bottom of the tube:

	Tube No.		
Additions (μL)	*1*	*2*	*3*
10 × *Eco*R1 buffer	2	2	2
pUC 9 (commercial) (1 μg)	2	–	–
λ DNA (commercial) (3 μg)	–	6	–
pUC 9 (isolated on day 1 from JM 101)	–	–	15
Sterile H_2O	14	10	1
*Eco*R1 enzyme	2	2	2
Total	20	20	20

2. Briefly spin the tubes in the microfuge to mix the solutions; then incubate the three tubes for 1 hr at 37°C.
3. Prepare an 0.8% agarose "minigel" while the tubes are incubating.
   a. Weigh out 0.32 g of agarose, and add 40 mL of 1× TBE buffer (TBE buffer contains 0.09 mM Tris, 0.09 mM boric acid, 2 mM EDTA, pH 8.0) in a 125-mL flask. This is sufficient for two gels.
   b. Heat in a microwave oven until agarose is dissolved, and allow the gel to cool to *ca* 60°C.
   c. Add 6 μL of a solution (10 mg/mL) of ethidium bromide, mix by gently swirling the flask, and put the comb in place in the slots provided in the tray. **WEAR GLOVES WHEN HANDLING ETHIDIUM BROMIDE; IT IS A MUTAGEN.**
   d. Pour *ca* 15 to 20 ml of the molten agarose solution into each of two trays, and allow the gels to cool.
   e. Remove the comb carefully when the gels have solidified, and fill the chamber with enough 1× TBE buffer to cover the gels.
   f. After the incubation period, the plasmids are checked for proper digestion. Using a P20 micropipettor, carefully load six wells as follows:
      i. Commercial λ cut with *Hind* III (0.1 mg/mL) 2 μL and 4μL + 1 μL of 5× loading dye in each of two wells.
      ii. 4 μL pUC9 from tube 1 + 1 μL of 5× loading dye.
      iii. 4 μL of λ from tube 2 + 1μL of 5× loading dye.
      iv. 4 μL of pUC9/JM 101 from tube 3 + 1 μL of 5× loading dye.
      v. 4 μL of uncut commercial pUC9 (0.2μg) + 1 μL of 5× loading dye.
   g. Attach the apparatus to a power supply, and allow the gel to run for 1 hr at 100 V. Be careful to check the orientation of the wells in relation to the flow of the current (the fragments must move toward the anode).
   h. Visualize the gel under UV light, and take a photograph.

4. Check for completion of digestion.

   a. When pUC9 has been cut to completion, one band of size *ca* 2.7 kb will be obtained.

   b. The λ *Eco*R1 digests should contain six fragments of sizes 3.5, 4.9, 5.6, 5.8, 7.4, and 21.2 kb. (Fig. 7.2).

   c. Construct a standard curve by plotting the distance of the bands from the point of sample application against the log kb for *Hind* III bands: see Figure 7.2 for their sizes. This standard is included in every gel, a new curve is established each time, and the sizes of the unknown fragments are determined from the curve.

   d. If digestion is complete, inactivate the *Eco*R1 enzyme by heating the samples (in tubes 1, 2, and 3) at 65°C for 10 min. Cool on ice immediately. The tubes with these digests can now be stored until the next laboratory period in a freezer.

E. Preparation of Competent Cells

Aseptic technique must be used throughout.

1. Grow an overnight culture of JM 101 to be transformed. Inoculate 100 mL of LB broth (see Appendix I for composition) in a 500 mL flask with 1 mL of overnight culture. This culture should be grown at 37°C with shaking until the culture has reached an absorbance of 0.4 to 0.5 at $A_{600\ nm}$ (which is in the logarithmic growth phase). If the culture medium is at 37°C at the time of inoculation, this will take 1.5 to 2.0 hr.

2. Transfer the cells to sterile 50 mL centrifuge tubes and cool on ice for 10 min. Centrifuge at 6000 rpm for 5 min.

3. Discard the supernatant and resuspend the cells in 7.5 mL of *ice cold* sterile CM buffer (0.05 M $MgCl_2$ and 0.05 M $CaCl_2$ in 10% glycerol). This exposure to Ca^{++} is thought to render intact bacterial cells more receptive to plasmid DNA.

4. Incubate on ice for 15 min.

5. Centrifuge at 6000 rpm for 10 min and resuspend in 3.5 mL of cold CM buffer.

6. Place on ice for 5 min, add 125 μL DMSO (dimethyl sulfoxide), and cool on ice again for 5 min.
7. Add another 125 μL DMSO, and hold on ice for 5 min.
8. Dispense 0.1-mL aliquots of cell suspension into *sterile, cold* 1.5-mL microfuge tubes and freeze at − 70°C. The cells are now ready to be transformed and may be stored at this temperature for several weeks.
9. To determine the transformation efficiency of these cells, the following procedure is used:
   a. To one vial containing 100 μL of competent cells, add 50 ng of commercial pUC9.
   b. Incubate on ice for 30 min.
   c. Heat shock the cells at 42°C for 90 sec.
   d. Return to the ice bath for 2 min.
   e. Add 0.8 mL of LB medium with 20 mM glucose (prewarmed to 37°C).
   f. Incubate in a 37°C water bath with vigorous shaking for 60 min.
   g. Plate 25 μL and 100 μL of a 1:10 dilution of the cells on LB plates containing X-gal, IPTG, and ampicillin.
   h. Incubate the plates for 18 to 24 hr at 37°C. Then count the colonies, and calculate the transformation frequency.

F. Ligation of Plasmid Vector and Insert DNA

After isolation and digestion of the plasmid DNA with restriction enzymes, the cleaved DNA is joined to a foreign DNA *in vitro*. This process is called *ligation* and requires the enzyme DNA ligase, which can be purchased.

In the following experiments, the pUC9 DNA fragments that contain the termini compatible with those also generated by *Eco* R1 digestion of the λ plasmid will be joined. The recombinant plasmids thus formed should be circular and will then be used to transform *E. coli* to ampicillin resistance.

Procedure for the Ligation

1. Estimate the amount of DNA present in the digests. Analyze the data obtained from the "minigel" as outlined in Section E above for pUC9, the vector (tube 3).
2. Calculate an amount of this preparation that will contain 200 ng of the DNA and the amount of commercial λ to provide 200 and 1000 ng of *insert* DNA (tube 2, 150 ng/μL), respectively.
3. Set up three sterile 1.5-mL microfuge tubes for the ligation procedure by combining vector and insert as follows:
   a. 1:1; 200 ng (vector), 200 ng (insert)
   b. 1:5; 200 ng (vector), 1000 ng (insert)
   c. Vector only (tube 1), 200 ng
4. *Preparation of DNA fragments for ligation.* To increase the chances for success, the digests to be ligated are precipitated together with EtOH, concentrated, and resuspended prior to treatment with DNA ligase. Proceed with the precipitation of the combined DNA fragments in these three tubes as follows.
   a. Add sterile TE buffer, pH 8.0, to a final volume of 50 μL.
   b. Add 5 μL of 3 M sodium acetate, pH 5.2.
   c. Add 110 μL of 100% EtOH.
   d. Mix by vortexing, and incubate the mixture on ice for 15 min.
   e. Centrifuge in the microfuge for 15 min at 4°C. Orient the cap joint facing outward.
   f. Pour off the EtOH supernatant, and gently rinse the pellet with 200 μL of 70% EtOH.
   g. Centrifuge again for 5 min, pour off the supernatant, and allow to drain on a Kimwipe to remove any remaining alcohol.
   h. Dry the pellet by vacuum desiccation for 10 min.
   i. Resuspend in 7 μL TE buffer, pH 8.0.
5. Ligation of pUC9 vector and insert(s).
   a. To the resuspended pellets add 2 μL of 5× ligation buffer (0.25 M Tris-HCl, pH 7.6; 50 mM $MgCl_2$; 5 mM

ATP; 5 mM dithiothreitol; 25% [w/v] polyethylene glycol) and 1 μL of T4 ligase (1 Unit).

b. Incubate overnight at 16°C. The ligated plasmids can now be used to transform competent host cells. Store at 4°C until the next laboratory period.

G. Transformation of *E. coli* cells

Following the ligations, the plasmids can be introduced into *E. coli* to carry out the screening. The strain that will be transformed, JM 101, ampicillin sensitive, β-gal negative. After transformation, the cells will be grown on ampicillin-containing medium to select for those colonies that are ampicillin resistant, containing either recombinant or nonrecombinant plasmids. The X-gal substrate and IPTG in the medium allow for the selection of colonies that will be analyzed later.

This series of steps allows for entry of the cloned plasmid DNA into the recipient cells. The mechanism for entry is believed to involve a chemically induced change in the permeability of the cell membrane. The cold incubation allows for the adsorption of the plasmid DNA to the cell surface. The 42°C heat treatment induces DNA uptake, and the 37°C incubation permits cell growth in a rich medium before plating onto a selective medium. Cells were made susceptible to transformation by exogenous DNA by prior treatment with Ca^{++}

Maintain aseptic technique.

1. Transfer 0.1 mL of competent *E. coli* JM 101 cells to five 1.5-mL microfuge tubes and to each one add:

   a. No DNA–*negative control.*

   b. 1 μL of commercial uncut pUC9 at a concentration of 5 ng/μL–*positive control* (this should yield blue colonies only, and the number of blue colonies is then used to calculate the transformation efficiency).

   c. 1 μL of the 1:1 ligation mix (from Section F.3.a).

   d. 1 μL of 1:5 ligation mix (from Section F.3.b).

   e. 1 μL of the religated vector–*no insert* (from Section F.3.c).

2. Incubate the tubes on ice for 30 min.
3. Incubate in a 42°C water bath for 90 sec. (This heat pulse is thought to increase the permeability of the membrane and therefore improve DNA uptake.)
4. Place on ice for 2 min.
5. Add to each tube 0.8 mL of LB medium containing 20 mM glucose prewarmed to 37°C, and incubate with shaking at 37°C for 60 min. (This permits the cells to recover and begin replicating the plasmid and expressing the ampicillin-resistance gene.)
6. Prepare culture plates, each containing LB agar and the following:
   a. Ampicillin (to a final volume of 100 μg/mL).
   b. X-gal (40 μg/mL).
   c. IPTG (40 μg/mL).
7. Mix the tubes by inverting them several times to obtain a uniform cell suspension, and make a 1:10 dilution of each in LB medium.
8. Deliver 100 μL of each of the diluted transformations onto five plates, and spread the cell suspension with a glass spreader which has been dipped in alcohol and flamed.
9. Allow the agar to absorb the liquid, invert the plates, and incubate at 37°C for 15-18 hr. (The plates should not be incubated for more than 24 hr as satellite colonies begin to appear—the plates can be sealed with parafilm or tape and refrigerated for a day or two if necessary.)

H. Analysis of Recombinant Plasmids

From the overnight cultures, a preparation of recombinant plasmids is made using the "boiling miniprep" method. The DNA is digested with *Eco*R1 and this time with a second endonuclease, *Bam*H1, which has restriction sites within most of the λ fragments produced by *Eco*R1 digestion (Fig. 7.2). The measurements of the bands are then used to calculate the sizes of the fragments resulting from the two digestions. From these are determined (1) the identity of the λ segment that was inserted and (2) the orientation of the insert in the recombinant plasmid.

1. Count or estimate the number of colonies of each phenotype that grew on the agar plates. Count only large blue colonies and white colonies of similar size.

   a. Calculate the transformation efficiency and the percent insertional inactivation from the ligation mixtures as follows:

      i. $\text{Transformation efficiency} = \dfrac{\text{Total no. colonies*}}{\mu\text{g of plasmid plated}}$

      ii. $\text{Insertional inactivation} = \dfrac{\text{No. recombinants (white)}}{\text{total (blue + white)}}$

   b. Pick two large white colonies off the agar plate using a heat-sterilized wire loop, and inoculate these into each of two tubes containing 5 mL of liquid LB medium and ampicillin.

2. Grow the cultures overnight at 37°C with vigorous shaking. The plasmids will be isolated this time using a "boiling miniprep" method and digested with *Eco*R1 and another restriction enzyme, *Bam*H1 (see Fig. 7.2).

3. Isolation of recombinant plasmids by a "boiling miniprep" method.

   a. Transfer 1.5 mL of one of the overnight cultures to a sterile microfuge tube. Spin 1 min in the microfuge. Discard the supernatant. Wash the cells with sterile saline. Spin, and discard the supernatant.

   b. Resuspend the bacterial pellet in 350 μL of STET solution (0.1 M NaCl; 10 mM Tris, pH 8.0; 1 mM EDTA, pH 8.0; and 5% Triton X-100).

   c. Add 25 μL of a freshly prepared solution of lysozyme (10 mg/mL in 10 mM Tris, pH 8.0). Mix by vortexing briefly.

   d. Place the tube in a boiling-water bath for exactly 40 sec.

   e. Centrifuge the bacterial lysate in the microfuge for 10 min.

---

*Blue colonies on the positive control plate (see Section G.1.b) derived from the transformation of competent cells with uncut pUC9.

f. Carefully transfer the supernatant to a clean 1.5-mL microfuge tube. Use a pipettor, and make sure that the tip does not touch the pellet.

g. Add to the supernatant 40 μL of 2.5 M sodium acetate, pH 5.2 and 420 μL of isopropanol. Mix by vortexing, and incubate the tube for 5 min at room temperature.

h. Recover the pellet of nucleic acids by centrifugation for 15 min at 4°C in a microfuge.

i. Remove the supernatant. Stand the tube in an inverted position on a paper towel to allow all of the fluid to drain away. Remove any drops of fluid adhering to the walls of the tube.

j. Add 1 mL of 70% EtOH and recentrifuge for 5 min at 4°C in a microfuge.

k. Discard the supernatant, drain the tube, and allow the pellet to dry in a vacuum desiccator for 10 min.

l. Resuspend the nucleic acid pellet in 30 μL of TE buffer, pH 8.0.

4. Digestion of recombinant plasmid with *Eco*R1 and *Bam*H1

a. Set up two digestion tubes as follows:

	Tube no.	
Additions (μL)	*1*	*2*
Plasmid isolated from transformed colonies	10	10
Sterile H_2O	6	6
10× buffer	2	2
*Eco*R1 enzyme	2	–
*Bam*HI enzyme	–	2
Total	20	20

b. Incubate for 1 hr at 37°C.

c. Add 5 μL of tracking dye (0.25% bromphenol-blue in 50% glycerol) to stop the reaction, and store the digests in the refrigerator until the next laboratory period.

5. Agarose gel electrophoresis of recombinant fragments and analysis of data.

   a. Prepare the 0.8% agarose solution by adding 2.4 g of powder to 300 mL of TBE buffer. Dissolve by bringing to a boil over a Bunsen burner or in a microwave oven. Allow solution to cool to 50°C.

   b. Add 15 μL of ethidium bromide from a stock solution (10 mg/mL) to obtain a concentration of 0.5 μg/mL **(WEAR GLOVES)**.

   c. A large horizontal slab gel will be used. The plastic form, which is open at both ends, must be taped securely to provide the mold into which the agarose is poured. Insert the comb, which will form the sample wells. (Be sure that the teeth do not cut through the bottom of the gel. There must be at least a 1-mm layer of agarose between the bottom of the teeth and the base of the gel.) Pour the cool agarose solution carefully to avoid air bubbles.

   d. Allow the gel to set completely, which takes about 30 min. ***CAREFULLY*** remove the comb and the two strips of tape.

   e. Transfer the slab gel to the electrophoresis tank, and add enough TBE buffer to cover the gel to a depth of at least 1 cm.

   f. Load the following samples into each of four wells:

      i. Commercial *Hind* III fragments, 5 μL + 1 μL tracking dye.

      ii. Uncut plasmid, 5 μL + 1 μL tracking dye.

      iii. 10 μL of the *Eco*R1 digested plasmid + 1 μL tracking dye.

      iv. 10 μL of the *Bam*H1 digested plasmid + 1 μL tracking dye.

   g. When all of the samples have been loaded, connect the electrophoretor to a power supply (be sure to connect the red lead to the bottom of the apparatus) and apply 150 V for about 3 hr.

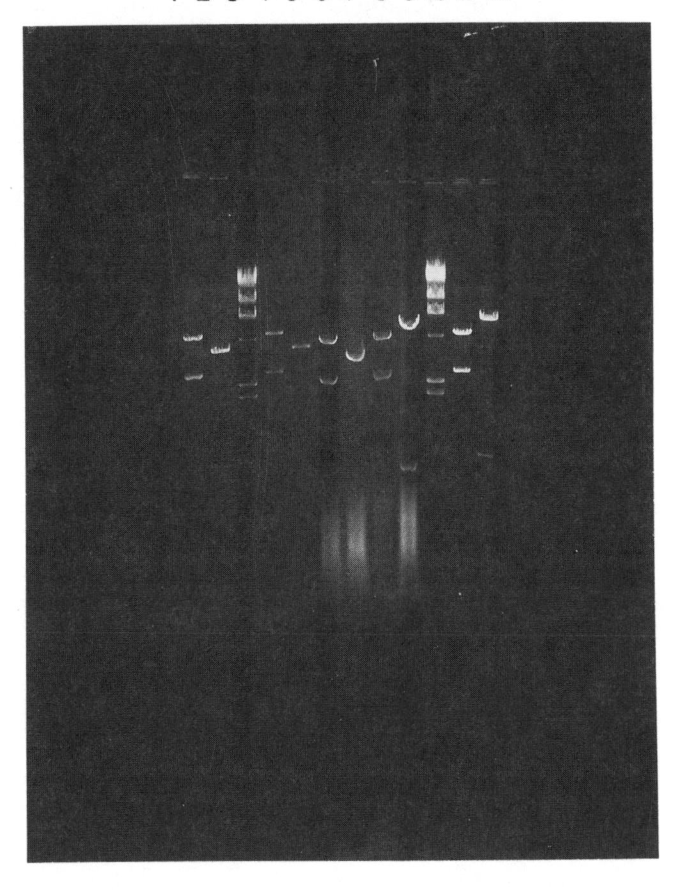

Fig. 7.3. Typical gel pattern for recombinants digested with *Eco*R1 and *Bam*H1 endonucleases.

Lane#
1 *Eco*R1 cut recombinant- boiling miniprep (+RNAse)
2 Same as 1, cut with *Bam*HI
3 λ *Hind* III ladder, 5μL (0.1 mg/mL)
4 *Eco*R1 cut recombinant- alkaline miniprep (+RNAse)
5 Same as 4, cut with *Bam*HI
6 *Eco*R1 cut recombinant- alkaline miniprep
7 Same as 6, cut with *Bam*HI
8 *Eco*R1 cut recombinant- alkaline miniprep
9 Same as 8, cut with *Bam*HI
10 λ *Hind* III ladder, 10μL (0.1 mg/mL)
11 *Eco*R1 cut recombinant- boiling miniprep (+RNAse)
12 Same as 11, cut with *Bam*HI

h. Turn off the power supply, and disconnect the apparatus carefully. Then, wearing gloves to avoid contact with ethidium bromide, remove the gel to a UV viewing box to note the fluorescing bands.

i. Photograph the gels, and make measurements of the bands from the photographs. (A typical picture of such a gel is shown in Fig. 7.3.)

j. Calculate the sizes of the fragments obtained from the *Eco*R1 and *Bam*H1 digestions based on the distances of the bands from the origin and the *Hind* III standard. Refer to the restriction map for λ DNA (Fig. 7.2), identify the insert that was ligated into the vector, and determine its orientation on pUC9.

CHAPTER 8

Lipids

Day 1

1. Safety in the laboratory.
2. Extraction and isolation of lipids.
3. Start of gravimetric analysis.

Day 2

1. Continuation of gravimetric analysis and calculation of *total* lipids.
2. Separation of lipids on a Sep-Pak column.
3. Concentration of *neutral* and *polar* lipids.

Day 3

1. Preparation of fractions for cholesterol determinations.
2. Cholesterol determinations.
3. Selection of developing solvents.

Day 4

1. Preparation of fractions for spotting TLC plates.
2. Spotting, development, and visualization of TLC plates.
3. Conclusion of gravimetric analysis.

Day 5

1. Saponification of *neutral* lipids and methylation of fatty acids.
2. Separation by argentation chromatography.

3. Preparation of tubes for the P_i determinations–bring solutions to dryness before the next laboratory period.

Day 6 Ashing of lipids and determination of inorganic phosphate in fractions.

REFERENCES: Dryer, R.L. and Lata, G.F. *Experimental Biochemistry*. Oxford University Press., New York, pp. 360-362 (1989).

Gebhardt, S. and Matthews, R. (eds) *Nutritive Value of Foods*, U.S. Government Printing Office, Washington, D.C. (1985).

Gurr, M.I., and James, A.T. *Lipid Biochemistry–An Introduction*, 2nd ed. Halsted Press, New York (1975).

Heftmann, E. *Chromatography*, 3rd ed. Van Nostrand, New York (1975).

Kates, M. *Techniques of Lipidology: Isolation, Analysis and Identification of Lipids*, 2nd ed. Elsevier, New York (1986).

Windholz, M. (ed) *Merck Index*, 10th ed. Merck Inc., Rahway, N.J. (1983).

Randerath, K. *Thin-layer Chromatography*, 2nd ed. Academic Press, New York (1968).

Rawn, Ch. 9.

Robyt and White, pp. 373-378.

Stryer, Ch. 23.

A. Introduction

Lipids are a heterogeneous group of biomolecules that are insoluble in water but highly soluble in organic solvents such as ether and chloroform. They are constituents of membranes, serve as energy storage molecules, and some, such as the hormones and vitamins, have biological activity. They are also found in combination with other molecules, such as proteins (lipoproteins) and carbohydrates (glycolipids). Although lipids vary greatly in their structures, upon alkaline hydrolysis most break down into some type of backbone structure and fatty acid moieties.

Two classes of lipids, based on their solubility characteristics, are studied here. The *neutral* lipid fraction will include the least polar triacylglycerols, cholesterol, and cholesterol esters as well as fatty acids. The *polar* fraction will consist mainly of the phosphate-containing phospholipids. The basic structure of the triacylglycerols is illustrated below:

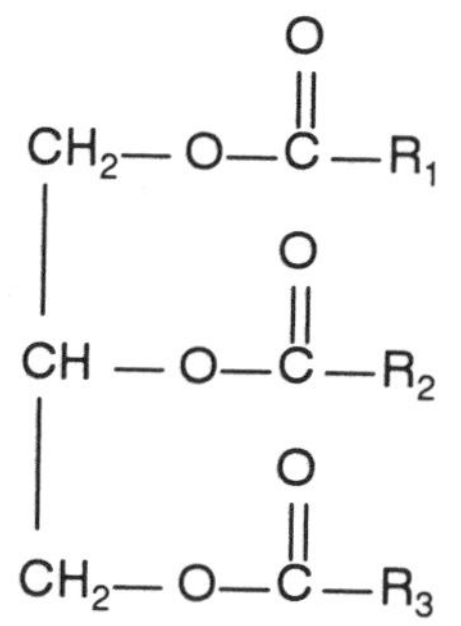

$$
\begin{array}{l}
CH_2-O-\overset{\overset{\displaystyle O}{\|}}{C}-R_1 \\
|\\
CH\ -O-\overset{\overset{\displaystyle O}{\|}}{C}-R_2 \\
|\\
CH_2-O-\overset{\overset{\displaystyle O}{\|}}{C}-R_3
\end{array}
$$

in which the glycerol backbone is attached by ester linkages to three fatty acid molecules, which may be of the same or different composition. Also found in nature are mono- and diacylglycerols, in which only one or two fatty acids are attached to the glycerol molecule. Most naturally occurring fatty acids are of C12 to C20 chain lengths and vary in extent of unsaturation, as will be discussed below in terms of the separation of fatty acids.

The most common sterol associated with the neutral lipids is cholesterol, which may occur free or esterified to a fatty acid at the site of the free hydroxyl group:

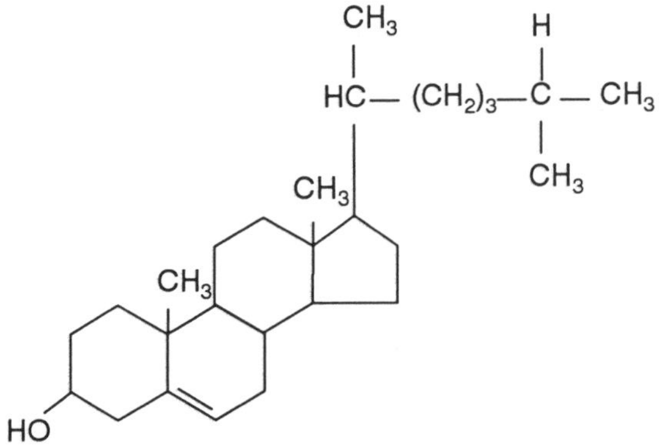

CH_3 H

HC— $(CH_2)_3$—C — CH_3

CH_3 CH_3

CH_3

HO

Cholesterol

The phosphoglycerides, which are commonly referred to as *polar* lipids, are constituents of membranes and also contain a glycerol backbone, but, whereas the hydroxyl groups of C1 and C2 are esterified to the -COOH groups of two fatty acids, the third is esterified to phosphoric acid. The common natural phosphoglycerides are then further esterified to an "alcohol." The most common phosphoglycerides contain serine, ethanolamine, or choline. They have a glycerol backbone and the headgroups in place of the (X) as indicated below:

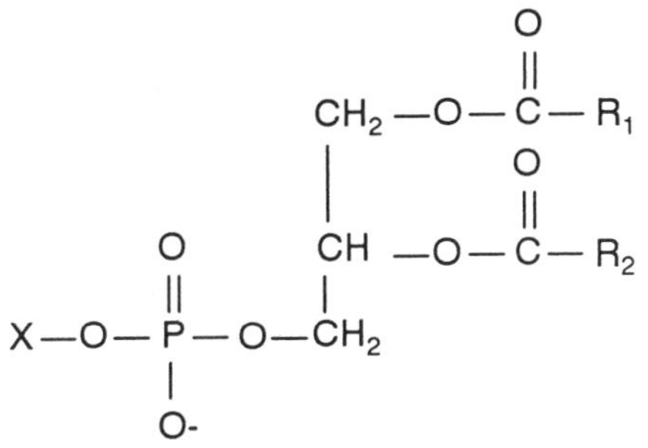

$$
\begin{array}{l}
\qquad\qquad\qquad CH_2-O-\overset{O}{\overset{\|}{C}}-R_1 \\
\qquad\qquad\qquad | \\
\qquad\qquad\qquad CH-O-\overset{O}{\overset{\|}{C}}-R_2 \\
\qquad\qquad\qquad | \\
X-O-\underset{O^-}{\underset{|}{\overset{O}{\overset{\|}{P}}}}-O-CH_2
\end{array}
$$

Phospholipid	*–X group*
Phosphatidic acid	-H
Phosphatidylserine	$-CH_2CH(\overset{+}{N}H_3)COOH$
Phosphatidylethanolamine	$-CH_2CH_2\overset{+}{N}H_3$
Phosphatidylcholine	$-CH_2-\overset{+}{N}(CH_3)_3$
Phosphatidylinositol	-Inositol
Phosphatidylglycerol	-Glycerol

Sphingomyelin, which is also found in membranes and is isolated in the polar fraction, does not have a glycerol moiety. It consists of a molecule of sphingosine, choline, and a fatty acid:

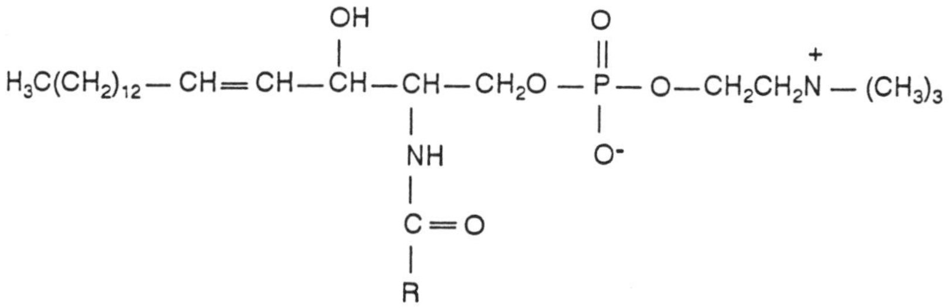

$$H_3C(CH_2)_{12}-CH{=}CH-CH(OH)-CH(NH-C(=O)-R)-CH_2O-P(=O)(O^-)-O-CH_2CH_2\overset{+}{N}-(CH_3)_3$$

Glycolipids are sugar-containing lipids present in some animal and plant cells. In animals, sphingolipids are most abundant in nerve cells. They consist of a sphingosine backbone as sphingomyelin, but the phosphoryl choline group is replaced by one or more sugar residues. In a cerebroside, the sugar residue may be either glucose or galactose. Plant tissues commonly contain mono- and/or digalactosyldiglycerides.

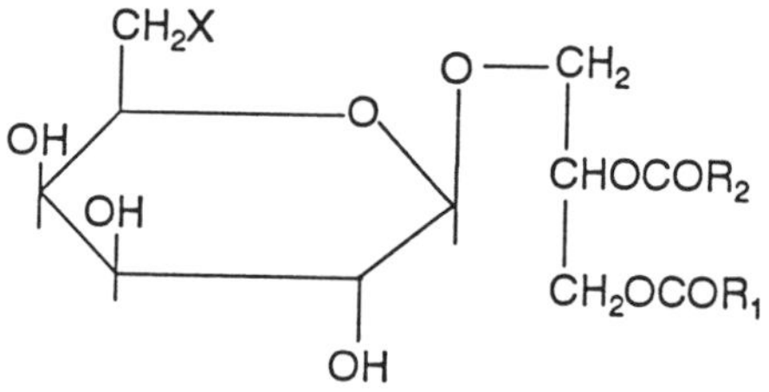

CH_2X; OH; OH; OH; $O-CH_2$; $CHOCOR_2$; CH_2OCOR_1

Glycolipid	*-X group*
Monogalactosyldiacylglycerol (MGDG)	-OH
Digalactosyldiacylglycerol (DGDG)	-Galactose

The lipids found in cell membranes serve a dual function. One end of the molecule, the head, is *polar*, and the longer portion, or tail, of the lipid is *nonpolar*. Thus the molecule has hydrophilic and hydrophobic groups, which confer solubility properties in either aqueous or organic solvents. For the most part, the nonpolar lipids are readily soluble in nonpolar solvents such as chloroform, ether, or hexane and only slightly soluble in methanol or water. A mixture of solvents is generally used in working with polar lipids, as is discussed in a subsequent section.

The nonpolar tails of the membrane lipids are fatty acids containing 14 to 20 carbon atoms. The fatty acids frequently are *saturated*, containing no double bonds, but they may be *unsaturated* and contain one or more double bonds. Unsaturated fatty acids have low melting points, as in oils that are rich in unsaturated fats. Having an abundance of unsaturated fatty acids confers on the membranes greater fluidity. The polar heads have a glycerol-3-phosphate backbone as exemplified by phosphatidylethanolamine, phosphatidylserine, and phosphatidylcholine. Interspersed among these molecules is *cholesterol*, which also has a polar hydroxyl head and a nonpolar tail. The heads and tails of the cholesterol molecules thus line up with the heads and tails of their neighbors. However, the rigid ring structure of cholesterol fits in between two adjacent phospholipid chains, and this combination imparts stiffness to the membrane.

Most animal membranes contain a mixture of phospholipids as well as cholesterol. Rat liver tissue, for example, has the following composition:

Type of phospholipid (phosphatidyl, p –)	*Percent of phospholipids*
p-choline	52.2
p-ethanolamine	25.2
p-inositol	7.6
p-serine	3.4
p-glycerol	4.8
sphingomyelin	4.2
other	2.6

From Rawn, p.223.

In the following experiments, students will isolate lipids from a sample of their choosing. Good lipid sources are eggs, nuts, avocado, seeds, fresh meat, or fish. The materials should be in their natural form, as processed foods may have been altered. Sufficient sample should be used for analysis to provide *ca* 1 g of lipid. *Nutritive Value of Foods* (1985) is a good source of information on the composition of many foods, and it lists values for the following: fats, calories, saturated fatty acids, unsaturated fatty acids (oleic and linoleic acids), phosphorus, vitamin A, and protein. Discussion of the latter two will also be useful for the nutritional studies in Chapter 9.

Poultry products are important sources of nutrients in the diet. Although the meat itself is relatively low in fats, other components such as the skin, egg yolks, and liver are rich in lipids. The data in Table 8.1 are illustrative of the distribution of the lipids found in animal tissue.

TABLE 8.1. Lipid Composition of Chicken Liver and Eggs[a]

		Percent	
		Eggs	Liver
Triglycerides		65	48.5
Fatty acids			
Myristic	14:0	0.4	–
Palmitic	16:0	27.8	29.0
Palmitoleic	16:1	5.3	–
Stearic	18:1	5.7	6.3
Oleic	18:1	46.6	48.7
Linoleic	18:2	13.7	9.6
Linolenic	18:3	0.5	–
Phospholipids		31.5	48.0
Sphingomyelin		2.3	5.5
p-Choline		76.3	54.7
p-Ethanolamine		17.4	29.3

[a]Christie, W.W., and Moore, J.H. The lipid composition and triglyceride structure of eggs from several avian species. *Comp. Biochem. Physiol* 41B:287–306 (1972).

B. Precautions During Lipid Procedures

Experimental work in lipid biochemistry requires some special precautions. **OBSERVE THESE SAFETY MEASURES**:

1. Use a hood whenever possible!
2. To pipette highly volatile solutions (*e.g.*, ether), use a propipette. Draw the solution up into the pipette, and let it drain down several times to saturate the space above the liquid; otherwise, the solution will tend to be expelled from the tip of the pipette. Be gentle when using either the large propipette or a Pasteur pipette bulb; when solvent reaches the bulb, contaminants are introduced, and the rubber of the bulb deteriorates. **NEVER** invert a pipette with fluid in it.
3. Since many of the solutions used in these experiments are highly volatile, solutions should always be kept stoppered (with aluminum foil or tightly fitting glass stoppers) when not in immediate use. This will prevent an accumulation of

toxic and/or highly flammable vapors in the air as well as keeping the concentration of the solutions reasonably constant. Store solutions, tightly capped, at the lowest temperature possible (*e.g.*, −20°C).

4. Lipids are highly prone to oxidation and polymerization, processes induced by both atmospheric oxygen and light. Solutions to be stored should be gassed briefly with nitrogen and stored in the dark.
5. **MOST IMPORTANT**: Solvents are highly flammable; do not use near open flames.
6. Dispose of any waste solvent properly. A container for this purpose is located in the hood. Never pour organic wastes down the sink.
7. Properties of some common organic solvents are shown in Table 8.2.

C. Isolation of Lipids From Natural Products and Gravimetric Analysis

Extraction of Lipids

The lipids are released from intact cells by grinding in methanol (MeOH) and extracted from the tissue with ether and MeOH. The debris is removed by centrifugation, and nonlipid contaminants that are water soluble are removed by extraction with a salt solution.

Experimental Procedure

1. Weigh out an appropriate amount of natural product to obtain approximately 1 g of fat. Record the weight to the nearest 0.1 g, and transfer the sample to a mortar or homogenizer. Add about an equal amount of acid-washed sand (for tender tissue, grind in a homogenizer), add 5 mL of MeOH, and grind to a smooth paste. Transfer to a 50-mL centrifuge tube, using another 5 mL of MeOH to rinse the mortar. Heat by immersing the preparation in a 60°C water bath for about 3 min. Allow it to cool to room temperature. (If the lipid source is a liquid, use a separatory funnel for the extraction.)
2. Transfer the material back to the mortar with about 5 mL of ethyl ether, and grind again. Return the suspension to the 50-mL centrifuge tube, and rinse the mortar with an-

TABLE 8.2. Properties of Common Organic Solvents

	Acetone	Chloroform	Ether (Ethyl ether)	Ethanol	Isopropanol	Methanol	Petroleum ether
Chemical formula	CH_3COCH_3	$CHCl_3$	$CH_3CH_2OCH_2CH_3$	CH_3CH_2OH	$CH_3CHOHCH_3$	CH_3OH	Mixed alkanes
Density	0.78	1.48	0.70	0.78	0.78	0.79	Varies but $<H_2O$
Miscible with H_2O	Yes	No	No	Yes	Yes	Yes	No
Boiling point	56°C	61°C	35°C	78.5°C	82°C	65°C	Varying ranges
Flammable	Flammable	Non-flammable	Explosive	Flammable	Flammable	Flammable	Highly flammable
Health hazard	Low hazard. Eye and skin irritation. Inhalation causes headaches.	Toxic. Carcino-genic. Irritating to skin.	Vapors produce anesthesia. Irritating to eyes and mucous membranes.	Low hazard.	Low hazard. Mildly irritating to eyes, nose, throat.	Ingestion can cause blindness.	Irritating to skin and when inhaled.
Miscellaneous properties	Pungent. Sweetish taste.	Character-istic odor. Sweetish taste. Light sensitive.	Decomposes on long exposure to air. Highly volatile.			Good solvent. Relatively high polarity.	Good all-purpose solvent. Low polarity.

other 5 mL of ethyl ether. Centrifuge in a clinical centrifuge for 5 min. Be sure opposite tubes are balanced. Decant the supernatant into a 50-mL Erlenmeyer flask. Reextract the pellet three times more with 5-mL portions of MeOH-ether (1:1), heating the preparation cautiously in the water bath for 3 min during each extraction, and repeat centrifugations. Combine the supernatants in the flask. (The centrifuge tubes should be covered with aluminum foil to minimize evaporation of ether and the centrifuge should be used in a hood if space permits.)

3. Carefully pour the combined extracts through a funnel fitted with fluted filter paper (Whatman No. 2) into a 100-mL graduated cylinder containing 40 mL of isotonic saline (0.9% NaC1) solution. Rinse the Erlenmeyer flask with 10 mL of ether, and add the ether to the filtering apparatus.
4. Mix the contents of the cylinder thoroughly with a flattened stirring rod, and allow the mixture to settle until two distinct layers have formed. Transfer the clear ether layer with the aid of a Pasteur pipette to a 25-mL graduated cylinder. Add 5 mL more of ether, mix again, and permit the layers to settle. Add the upper layer to the first extract in the graduated cylinder. Repeat this extraction of the saline solution two more times, and then adjust the volume in the graduated cylinder to 25 mL either by evaporation with N_2 or by addition of ethyl ether.
5. Pipette 5 mL onto the preweighed boat No. 1 for the gravimetric analysis (p. 165).
6. Transfer the remaining 20 mL of lipid extract to a 50-mL glass tube fitted with a ground glass stopper. Label it "*Total Lipids*," and store it in the refrigerator for later use.

Column Fractionation of Lipids

Silicic acid column chromatography separates the less polar neutral lipids from the more polar phospholipids on the basis of differential adsorption to an adsorbent. The compounds to be separated are put on the column and then washed through the adsorbent by a steady but slow flow of eluant. The least strongly adsorbed compound will be washed off the column first, and the most strongly will be recovered last. The rate at which the column is developed depends on the polarity of the eluant. By starting the elution with ether, the less polar "neutral" fraction will be collected first. Then the polarity of the eluant will be increased by adding MeOH, and the "polar" fraction will be recovered.

Experimental Procedure

CARRY OUT THIS PROCEDURE IN A FUME HOOD!

A commercial product, the Sep-Pak cartridge, can be used for convenience, but a silicic acid column can be used as well, although it results in a somewhat slower separation. For the silicic acid separation, a glass column, 1 × 25 cm in size, is attached to a ring stand, and the following procedure is followed:

1. Plug the column with a small amount of glass wool, and cover this plug with 1 cm of acid-washed sand.
2. Fill the column approximately one-half full with hexane, and pour into it a well-mixed slurry of 4 g of silicic acid and 2 g celite in 30 mL of hexane.
3. Allow the column to settle, but do not let the adsorbent run dry.
4. Bring the level of the hexane to the top of the packing material and carefully layer on the sample of "total lipids." (The amount of lipid in this sample should not exceed 50 mg.)

The "Sep-Pak" cartridge is a convenient silica column that is used to separate compounds on the basis of their polarity. (For a discussion of Sep-Pak columns, see Chapter 4). The sample to be applied should be suspended in a nonpolar solvent, and the compounds of interest are removed by washing the cartridge with solvents of increasing polarity. The cartridges are 1-mL columns made of silica and an aqueous slurry, and they have a binding capacity of 100 mg of lipids.

1. Calculate the volume of the "total lipids" sample containing 70 to 80 mg of lipids from the gravimetric analysis.
2. Evaporate the appropriate amount of "total lipids" to dryness under N_2, and dissolve these lipids in approximately 2 mL of ether:MeOH (100:1; v/v).

3. a. Clamp a syringe barrel in the vertical position, Luer end down.
   b. Before attaching the Sep-Pak, rinse the syringe first with acetone and then with ether:MeOH (100:1; v/v) to ensure that the syringe is clean. Allow to dry.
   c. Attach the longer end of the Sep-Pak cartridge to the Luer end of the syringe. The syringe barrel will act as the solvent reservoir.
   d. Set up a pressure bulb that will fit snugly on the syringe barrel top (see Fig. 8.1).

THE COLUMN MUST NOT RUN DRY. Therefore, it is important to have all the solutions available before starting this procedure. Set up ahead of time the following solutions in seven 50-mL ground glass-stoppered tubes. Then transfer the entire contents of each tube with a Pasteur pipette.

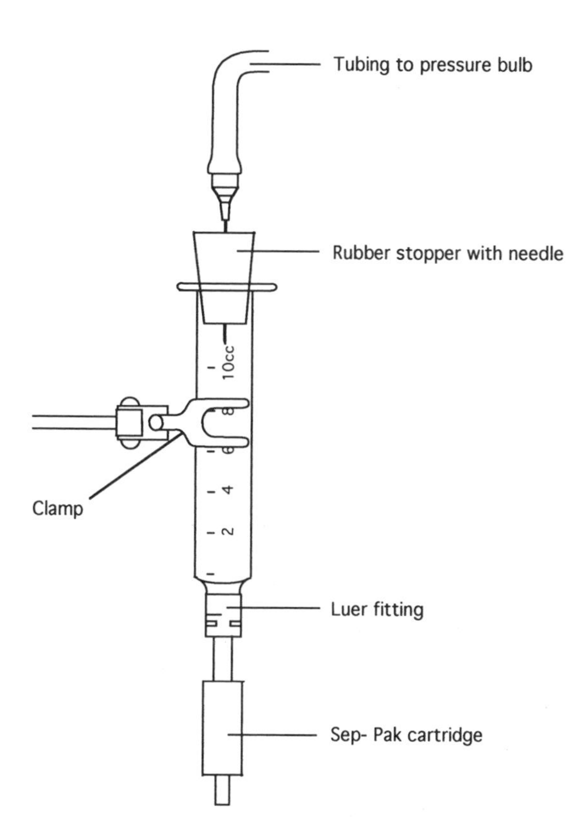

Fig. 8.1. Sep-Pak cartridge attached to a glass syringe with Luer fitting.

Tubes with samples to be applied	*Tubes with samples to be collected*
1. 2 ml "total lipids" sample	(1) Lipid fraction
2. 2 ml ether: MeOH (100:1)	
3. 10 ml ether: MeOH (100:1)	(2 + 3) Neutral fraction
4. 8 ml acetone	
5. 8 ml acetone: acetic acid (100:1)	(4 + 5) Glycolipid fraction
6. 5 ml ether: MeOH (1:1)	
7. 5 ml MeOH	(6 + 7) Polar fraction

4. Load the sample on the column, and allow it to enter the gel. Then rinse the tube with approximately 2 mL ether:methanol (100:1; v/v), and add this wash to the column. Begin collecting the "neutral lipid" fraction in a 50-mL glass-stoppered tube. Label this tube "*Neutral Lipids*"
5. Add another 10 mL of ether:MeOH (100:1; v/v) to the column, and continue collecting eluant in the same "neutral lipids" fraction tube.
6. If glycolipids are contained in the sample, add approximately 8 mL of acetone. Begin collecting the "*Glycolipids*" fraction in a second 50-mL glass-stoppered tube once the first milliliter of the acetone has entered the column. Add another 8 mL of acetone:acetic acid (100: 1; v/v), and add this wash to the "glycolipids" fraction tube.
7. Follow with 5 mL of ether:MeOH (1:1), and begin collecting the "*Polar lipids*" fraction in a third glass-stoppered tube. Add 5 mL of MeOH, and add the eluant to the "polar lipids" fraction tube.
8. Bring the volumes of the "neutral," "glycolipid," and "polar" fractions to 20 mL in a graduated cylinder. Remove 5 mL of each fraction to preweighed aluminum weighing boats for gravimetric analysis, and transfer the remaining fractions to 50-mL glass stoppered tubes.
9. Bring the remaining samples to dryness using a stream of N_2 and a warm water bath to facilitate evaporation. Resuspend the neutral fraction in 5 mL of ether:MeOH 100:1; v/v) and the polar fraction in 5 mL of ether:MeOH (1:3;

v/v). *Record the volume of each fraction* after storage, because some solvent will evaporate over time even in the cold. The change in volume can be accounted for in making the calculations, or the volume can be restored by adding some of the solvent mixture in which the fraction was suspended. Store fractions at 4°C.

10. Proceed with the gravimetric analysis.

Gravimetric Analysis

The gravimetric analysis will permit determination of yield and concentration of lipid in the extract and in the "neutral" and "polar" fractions obtained from the separation on the column.

1. Obtain three aluminum weighing boats, and label them No. 1, 2, and 3 (add a fourth if glycolipids were isolated). Weigh each boat at least three times. Using forceps only, tare three aluminum foil weighing boats to the nearest 0.1 mg on an analytical balance. Place the boats on glassine weighing paper at all times. It is important not to handle boats directly, as the weights may be affected appreciably.
2. Add *exactly* 5.0 mL of each of "total lipids," "neutral," and "polar" fractions to the preweighed boats, and cover them with watch glasses.
3. Place these boats on weighing papers, and store them in the hood to dry until the next laboratory period.
4. Determine the concentration of lipids in the original material.
   a. Report the data in mg of lipid/g of tissue.
   b. Calculate the percent lipid in the tissue.
   c. Calculate the amount of "neutral" and "polar" lipids present per gram of tissue (and "glycolipids" if applicable).
   d. Calculate the percent recovery from the Sep-Pak column.

D. Determination of Cholesterol in the Lipid Fractions

The presence of cholesterol and cholesterol esters in the lipids extracted from the original material will be demonstrated by the thin-layer chromatography method. In this experiment, the amount of total cholesterol in the three lipid fractions will be determined.

This method may be useful in determining the cholesterol content of various tissues. Regardless of source material, it is necessary first to extract the cholesterol in absolute ethanol. The samples will be evaporated to dryness under N_2 and the residues resuspended in ethanol. The procedure is given for bovine liver, which has an approximate cholesterol concentration of 0.26%. The amounts used in the assay can be adjusted when other materials are examined.

Preparation of the Samples for the Assay

Experimental Procedures

1. Carefully check the volumes of the three fractions, because some evaporation may have occurred during storage. Transfer the *total* fraction to a 25-mL graduated cylinder and the *neutral* and *polar* fractions to 10-mL graduated cylinders. Bring these to 20, 5, and 5 mL, respectively, with the appropriate solvents.
2. See Table 8.3 for volumes of each sample to assay, and transfer these amounts to 13 × 100 mm test tubes. Evaporate to dryness under N_2; then resuspend the samples in the amounts of absolute alcohol indicated. While the tubes are drying, proceed with the standard curve.
3. Construct a flow chart giving the amount of lipid in the fractions that were evaporated, the amounts present in the samples resuspended in alcohol, and the concentration in the aliquots delivered into the assay tubes. This

TABLE 8.3. Volumes of Each Fraction to Analyze for Cholesterol Determinations

Fraction	Volume (mL)	Source	Volume of alcohol added to resuspend (mL)
Total lipids	1.0	Original extract	1.0
Neutral	1.0	Concentrated column fraction	0.2
Polar	2.0	Concentrated column fraction	0.2

diagram will facilitate the calculations to be made from the data.

Construction of a Standard Curve for Cholesterol

1. Set up a series of seven test tubes, and pipette varying amounts of a standard solution (0.1 mg/mL) so that the tubes contain 0, 20, 40, 80, 120, 160, and 200 mg of cholesterol.
2. Add sufficient absolute alcohol to the tubes to bring the volume to 2.0 mL.
3. Add 2.0 mL of ferric chloride reagent, and mix *carefully* at the lowest setting of the Vortex mixer. **THIS REAGENT IS STRONGLY ACIDIC AND MUST BE HANDLED WITH EXTREME CARE**. (Ferric chloride reagent consists of 80 mL of a solution of 2.5% $FeC1_3 \cdot 6H_2O$ in conc. H_3PO_4 and 920 ml of conc. H_2SO_4 per L. It is best to transfer this potent reagent with an automatic dispenser.)
4. Incubate the reaction tubes for 30 min at room temperature, and read at 550 nm. Transfer to cuvettes very carefully. Do not allow any liquid to drip along the outside of the tubes. Wash any spills immediately with a solution of bicarbonate followed by a generous amount of water.

Determination of Cholesterol Content of Lipid Fractions

1. Pipette the solutions indicated in Table 8.3 for each of the three lipid fractions, and evaporate them to dryness in small test tubes under N_2. Redissolve the dried samples in absolute ethanol. If the samples do not readily go into solution, sonication may be necessary.
2. Set up a series of 10 test tubes, and deliver 0.1, 0.05, and 0.02 mL of each fraction using an automatic micropipettor since the amount of sample is small.
3. Add sufficient absolute ethanol to bring the volume to 2.0 mL. To tube No. 10 add 2.0 mL of ethanol only for the blank.
4. Add the cholesterol reagent, and proceed as with the standard curve.
5. Determine the amount of cholesterol (in mg/g of tissue) found in each fraction, and calculate the percent cholesterol in the tissue analyzed.

E. Thin-Layer Chromatography

Preparation of Silica Gel Plates (Optional)

REFERENCES: Randerath (1968).
Wharton, D.C., and McCarty R.E. *Experiments and Methods in Biochemistry*. Macmillan, New York, pp. 111–115 (1972).

Thin-layer chromatography (TLC) plates can be prepared as indicated here, or 20 × 20 cm silica gel chromatogram sheets for TLC can be used. (These plates are precoated on a plastic backing and can be purchased from Eastman Kodak Co., Rochester, NY).* The plates are activated before use by briefly heating them in an oven at 110°C. Care must be taken to keep the plastic backing from curling when commerical plates are used.

Experimental Procedure

1. The thickness of the adsorbent will determine how much material can be chromatographed. Masking tape will be used to regulate the adsorbent thickness. Clean three 20 × 20 cm glass plates with 95% EtOH to ensure adhesion of the adsorbent. Place the plates on a large sheet of clean paper, and attach masking tape (a single layer) along the two vertical edges of the plates.
2. For the determination of the solvent system of choice, microscope slides will be used as the plates, and beakers will be used as the developing chambers. The procedure is similar to that used for the larger plates, but six microscope slides are aligned and held together on a sheet of clean paper by placing masking tape along the free edges of the slides.
3. Suspend 4 g of silica gel H in 8 mL of water in a 125-mL Erlenmeyer flask for the slides or 5 g of silica gel H in about 12 mL of water for each of the large plates. Quickly mix the slurry by shaking to give a uniform suspension. Pour the suspension onto the large glass plate or across the small plates. Spread the slurry evenly across the plates with a

---

*One of the advantages of preparing plates rather than using the commercial ones is that the properties of the adsorbent can be modified. For example, silver nitrate can be added to the silica gel to permit differentiation between varying levels of unsaturation in lipids. Ag-TLC chromatography will be discussed in a subsequent section.

glass rod. Spreading must be carried out within a few minutes after mixing the water and silica gel, because the calcium sulfate binder sets quickly.

4. Allow the plates to air dry, remove the tape, and store the plates in a storage cabinet with desiccant. The silica gel coating should appear uniform.
5. Before spotting the plates, they should be dried in an oven at 110°C for 15 min.

Selection of Developing Solvent Systems

Many different proportions of solvents for separation of identical compounds have been reported in the literature. One of the major differences with TLC solvent systems for lipids is the amount, if any, of water or other polar solvent present. In this experiment, the proper amount of water or acetic acid (HOAc) for best separation will be determined.

Experimental Procedure

1. Set up eight developing chambers (250-mL beakers) with the solvent systems listed below (20 mL of each should be sufficient). Cover each beaker with aluminum foil and mix solvents well. **WORK IN THE HOOD!**

   a. Polar lipids

$CHCl_3$:	CH_3OH:	H_2O
25	15	0, 1, 2, 4

   b. Neutral lipids

hexane:	ether:	HOAc
40	10	0, 0.5, 1, 2.5

   c. Glycolipids

$CHCl_3$:	CH_3OH:	H_2O
25	10	0, 1, 2, 5

2. Spot eight silica gel-coated microscope slides or commercially prepared plastic sheets cut to the same size with 5 μL of "total lipid" fraction and with 10 μL of a mixture of standards.
3. When the solvent front has risen to within 1 cm of the top of each slide, remove the slide from the chamber, allow the solvents to evaporate in the hood, and "visualize" the spots after placing slides in an iodine chamber for a few minutes. A beaker covered with aluminum foil makes a suitable chamber for this procedure.

TABLE 8.4. Unknowns and Standards Used for Application

(1) Sample	(2) Concentration	(3) Solvent	(4) Amount to spot (μL)
A. Unknowns[a]			
1. Total lipids	As originally isolated (1 mL of 20)		
a. Animal source		0.1 mL	5
b. Plant source		0.1 mL	5
2. Polar extract	Concentrated Sep-Pak extract (1 mL of 5)		
a. Animal source		0.05 mL	2 and 5
b. Plant source		0.05 mL	5 and 10
3. Neutral extract	Concentrated Sep-Pak extract (1 mL of 5)		
a. Animal source		0.05 mL	2 and 5
b. Plant source		0.05 mL	5 and 10
B. Polar lipid standards			
1. Phosphatidylserine	25 mg/mL	$CHCl_3$:MeOH	2

2. Phosphatidylcholine	25 mg/mL	$CHCl_3$:MeOH	2
3. Phosphatidylethanolamine	25 mg/mL	$CHCl_3$:MeOH	2
4. Sphingomyelin	25 mg/mL	$CHCl_3$:MeOH	2
5. Cardiolipin	12.5 mg/mL	Abs. EtOH	5
6. Standards for brain tissue			
a. Gangliosides	25 mg/mL	MeOH	2
b. Cerebrosides	25 mg/mL	MeOH	2
c. Sphingosine	25 mg/mL	MeOH	2
7. Mixture of 1, 2, 3, and 4	(1:1:1:1)		10
C. Neutral lipid standards			
1. Monoolein	25 mg/mL	$CHCl_3$:MeOH	2
2. Cholesterol	25 mg/mL	$CHCl_3$:MeOH	2
3. Cholesterol esters (cholesterol stearate)	25 mg/mL	$CHCl_3$:MeOH	5
4. Triolein	25 mg/mL	$CHCl_3$:MeOH	2
5. 1,3 Di-olein	25 mg/mL	$CHCl_3$:MeOH	2
6. Mixture of 1 through 5	(4:1:2:2:2)	$CHCl_3$:MeOH	10
D. Glycolipid standards			
1. Monogalactosyl-diglyceride	10 mg/mL	MeOH	5
2. Digalatosyldiglyceride	10 mg/mL	MeOH	5
3. Mixture of 1 and 2	(2:1)	MeOH	10

[a]Bring 1 mL each of total, polar, and neutral extracts to dryness under N_2 and resuspend in $CHCl_3$:MeOH (1:1) as indicated in column 3.

Thin-Layer Chromatography of Lipids

TLC (for a discussion of TLC, see Chapter 4) will be used to fractionate neutral lipids, phospholipids, or glycolipids (plants) into their constituent classes (cholesterol, triacylglycerols, cholesterol esters, etc., or phosphatidylcholine, phosphatidylserine, etc.).

Experimental Procedure

1. By heating at 110°C for *ca* 10 min, activate two previously prepared glass plates or two 20 × 20 cm silica gel-coated sheets for 2 to 3 min, one for polar lipids (or glycolipids) and one for the neutral lipids.
2. Bring 1 mL each of total, polar (glycolipid), and neutral extracts to dryness under N_2 and resuspend in $CHCl_3$:MeOH (1:1) as indicated in column 3 of Table 8.4.
3. Prepare the solvents that were previously selected for the developing tanks. Allow about 150 to 200 mL per tank. Mix the solvents well, and line the tanks with blotting paper to ensure saturation of the atmosphere.
4. Start spotting and developing the polar plate first, as it takes longer to run.
5. With black pencil, draw a light line 2 cm from the bottom of the silica-coated plate. Place nine dots on the line, 2 cm apart, starting 2 cm from the left edge according to the diagram shown below. Record which solution will be spotted on each dot as indicated below. It is advisable that the standards and the unknowns be alternated. In case some area of the plate does not develop properly, the remaining spots can be analyzed.

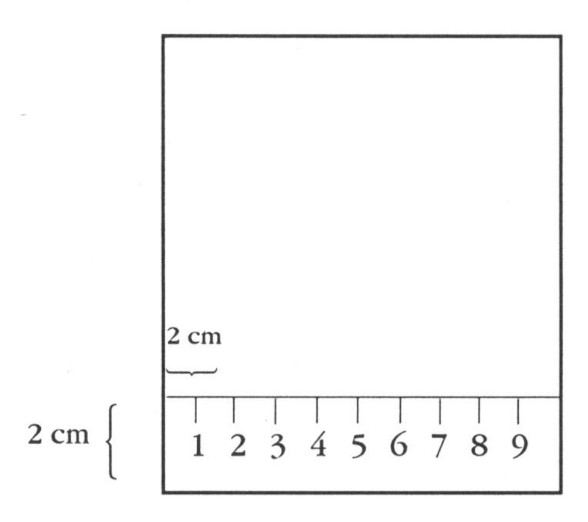

6. Apply the samples in the following order on both plates
   a. Standard.
   b. Standard.
   c. Sample–neutral, polar, or glycolipid.
   d. Standard.
   e. Standard.
   f. Sample–neutral, polar, or glycolipid.
   g. Sample–total lipids.
   h. Standard–mixture.
   i. Standard.
7. Using disposable microcaps, apply the samples of the unknowns in the places indicated on the chart shown above, and deliver the quantities designated in column 4 in Table 8.4 for the appropriate standards.

Visualization of Chromatograms

1. After the chromatograms have been removed (*ca* 1.5 to 2 hr for the polar or glycolipids and 45 min for the neutral lipids) and air dried, place them, solvent front downward, in iodine vapor in a tank in the hood. Leave for 1 to 3 min, and trace the spots on transparent paper. After allowing reversibly bound iodine to evaporate in the hood, spray the chromatogram with one of the solutions listed in Table 8.5. Again trace all spots on the transparent paper, and record the colors observed.
2. Calculate the relative mobilities (R_f) of all spots, using the midpoint of the spots to take the measurements. Identify the lipids present in each fraction from the R_f values and color development.
3. Tabulate the results and identify the unknowns.

F. Extraction and Separation of Fatty Acids From the Total Lipid Fraction

REFERENCES: Dryer and Lata, pp. 360–362.

Mangold, H.K. Thin layer chromatography of lipids. *J. Am. Oil. Chem. Soc.* 38: 708–727 (1961).

Robyt and White, pp. 372–380.

Wynne, E.B., Schmidt, J.A., and Umbreit, G.R. Perchloric acid and catalyzed reactions. I. Methylation of fatty acids, in *Biomedical G.C. Notes*. F. & M. Scientific, Avondale, PA, (1965).

TABLE 8.5. Detection Reagents for Thin-layer Chromatography[a]

Iodine vapor

Dry crystals of iodine are placed in a dry tank. As the iodine vaporizes, a purple vapor is produced. This method detects unsaturated fatty acids well but reacts reversibly with any–HC=CH– linkages.

Results: Brown to yellow spots that will fade as the iodine evaporates. Then a second reagent may be used on the same plate.
Use: All plates.

Sulfuric acid

A solution of 10% H_2SO_4 is sprayed in a fine mist, and followed by heating at 110°C for a few minutes.

Results: Cholesterol and cholesterol esters yield red to purple spots.
Use: Neutral plates.

Molybdenum blue reagent (optional)

An acidified solution of molybdenum is sprayed in a fine mist, and the color is allowed to develop in the hood.

Results: All phosphorus-containing compounds will form blue spots.
Use: Polar plates.

Ninhydrin

The plates are sprayed with ninhydrin reagent and heated at 110°C for 20 min.

Results: Lipids containing primary and secondary amines will yield red-purple spots.
Use: Polar and glycolipid plates

[a]Zweig, G. and Sherman, J. *CRC Handbook of Chromatography*, Vol. 1., CRC Press, Cleveland (1972).

Fatty acids are major components of the storage lipids. The fatty acids comprise the R groups of the triacylglycerols, and in membranes they are the nonpolar "tail" moieties of the structural lipids. The most common fatty acid chains contain from 12 to 24 carbon atoms and have systematic names that are indicative of their length. The fatty acids often are referred to by common names.

Fatty acids in animal and plant tissues are *saturated* (*i.e.,* contain no double bonds) or they are *unsaturated* (*i.e.,* have one or more double bonds). Table 8.6 lists some examples of fatty acids and shows the location of the double bonds.

[illegible] d the degree of saturation [illegible] atty acid. The unsaturated [illegible] ngly more rigid, and their [illegible] as the number of double [illegible]

[illegible] cids from the "total lipids" [illegible] ed, and separation will be [illegible] aphy. Then the fatty acid [illegible] ed under UV light.

[illegible] r potassium salts from lipid [illegible] own in *equation 1*. Methyl [illegible] e fatty acid salts with strong [illegible] *ation 2*). The methyl esters [illegible] ether and concentrated for [illegible]

[illegible] a TLC method that dis- [illegible] acids of differing degrees [illegible] of silica gel to which $AgNO_3$ [illegible] kyl groups of free fatty acids [illegible] gel, it is necessary first to [illegible] ethyl esters. The esters are [illegible] ted plates. The interactions [illegible] ls of the fatty acids result in [illegible] d by this method. As shown [illegible] movement of the free fatty

[illegible] nds in Fatty Acids

No. of carbons and double bonds[a]	Common name	Melting point (°C)
16:0	Palmitate	63.1
16:1 (*cis* 9)	Palmitoleate	−0.5
18:0	Stearate	69.6
18:1 (*cis* 9)	Oleate	13.4
18:2 (*cis* 9, 12)	Linoleate	−5.0
18:3 (*cis* 9, 12, 15)	Linolenate	−11.0
20:0	Arachidate	76.5
20:4 (*cis* 5, 8, 11, 14)	Arachidonate	−49.5

[a]No. of double bonds given after the semicolon, and their location is given in parentheses.

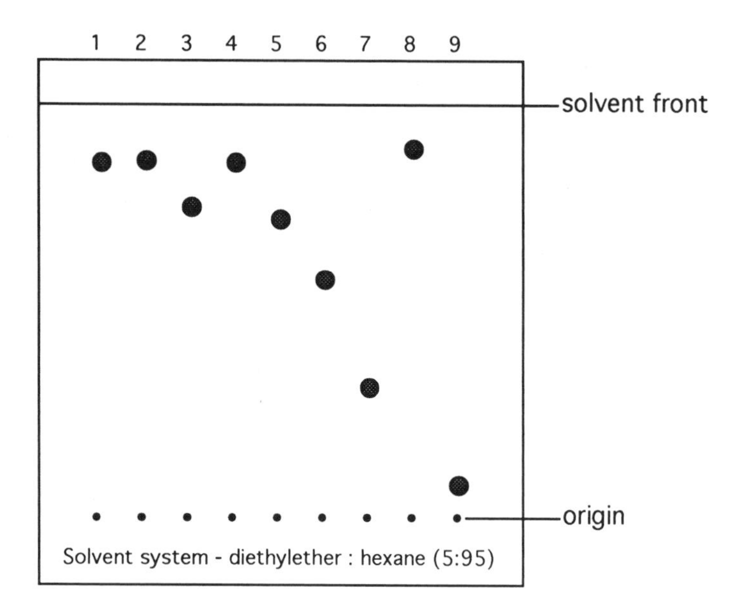

Fig. 8.2. Separation of fatty acid esters by argentation chromatography (the plate was sprayed with 0.025% dichlorofluorescein and visualized under UV light). 1, methyl caprylate (C8:0), 2, methyl palmitate (C16:0), 3, methyl palmitoleate (C16:1), 4, methyl stearate (C18:0), 5, methyl oleate (C18:1), 6, methyl linoleate (C18:2), 7, methyl linolenate (C18:3), 8, methyl arachidate (C 20:0), 9, Free fatty acids.

acids, but methyl esters of saturated fatty acids with various chain lengths move rapidly. Varying mobilities are observed between fatty acid esters having one, two, or three double bonds, but there is increasing retardation as more double bonds are available for interaction with Ag^+.

Equation 1: Saponification

$$\begin{array}{l} H_2C-O-\overset{O}{\overset{\|}{C}}-R \\ \;\;| \\ HC-O-\overset{O}{\overset{\|}{C}}-R \\ \;\;| \\ H_2C-O-\overset{O}{\overset{\|}{C}}-R \end{array} \xrightarrow{KOH,\ \Delta} \begin{array}{l} CH_2OH \\ \;\;| \\ CHOH \\ \;\;| \\ CH_2OH \end{array} + 3\,R-\overset{O}{\overset{\|}{C}}-OK$$

Triacylglycerol — Glycerol — Potassium salt of fatty acids (soaps)

Equation 2: Methylation of free fatty acids

$$RC(=O)OK \xrightarrow{H^+} RC(=O)OH \xrightarrow[CH_3OH]{H^+,\ \Delta} RC(=O)OCH_3$$

Experimental Procedure

1. Transfer 2 mL of the "total lipids" fraction to a 50-mL glass-stoppered test tube, and bring the liquid to dryness under N_2.

2. Add 5 mL of 0.5 N KOH in ethanol to the residue, and incubate the sample at 55°C for 20 min. This is the saponification reaction. Cool the tube in an ice bath.

3. During this incubation period, mark the silver nitrate–impregnated plate with a line at the origin and with 10 equally spaced dots for sample applications. Activate the plate by heating it at 110°C for *ca* 10 min. (These plates are fragile and expensive; *handle them with care*).

4. Add 2.5 mL of conc. HCl, and extract the fatty acids with 5 mL of petroleum ether. Return the tube to the 55°C bath and incubate for 2 min. Mix the tube thoroughly, and collect the extract (top layer) in a glass-stoppered tube. Repeat the extraction with another 5 ml of petroleum ether, and add the second extract to the same tube.

5. Filter the combined extracts through anhydrous sodium sulfate to remove any water that may be present. Use a funnel fitted with a plug of glass wool, and add the salt to a depth of about 5 cm. Add sufficient petroleum ether to wet the salt. Pass the extract from the glass-stoppered tube through this salt layer using a Pasteur pipette to make the transfer, and rinse the tube with 2 to 3 mL of the same solvent. Add this rinse to the salt in the funnel. Continue by adding another 5 mL of petroleum ether to wash any residual product out of the salt. Collect these extracts and washes together in a clean tube.

6. Take the filtrate to dryness under N_2 and dissolve the residue in 10 mL of perchloric acid: MeOH mixture (5:95). Incubate at 55°C for 10 min.

7. Extract the esters twice with 5 mL of petroleum ether while mixing on the Vortex mixer, and transfer the combined extracts into a 15-mL glass-stoppered test tube.

8. Again filter the esters through another funnel containing anhydrous sodium sulfate, and take the extract to dryness under N_2. Set up the developing tank at this time. Prepare 150 to 200 mL of diethyl ether:hexane (5:95) solvent.

9. Resuspend the fraction of methyl esters in 0.2 mL of petroleum ether, and transfer to a microfuge tube. Spot 10 and 20 μL of the methyl esters derived from the "total lipids" and 5 μL of the following fatty acid ester standards (10 mg/mL): methyl palmitate, methyl palmitoleate, methyl stearate, methyl oleate, methyl linoleate, methyl linolenate, methyl arachidate, and a sample of free fatty acids.
10. Develop the chromatogram until the solvent front reaches 5 to 8 cm from the top. Allow the plate to air-dry in the hood, and spray it with 0.025% dichlorofluorescein in 95% EtOH (in the hood). Allow the plate to air dry, and "visualize" it under UV light.
11. Gently circle the spots with a soft pencil while the plate is under the UV light. Then reproduce the appearance of the plate onto a transparency by tracing the points of application, the solvent front line, and the spots. Calculate the R_f values, and identify the fatty acids.

G. Determination of Inorganic Phosphate in the Lipid Fractions

Inorganic phosphate (P_i) will be determined and the amounts of phospholipids in the fractions will be calculated after they have been treated with strong acid and heat. The ashing procedure will be carried out in a dry heating block at 140°C in the hood.

Experimental Procedure

1. Preparation of materials.
   a. Construct a flow chart similar to the one made for the cholesterol determinations. Indicate the amount of lipid material in the samples that were ashed, the aliquots used in the assay tubes, and the dilutions made if applicable.
   b. Throughout this experiment, *new* or disposable glassware should be used to avoid phosphate contamination from detergents used in previous washing of the glassware. If previously washed vessels must be used, rinse at least three times with Milli-Q H_2O of high purity. This high-purity water is relatively free of traces of P_i that might be present in regularly deionized water.
2. Construct a standard curve for determining P_i.

a. Set up a series of *new* disposable tubes (10 × 75 mm), and add a P_i solution as follows, using a *new* plastic disposable 1-mL graduated pipette (be sure to deliver the sample *directly* to the bottom of each tube):

Tube no.	*0.2 mM KH_2PO_4 (mL)*	*nmol P_i*
1 (Blank)	0	0
2	0.05	10
3	0.10	20
4	0.15	30
5	0.20	40
6	0.25	50
7	0.30	60

b. Place the tubes in an oven at 110°C until they are dry. (This may take overnight and should be done prior to this laboratory period.)

3. Preparation of lipid fractions for ashing and determination of P_i concentrations in the lipid fractions.

a. Make a 10-fold dilution of the "total lipids" fraction by adding 0.9 mL of $CHCl_3$:MeOH (1:1) to 0.1 mL of sample in a 1.5-mL microfuge tube.

b. Set up tubes 8 through 16 (use disposable 10 × 75 mm tubes) as indicated below. Use an automatic pipettor with new plastic tips to make each addition.

Tube no.	*Volume of sample (mL)*	*Lipid fraction*
8	0.01	Total (Diluted 1:9)
9	0.05	Total "
10	0.10	Total "
11	0.01	Neutral (Undiluted)
12	0.05	Neutral "
13	0.10	Neutral "
14	0.01	Polar (Undiluted)
15	0.05	Polar "
16	0.10	Polar "

c. Place these tubes in a heating block (140°C), and allow the tubes to come to dryness. (This will not take long since the solvent is quite volatile.) Allow the tubes to cool before proceeding.

4. Ashing of lipids and assay for P_i concentrations.
   a. After the tubes have cooled to room temperature, add 0.2 mL of 10% H_2SO_4 to each tube, and incubate the tubes at 140°C for 1 hr.
   b. Add 0.05 mL of 30% H_2O_2 to each tube, and incubate the tubes again at 140°C for 40 min.
   c. Allow the tubes to come to room temperature. Add 0.5 mL of Milli-Q H_2O to each tube, and mix the contents well on a Vortex mixer.
   d. Add 0.5 mL of molybdate reagent* to each tube, and again mix the tubes thoroughly. (This reagent should be colorless and should be prepared just prior to use. It deteriorates rapidly and should be discarded if a greenish color develops).
   e. Incubate the tubes for exactly 15 min at 50°C.
   f. Read all the tubes at $A_{820\ nm}$ against the blank in a spectrophotometer using 1-mL disposable plastic cuvettes. Plot A *vs.* nmol P_i for the standard. Determine the amounts of P_i in the samples assayed.
   g. Determination of phospholipids in the lipid fractions
      i. Assuming an average MW for phospholipids of 800, calculate the amount of phospholipids in mg/g of tissue for each fraction.
      ii. What percentage of the total lipids is phospholipid?
      iii. Based on the P_i determinations, calculate the percentage of each lipid fraction recovered from the column. How do these results compare with the values obtained from the gravimetric analysis?

---

*For 20 tubes of molybdate reagent (be sure to use new glassware): (1) Make up 15% ascorbic acid solution (1.5 g of ascorbic acid + 10 mL Milli-Q H_2O. (2) Combine 1.0 mL of 5% ammonium molybdate + 3.0 mL of 15% ascorbic acid + 6 mL of Milli-Q H_2O. *The molybdate reagent is prepared just prior to use and immediately protected from light by wrapping the container with aluminum foil.*

CHAPTER 9

Clinical/Nutritional Biochemistry

A. Analysis of Blood

Day 1

1. Preparation of protein-free filtrate and glucose analysis.
2. BUN.
3. Hemoglobin and hematocrit determinations.
4. Blood typing.

Day 2

1. Cholesterol determination.
2. Electrophoresis of serum proteins–allow to air dry until next laboratory period.
3. Triglycerides in serum.
4. ALT in serum.
5. Total proteins–use of the refractometer.
6. Collection of a urine sample.

B. Urinalysis and Tay-Sachs Assay

Day 3

1. Incubation of serum for Tay-Sachs assay.
2. Record volume of urine.
3. Albumin determination–biuret method.
4. Total solids.
5. Coproporphyrin determination.
6. Urinalysis–Ames Multistix method.

7. Save 100 mL of urine and add 1.5 mL of 6 N HCl. Store at 4°C for catecholamine extraction.
8. Staining and scanning of serum protein gel.

Day 4

1. Tay-Sachs assay.
2. Catecholamines determination.

C. *Nutritional Studies*

Day 5

1. Carotene standard curve.
2. Carotene isolation and quantitation from a green vegetable.
3. Extraction and hydrolysis of foods for protein analysis.
4. Electrophoresis of serum cholesterol.

Day 6

1. Nutritional evaluation of foods as protein sources.
   a. Microbiological assay for tryptophan.
   b. Fluorimetric assay for phenylalanine.
2. Staining and scanning of serum cholesterol gel.
3. Carotenes in serum.

Introduction

In this series of experiments, some of the standard procedures followed in the clinical laboratory will be introduced. The blood and urine analyses will be carried out on your own specimens. Two samples of about 5 mL of blood should be drawn on the first day of the module, just prior to the laboratory session, one in a red-capped tube (regular) and the other in a green-capped tube (heparinized). On that day, do not eat for a period of 3 hr before the blood collection.

Quantitative clinical chemistry permits the study of body functions in health and disease because many chemical constituents of body fluids change in concentration when the body passes from the normal to a pathological state. A large number of tests are now available in the clinical laboratory. Several analyses will be performed that were chosen because they are representative of the methodology available. For some of the tests described here, there are alternative methods available; the choice of method is dependent on equipment available, time required to carry out a given procedure, and cost. Many of the assays that will be performed on serum are now done routinely by automated equipment. However, the principles involved are basic and should be understood to assess properly the data generated by the more sophisticated technology. In particular, the computer printouts generated in the clinical laboratory are difficult for the uninitiated to interpret, and the values obtained from different laboratories can vary depending on the methods of analysis used. Furthermore, complex equipment can break down, and it is not unlikely that a test will need to be performed manually at some time.

The urine analyses must be performed on samples collected over a period of 24 hr because the concentration of metabolites excreted varies during the day. The assays that will be performed were selected to illustrate a variety of techniques, including methods of isolation of the desired constituent from urine and the use of the fluorimeter as well as the more rapid, though less quantitative, commercial *Multistix*.

BLOOD ANALYSIS

Since changes in blood chemistry are associated with many diseases, quantitative clinical chemistry plays an important part in the study of the functions of the body. Many of the analyses that will be done in this section are now automated in the clinical laboratory, but knowing the principles involved in sampling, in comparing spectrophotometric values of unknowns with those of standards, and in interpretating the results will permit a better understanding of the clinical data. Note that the values for blood components are reported in g or mg/100 mL, also termed dL.

REFERENCES: Oser, B.L. *Hawk's Physiological Chemistry*, 14th ed. McGraw-Hill, New York (1965).

Taylor, H.E. *Clinical Chemistry*. John Wiley & Sons, New York (1989).

Tietz, N.W. *Fundamentals of Clinical Chemistry*, 2nd ed. W.B. Saunders, Philadelphia (1976).

A. Blood Collection

Two samples will be collected for this study as indicated in the diagram below. The green-capped tube contains an anticoagulant that prevents clotting and a red-capped tube from which serum will be recovered.

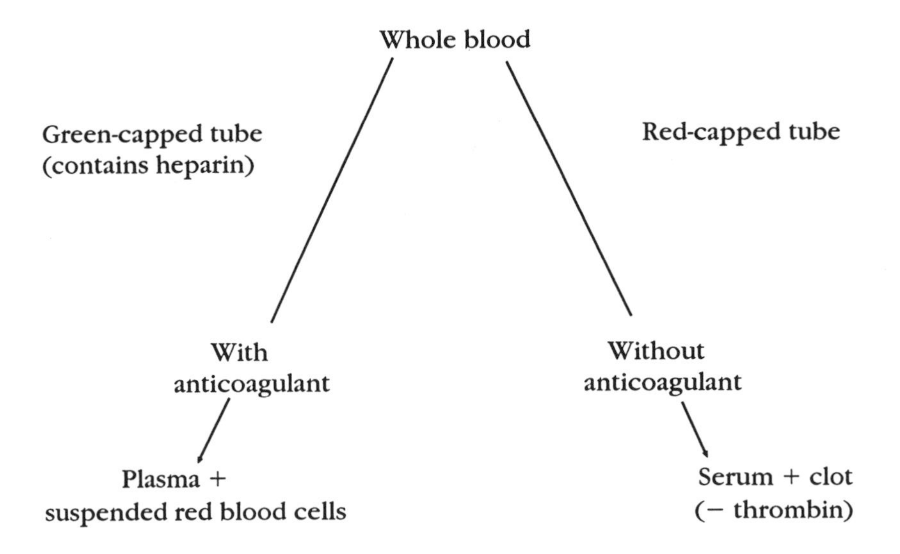

Special Handling of Blood Samples

When working with blood samples in the laboratory, special precautions should be taken to prevent accidental infection by contagious agents:

1. Samples should be taken by a licensed phlebotomist in a clinical laboratory.
2. Each person should work with his or her sample only.
3. Wear protective gloves and safety glasses at all times.
4. Clean any glassware that has come in contact with blood by thoroughly rinsing with a Clorox solution. It can then be washed in the usual manner.
5. Disposable materials are deposited in an autoclavable bag and sterilized before discarding.

B. Preparation of Protein-Free Filtrates

Proteins are removed from blood before analyses involving reduction or oxidation of a metal ion can be made, because reducing compounds other than glucose would also react with the reagents. In the method of Folin and Wu, which is used here, the proteins are precipitated by tungstic acid and the precipitate removed by filtration.

Procedure for Preparing Protein-Free Filtrates

1. Pipette 2.5 mL of whole blood after gentle mixing (green-capped tube containing the anticoagulant) into a 125-mL Erlenmeyer flask.
2. Add 17.5 mL H_2O and swirl flask gently to mix.
3. Add 2.5 mL of 10% Na tungstate solution. Again, mix by swirling the flask.
4. Add 2.5 mL of 0.67 N H_2SO_4 dropwise with continuous swirling.
5. Stopper flask, shake it gently, and allow it to stand for 10 min.
6. Pass the suspension through Whatman filter paper (No. 2) and collect the clear filtrate.

C. Determination of Glucose by the Method of Folin and Wu

In this procedure, the protein-free filtrate is heated in an alkaline copper solution, and cuprous oxide is formed. The Cu_2O is reacted with phosphomolybdic acid, resulting in a blue color that is read in a spectrophotometer at 420 nm against a blank. The value is compared with a standard of known concentration. Normal values range from 90 to 120 mg glucose/dL of blood, which is somewhat higher than the actual glucose content because of the presence of small amounts of other reducing substances.

Set up eight test tubes, follow the protocol shown in Table 9.1, and record the average value in the summary table (Table 9.12).

D. Preparation of Serum

1. Allow the second blood sample (red top) to stand for at least 1 hr.
2. Spin in a clinical centrifuge for 10 min.
3. Carefully remove the clear serum with a Pasteur pipette (do not disturb the clot).
4. Transfer the serum to three tubes; to each of two tubes add 0.5mL, and transfer the remainder to the third tube. The three tubes can be frozen and saved for later analyses.

TABLE 9.1. Procedure for Determination of Blood Glucose[a]

Additions (mL)	Tube no. 1 (Blank)	2	3	4	5	6	7	8
Blood filtrate	0	1.0	1.0	1.0	—	—	—	—
Glucose standard (0.2 mg/mL)					0.1	0.2	0.4	0.6
	← Bring volume to 1 mL with H_2O. →							
Alkaline copper reagent	1.0 →							
	Heat in a boiling H_2O bath for 8 min. Cool on ice.							
Phosphomolybdic acid	1.0 →							
	Add 7 mL of H_2O, mix, and read in the spectrophotometer against the blank.							
% $T_{420\ nm}$								
$A_{420\ nm}$								

[a]Note that a 10-fold dilution in preparing the protein-free filtrate was made. (Be sure to account for this in your calculations). Construct a standard curve, and use the average of the three determinations to report the data in mg/dL of glucose in blood.

E. Blood Urea Nitrogen

Urea is the end product of nitrogen metabolism in humans and is normally excreted by the kidneys. In serum, normal values range from 8 to 20 mg urea/dL of blood.

In this analysis, urease catalyzes the hydrolysis of urea to NH_3 and CO_2. The ammonia is converted to indophenol blue in the presence of sodium nitroferricyanide-phenol and hypochlorite reagents. The color of the unknown is read against a blank at $A_{625\ nm}$ and calculations are made by comparison with absorbance values of a standard of known concentration. Follow the procedure shown in Table 9.2 and record the results in Table 9.12.

F. Hematocrit

The method permits the determination of the volume occupied by the red blood cells (RBCs) in the blood. A capillary tube is filled with whole blood and spun in a centrifuge to pack the

TABLE 9.2. Procedure for Determination of Blood Urea Nitrogen (BUN)[a]

Additions (μL)	Tube no. 1 (Blank)	2	3	4	5	6	7
Standard (0.2 mg/mL)	—	10	10	10	—	—	—
Serum	—	—	—	—	10	10	10
H_2O	20	10 →					
Urease	20	→					
		Incubate at 37°C for 30 min.					
Phenol reagent (mL)	0.5	→					
Hypochlorite reagent (mL)	0.5	→					
		Mix and incubate at room temperature for 40 min. Add 4 mL of H_2O, mix, and read in a spectrophotometer against the blank.					
% $T_{625\ nm}$							
$A_{625\ nm}$							

[a]Use 13 × 100 mm test tubes for this assay.

RBCs. The hematocrit value is calculated as the ratio of the height of cells over the total height of fluid in the tube. Abnormal values are indicative of anemic conditions; normal values range from 35% to 50%.

Procedure for Hematocrit Determination

1. Using a hematocrit tube, fill it by capillary action to the mark with whole blood (green-topped tube).
2. Pack bottom end of the capillary tube with clay. Repeat this procedure with a second hematocrit tube.
3. Centrifuge for 5 min while balancing the two tubes against each other.
4. Read the percent cell volume by sliding the tube along a "Critocap" chart until the meniscus of the plasma intersects the 100% line.
5. Record the average value in the summary table (Table 9.12).

G. Hemoglobin Determination by the Cyanmethemoglobin Method

REFERENCES: Hainline, A. *Hemoglobin in Standard Methods of Clinical Chemistry*, vol. 2. Academic Press, New York, p. 49 (1958)

Stryer, pp. 150–171.

In this determination, whole blood is diluted with a potassium ferricyanide solution that oxidizes hemoglobin to methemoglobin, which in turn is converted to cyanmethemoglobin. The intensity of the color is measured photometrically. The procedure is standardized with blood that has been analyzed for its iron content. The standard curve (Fig. 9.1) has been obtained from the method of the Walter Reed Army Institute of Research. Read the tubes against this curve. Normal values for hemoglobin concentration range from 12 to 20 g/dL. Follow the procedure shown in Table 9.3 and calculate the average amount of hemoglobin in gram percent (g/dL). Record the value in the summary table (Table 9.12).

H. Cholesterol Determination in Serum

REFERENCES: Brown, M.S., and Goldstein, J.L. How LDL receptors influence cholesterol and atherosclerosis. *Sci. Am.* 251:58-66 (1984).

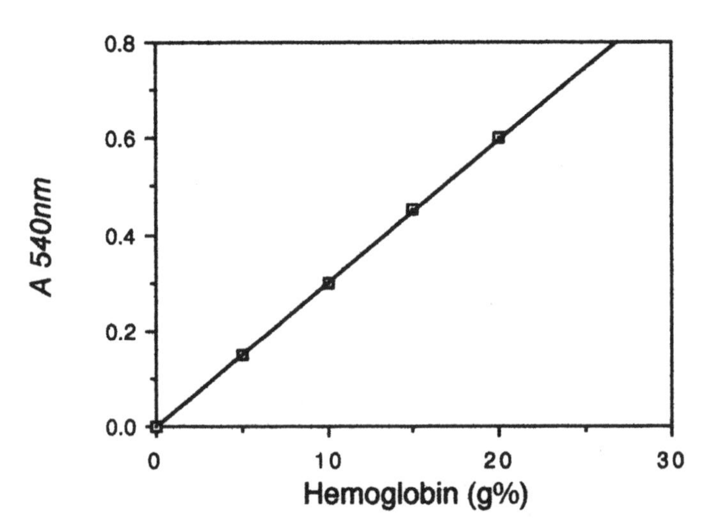

Fig. 9.1. Standard curve for the determination of homoglobin.

TABLE 9.3. Procedure for Hemoglobin Determination

Additions	Tube no. 1 (Blank)	2	3	4
Whole blood (μL)	–	20	20	20
H_2O (μL)	20	–	–	
K ferricyanide (mL)	5.0 ⟶			
	Cover with parafilm. Gently mix, incubate for 10 min at room temperature, and read against the blank.			
% $T_{540\text{ nm}}$				
$A_{540\text{ nm}}$				

HDL Cholesterol Electrophoresis. Helena Laboratories, Beaumont, TX (1984).

Zubay, G. *Biochemistry* 2nd ed. Macmillan, New York, pp. 733-746 (1988).

Cholesterol is an important blood component and is included in most serum profile analyses. Its role in the onset of heart disease has been well established, and it is a major risk factor that can be brought under control through proper diet and, if necessary, by use of one of a variety of drugs. In this study, total cholesterol values will be determined, and three lipoprotein fractions will be separated and analyzed.

Lipids are transported in serum in association with one of several proteins. The amounts and types of lipoproteins vary with diet and metabolic state. Normal concentrations (in mg/dL) of the major classes of lipids are as follows:

Total lipid	360–680
Total cholesterol	100–250
Triglycerides	80–240
Phospholipid	150–250

The lipoproteins that circulate in the vascular system have been classified into four major types, depending on sedimentation rate in an ultracentrifuge. High-density lipoprotein (HDL), very-low-density lipoprotein (VLDL), and low-density lipoprotein (LDL) are characterized by decreasing amounts of protein with concomitant increases in cholesterol and tryglyceride content as seen in the following table:

Method of separation of the fractions		**Composition (%)**			
Electro-phoresis	*Ultracentri-fugation*	*Protein*	*Tri-glycerides*	*Choles-terol*	*Phospho-lipids*
Chylo-microns	(Do not sediment readily)	2	98	–	–
β	LDL	21	12	45	22
Pre-β	VLDL	10	55	13	22
α	HDL	50	6	18	26

It has been well documented that abnormal plasma lipid levels are associated with coronary artery disease. Measurement of lipoprotein content is therefore an important tool in diagnosis as well as in disease prevention, as it is known that increased levels of HDL cholesterol (therefore decreased [total cholesterol:HDL] ratios) are directly related to lower risk of atherosclerosis and the development of coronary heart disease. It has also been shown that exercise, a low fat diet, and avoidance of cigarette smoking are factors responsible for increasing HDL cholesterol concentrations. Therefore, a combination of behavior modification factors and various medications can significantly decrease one's risk of developing heart disease.

Procedure for Total Cholesterol Determination

The method, a modification of the Zak reaction, requires that the cholesterol be extracted from the serum with ethanol. The extract is then reacted with a solution of $FeCl_3$ dissolved in phosphoric acid, and the resulting color is read in a spectrophotometer at 550 nm against a standard of known concentration that was treated in the same manner. Normal values for total cholesterol tend to increase with age, but the average values range from 100 to 250 mg/dL.

1. Determinations will be made in triplicate. To each of three screw-capped test tubes, transfer 0.1 mL of serum. Add 10 mL of absolute ethanol to each tube and mix rapidly on a Vortex mixer for 10 sec.
2. Centrifuge the tubes for 5 min at full speed in a clinical centrifuge, and transfer the extracts carefully to clean test

tubes. Measure 2.0 mL of extract into each of three new test tubes. It is important to have a clear solution; if some precipitate is carried over, spin the tube again.

3. The standard is prepared by adding 2.0 mL of a 0.02 mg/mL solution of cholesterol to a fourth tube.
4. Prepare a blank by delivering 2.0 mL of EtOH to a fifth tube.
5. Slowly add 2.0 mL of "color reagent" to each tube and mix by gentle swirling **(VERY STRONG ACID—HANDLE WITH CARE!)**
6. Cover the tubes with parafilm, and allow them to stand at room temperature for 30 min. The color that develops is stable for about 1 hr only.
7. Read the absorbance of the standard and of the unknowns at 550 nm against the blank, and calculate the average mg/dL value for total cholesterol in serum. Again, set up the ratio: [$A_{550\text{ nm}}$ unknown: $A_{550\text{ nm}}$ standard $\times$ concentration of the standard $\times$ 100] to calculate the concentration of unknown which is relative to that of the standard under the same conditions. Record this value in the summary table (Table 9.12).

Procedure for HDL and LDL Cholesterol Separation and Quantitation

In this experiment, a sample of serum is applied to an agarose gel plate, and the lipoprotein fractions are separated by electrophoresis. The resulting bands are stained in a solution of Fat Red dye and scanned in a densitometer at 525 nm. The equipment and the reagents are purchased from Helena Laboratories (Beaumont, TX).

1. Preparation of electrophoresis chamber.
   a. Dissolve a packet of sodium barbital buffer, pH 8.4, in 1.5 L of water.
   b. Pour 25 mL of buffer into the two inner chambers of the apparatus, and connect it to a power supply.
2. Application of serum to the agarose gel plate.
   a. Remove the plate from the sealed package and place it on a clean surface so that the numbers appear on the top of the plate.

b. Gently blot the area of the plate designed for sample application.

c. Place the template from the kit in this area, and gently press down to ensure that there are no air bubbles under it.

d. Place 2 μL of each sample on the template slots while spreading the serum over the entire slot area. Make sure not to puncture the gel while applying the sample.

e. Allow 8 min for the samples to diffuse into the gel after the last sample has been applied.

f. Carefully remove the template without disturbing the surface of the gel.

3. Electrophoresis.

   a. Gently squeeze the plate into the inner section of the chamber, agarose side down. (Two plates may be run at the same time.) The plate should be positioned so that the edges are firmly set in the buffer and the application is on the cathodic (−) side.

   b. Close the chamber and allow the plate(s) to equilibrate for 1 min; then turn on the power and allow the electrophoresis to proceed for 45 min at 80 V.

   c. Remove the plate(s) and allow to *air dry*.

4. Visualization of the bands.

   a. Place the plate **agarose side up** into a staining dish containing 0.1 % (w/v) Fat Red 7B stain in 95% methanol. The stain must be diluted just prior to use by slowly adding 15 mL of H_2O to 75 mL of prepared stain.

   b. Stain for 15 to 20 min.

   c. Remove the plate gently with tweezers, and wash it twice in destaining solution (3:1, MeOH:H_2O). Allow the plate to remain in each wash for at least 1 min and swirl gently to destain. Repeat if necessary to obtain a clear background.

   d. Follow with a wash in water.

   e. Remove from the wash and allow the plate to dry either at room temperature or by gently passing back and forth in front of a hair dryer *at a safe distance*.

5. Quantitation of lipoprotein fractions.

a. Scan the plate in a Cliniscan 2 densitometer (see Fig. 9.2) at 525 nm. The Cliniscan 2 densitometer is an automated instrument that reads the intensity of dye deposited on the agarose gel plate for each band. It is connected to a printer that generates a curve and provides readings of percentage values and unit calculations for the peaks that correspond to the bands that were scanned. (The setting for this instrument should be at F4). A sample plate stained and ready for scanning is shown in Figure 9.3A.

b. The HDL band is the fastest moving, as it has the most protein and is therefore more highly charged. It will be closest to the anode. The LDL band is the most prominent and migrates only slightly from the point of application, and the VLDL will be located between the two. The graph that is generated by the Cliniscan instrument is shown in Figure 9.3B.

c. Calculate the mg/dL HDL and LDL and the LDL/HDL ratio. Using the total serum cholesterol determined earlier, calculate the [total cholesterol: HDL ratio] and evaluate the risk by comparing these data with those summarized in the following table. Enter the data in the summary table (Table 9.12).

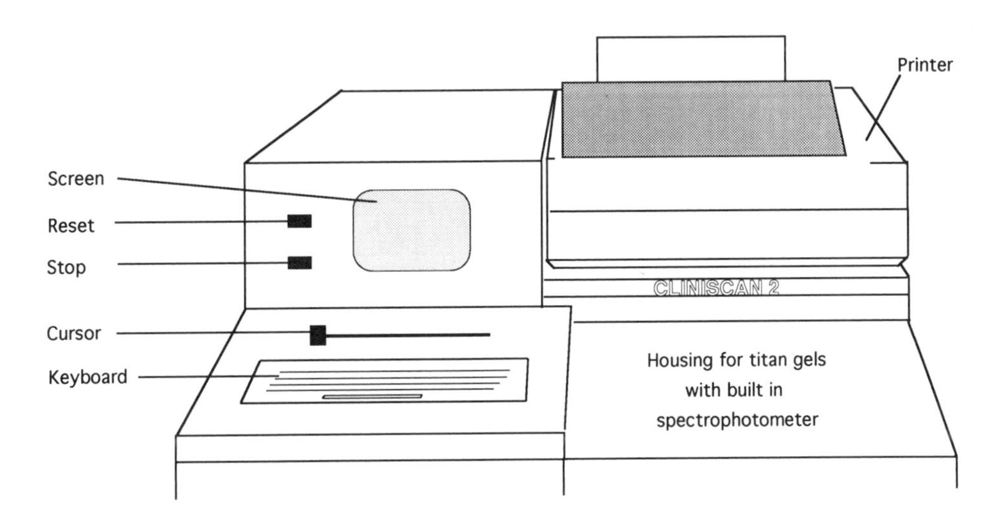

Fig. 9.2. The Cliniscan II–The instrument consists of a spectrophotometer connected to a computer and a printer. It scans a stained gel plate at a designated wavelength.

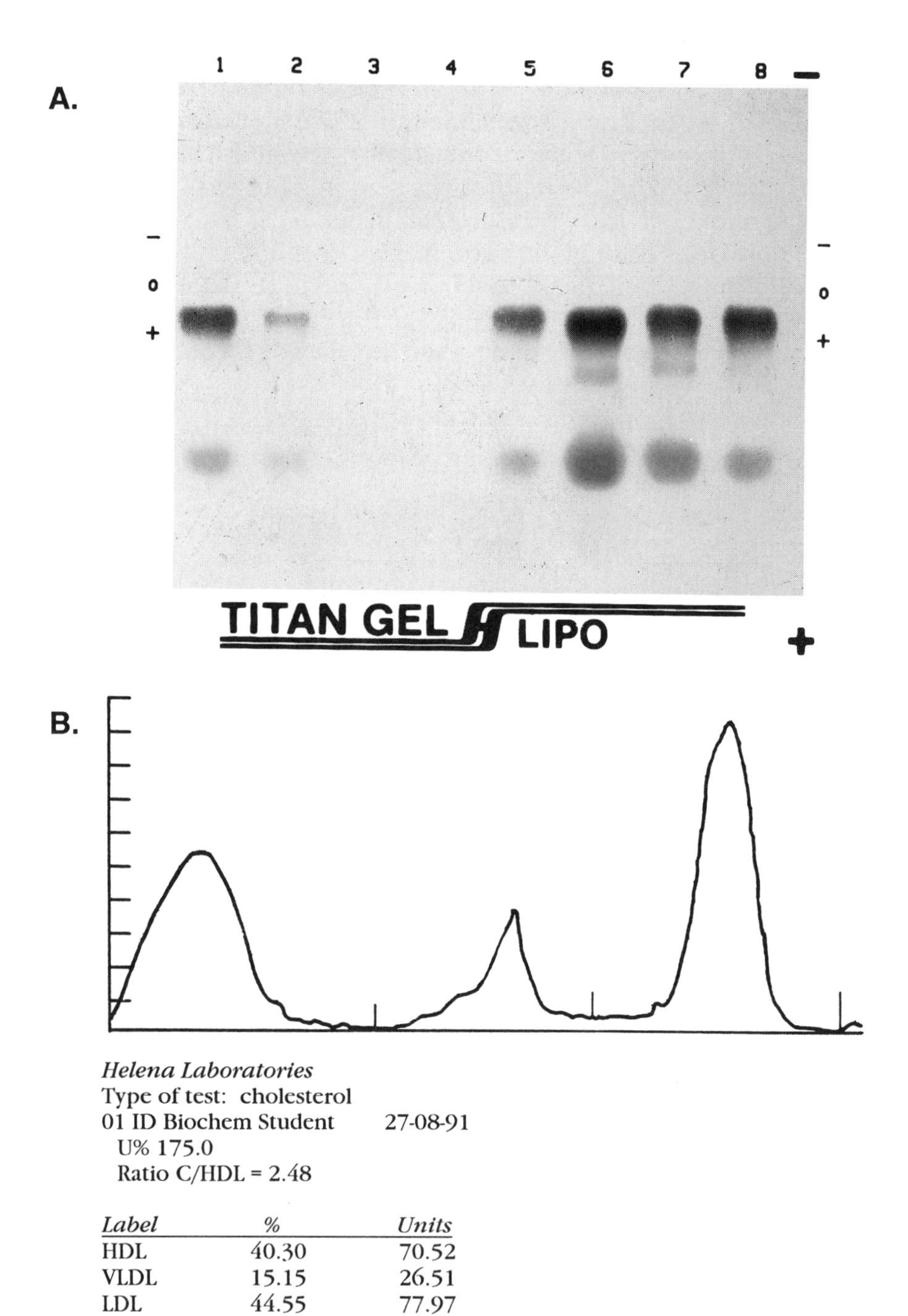

Helena Laboratories
Type of test: cholesterol
01 ID Biochem Student 27-08-91
U% 175.0
Ratio C/HDL = 2.48

Label	%	*Units*
HDL	40.30	70.52
VLDL	15.15	26.51
LDL	44.55	77.97

Fig. 9.3. Lipoprotein electrophoresis patterns. **A:** Titan gel lipoprotein electrophoresis plate that has been stained with Fat Red 7B. **B:** Scanning pattern obtained from the plate shown in A. Note that the instrument calculates both the ratio of C/HDL as well as % of lipoprotein fractions when the total cholesterol value (U%) is entered.

Atherosclerosis Risk Ratios (From the Framingham Study)

Risk	*LDL/HDL*	*Total C/HDL*
Men		
1/2 Average	1.00	3.43
Average	3.55	4.97
2× Average	6.25	9.55
3× Average	7.99	23.40
Women		
1/2 Average	1.47	3.27
Average	3.22	4.44
2× Average	5.03	7.05
3× Average	6.14	11.00

Reagents and Solutions used (purchased from Helena Laboratories, Beaumont, TX):

1. Agarose gel plates: 0.6% agarose with 5% sucrose and a preservative.
2. Sodium barbital buffer with preservative, pH 8.4 to 8.8.
3. Lipoprotein stain 0.1% (w/v) Fat Red 7B stain in 95% methanol.

I. Quantitative Determination of Serum Triglycerides

The method involves the complete enzymatic hydrolysis of triglycerides (also known as triacylglycerols) to glycerol and fatty acids. Although triglyceride levels vary considerably–depending on diet, amount of exercise, and the general metabolic state of the individual–the determinations are useful in following the course of atherosclerosis, diabetes mellitus, nephrosis, biliary obstruction, and endocrine malfunction.

The determination is made with a reagent kit purchased from Seragen Diagnostics, Inc., Indianapolis, IN. The reagent vials contain, after reconstitution, lipases, ATP, NAD^+, diaphorase, glycerol kinase, glycerol-1-phosphate dehydrogenase, and nitrophenyl-tetrazolium dye, which, on reduction, produces a colored formazan. The absorbance of this colored compound is read at 500 nm and is directly proportional to the amount of glycerol produced in reaction 1.

The reactions occurring in the determination are as follows:

1. Triglycerides $\xrightarrow{\text{lipases}}$ glycerol + fatty acids
2. Glycerol + ATP $\xrightarrow{\text{glycerol kinase}}$ glycerol-1-phosphate + ADP
3. Glycerol-1-phosphate + NAD^+ $\xrightarrow[\text{dehydrogenase}]{\text{glycerol-1-phosphate}}$ dihydroxy-acetone-phosphate + NADH
4. NADH+(indophenyl-nitrophenyl-tetrazolium) (colorless) $\xrightarrow{\text{diaphorase}}$ NAD^+ + formazan (pink)

Normal serum value range from 36 to 165 mg/dL

Procedure for Triglyceride Determination

1. Add "reconstituting reagent" according to directions given on the vial. This reagent is stable for 8 hr at room temperature and for 2 to 3 days at 4°C.
2. Transfer 1 mL of reconstituted substrate to each of four test tubes (13 × 100 mm), and incubate these tubes at 37°C for 10 min.
3. To one tube add 20 μL of water, to the second tube add 20 μL of standard (200 mg/dL), and to the other two add 20μL of serum.
4. Incubate the tubes again at 37°C for 10 min, and stop the reaction by adding 2 mL of acid (0.03 M HCl). Allow the tubes to stand at room temperature for 5 min.
5. Set the instrument to zero with the blank. Read the absorbance for both the standard and the serum tubes at 500 nm against the blank.
6. Calculate mg/dL triglycerides in the serum and record the data in Table 9.12.

J. Serum Protein Electrophoresis

REFERENCE: *Titan Gel Serum Protein System.* Helena Laboratories, Beaumont, TX (1984).

Although serum contains many individual proteins, each with a given function, it is possible to fractionate serum into five distinct categories by electrophoresis. The fractions are albumin, α_1, α_2, β, and γ. Albumin is important in maintaining the osmotic pressure of the serum and in the regulation of vascular fluid exchange. α_1 has antitrypsin activity and is markedly decreased in patients suffering from emphysema. α_2 is elevated in patients with cirrhosis of the liver and with diabetes, but it is lowered in patients with hemolytic anemia. The β fraction consists of transferrin, which is important in iron transport and regulation, and β-lipoprotein, which functions in the transport of serum lipids. The gamma fraction consists of globulins IgA, IgM, and IgG, the last accounting for about 85% of the total fraction. It also contains antibodies.

In this experiment, serum will be applied to an agarose gel, and the five fractions that will be obtained following electrophoresis will be stained with Ponceau S. The resulting bands will be scanned in a densitometer, and the relative protein concentrations will be calculated.

Procedure for Electrophoresis

The five major serum proteins will be separated by electrophoresis in a chamber containing 0.05 N Na barbital buffer at pH 8.8. Following electrophoresis, the protein bands will be stained in a solution of Ponceau S, and the relative dye intensities will be measured by scanning the stained strips in a densitometer. Analysis of the data will then permit identification and determination of the relative quantities of each of the proteins. Figure 9.4A, B illustrates typical results obtained by this method. The equipment and the reagents used in this study were purchased from Helena Laboratories. Each agarose gel plate can accommodate 10 serum samples.

1. Preparation of the electrophoresis chamber.
   a. A packet of serum protein buffer (sodium barbital, pH 8.4 to 8.8) is dissolved in 1500 mL H_2O.
   b. A volume of 25 mL of buffer is poured into each of the two inner chambers and the apparatus is connected to a power supply.
2. Sample application.
   a. There are 10 wells on each prepackaged agarose gel plate, and each student will apply a sample into one of the numbered slots.

A.

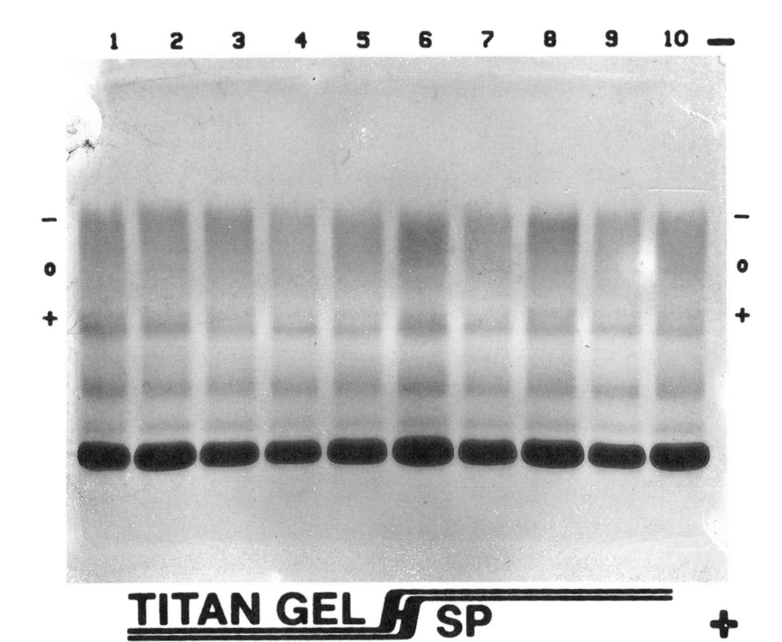

B.

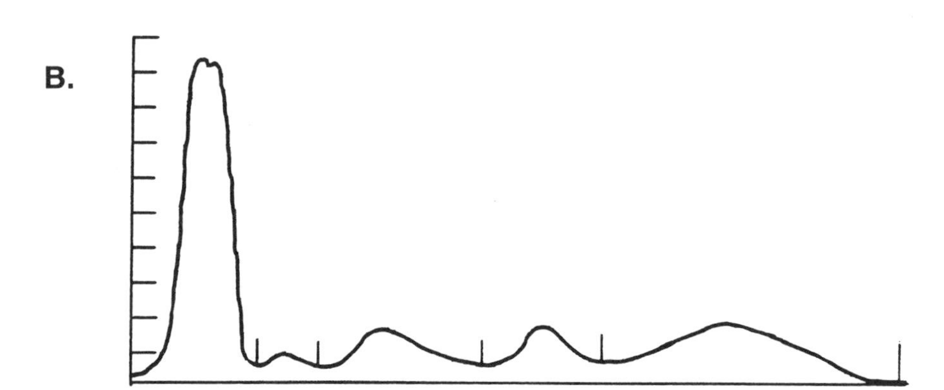

Helena Laboratories
Type of test: serum protein
01 ID Biochem student 29-08-91
TP GM% 7.500
Total value range: 7.13–8.09

Label	%	Units	Normal limits (units)
Albumin	48.46	3.63	3.49–4.11
α_1	3.22	0.24	0.21–0.31
α_2	13.50	1.01	0.91–1.29
β	10.55	0.79	0.91–1.29
γ	24.26	1.82	1.42–2.34

Fig. 9.4. Serum protein electrophoresis pattern. **A:** Titan gel serum protein plate showing bands obtained following electrophoresis and staining with Ponceau S. **B:** Scanning pattern obtained from the plate shown in A. The five major protein peaks were obtained on scanning at 525 nm, and their relative areas were calculated and converted to units (g/dL).

b. Gently blot the indicated area of the agarose plate, remove the blotter carefully, and align the template provided with the -O- signs located on the sides of the plate. Apply slight finger pressure to make sure that there are no air bubbles under the template. Dilute the serum with an equal volume of buffer by mixing 3 μL of each on a piece of parafilm.

c. Apply 3.0 μL of diluted samples on the template slots, and wait 5 min after the last sample has been applied to allow the serum samples to diffuse into the agarose.

d. Blot off any excess serum, and carefully remove the template.

3. Electrophoresis of sample plate.

   a. Place the plate into the inner sections of the chamber, agarose side down, by gently squeezing the plate into place. Be sure that the edges of the agarose are in the buffer and the application point is on the cathodic (−) side. Two plates may be accommodated at one time.

   b. Securely fasten the cover on the chamber, and apply 120 V for 15 min.

4. Visualization of the protein band.

   a. At the end of the electrophoresis period, remove the plate from the chamber and transfer to a Fixative/Destain solution for 10 min.

   b. Remove the plate from this solution and allow it to air dry. (A fan or hairdryer may be used if necessary but with great caution, as the plastic backing tends to buckle if the plate is overheated.)

   c. When the plate is completely dry, immerse it in the Ponceau S stain for 5 min.

   d. Remove the plate, drain it onto a paper towel, and destain by rinsing it in several consecutive washes of destaining solution. Allow the plate to remain in each wash for 1 min. The plate background should be clear.

   e. Allow the plate to dry completely once again, and scan it in a Cliniscan 2 densitometer at 525 nm as done for the lipoprotein separations.

5. Determination of total proteins in serum.

USE OF THE REFRACTOMETER

a. The Goldberg refractometer (American Optical Co.) can be used to determine total solids in urine as well as total proteins in serum.

b. To load the refractometer, hold the instrument in a horizontal position with the cover plate down on the prism. With a Pasteur pipette, deliver several drops of fluid to the semicircular area on the prism that is not covered by the plastic cover.

c. The liquid will spread by capillary action and form a thin, even layer over the prism.

d. Expose the refractometer to a bright light, and bring the eyepiece into focus by rotating the lens-holding sleeve.

e. Take a protein measurement on the appropriate scale at the point where the dividing line between bright and dark areas crosses the scale (Fig. 9.5).

f. Clean the prism carefully with a moistened Kimwipe. Be sure to handle this instrument with caution, and avoid scratching the surface of the prism.

K. Quantitative Determination of Alanine Aminotransferase (ALT) in Serum

REFERENCES: Teitz, pp. 673–682.

Technicon RA Systems, Technicon Corp. Bulletin SM4-013G85, Tarrytown, NY, July 1985.

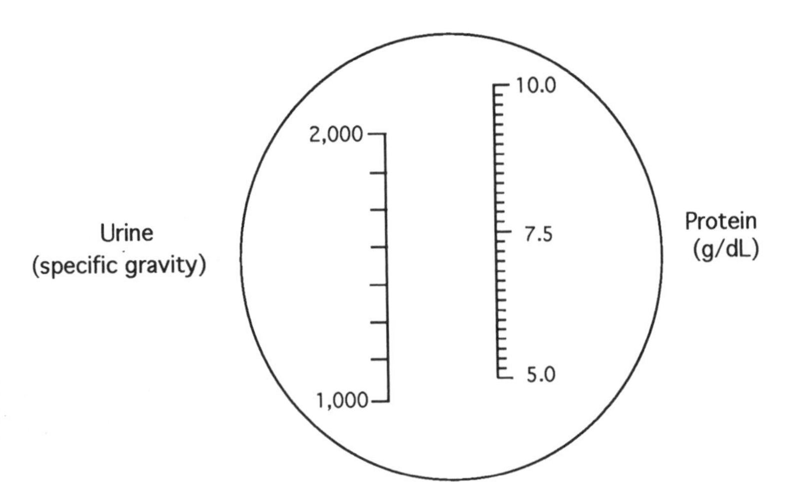

Fig. 9.5. Field seen in an AO refractometer with scales shown to read total serum protein in g/dL or specific gravity of urine.

The ALT method is used in the clinical laboratory, often in conjunction with the aspartate aminotransferase (AST) method, to detect tissue damage that caused increased enzyme activity in the serum. ALT enzyme activity is 450 times greater in heart cells, 2850 times in liver cells, and 1200 times in kidney cells than it is in serum. The enzyme is localized in the mitochondria of these tissues and is released into the serum when severe tissue damage occurs.

The ALT assay method is commonly used to aid in diagnosis of diseases related to liver malfunction. Because elevated levels appear before the clinical symptoms of diseases such as jaundice appear, early treatment is possible.

The assay system consists of a kit manufactured by Technicon Instruments Corp. (Tarrytown, N Y) and is based on determining the rate of NADH oxidation photometrically at 340 nm. The reagents contained in the kit include: L-alanine, 546 mM; NADH, 0.28 mM; α-ketoglutarate, 16.3 mM; *ca* 2400 units of lactate dehydrogenase; and pyridoxal 5-phosphate, which is a cofactor of ALT.

In the presence of ALT in serum, the following reactions occurs:

$$\text{1) L-alanine} + \alpha\text{-ketoglutarate} \xrightarrow{\text{ALT}} \text{pyruvate} + \text{L-glutamate}$$

$$\text{2) Pyruvate} + \text{NADH} \xrightarrow[\text{dehydrogenase}]{\text{lactate}} \text{lactate} + \text{NAD}^+$$

The rate of NADH oxidation is therefore directly proportional to the ALT in the sample. From the rate of *decrease* in absorbance of NADH, the activity in International Units (IU) per liter of serum can be calculated. Normal values range from 5 to 35 IU/L at 30°C.

Procedure for ALT Determination

1. *Preparation of reagent.* Reconstitute the reagents in the vial by adding 10 mL of water, and swirl gently until the powder is dissolved. The reagents are stable for 12 hr at room temperature and for 7 days at 4°C.

2. Allow both reagents and serum to equilibrate to room temperature.

3. Add 1 mL of reagent to a quartz cuvette, and read the absorbance at 340 nm in the spectrophotometer using a UV light source.

4. Prior to adding the serum to the substrate, the spectrophotometer must be calibrated. One method is to set the absorbance of the unreacted reagent to an initial reading of 0.500; the other is to use the reagent as a "blank" by setting the absorbance at zero. In the latter instance, negative values will be obtained, but this does not present a problem since it is the decrease in rate and not the absolute values that are of interest.
5. Add 0.1 mL of serum, mix well, and record the absorbance at 1-min intervals for 6 min.
6. Plot a graph of $A_{340\ nm}$ *vs.* time. From this graph, calculate the ΔA/min. Discard the first minute's reading, since the change in absorbance may be due to a mixing effect rather than to enzymatic activity.
7. The results are reported in International Units (IU) per L of serum and calculated as follows:

$$\mathrm{IU/L} = \frac{\Delta \mathrm{A/min} \times 1000}{6.22 \times 0.1}$$

where ΔA/min = change in absorbance observed, 1000 = conversion from mL to liter since the reaction takes place in a volume of 1 mL, 6.22 $mM^{-1}\,cm^{-1}$ = extinction coefficient for NADH at 340 nm and 0.1 = volume (in mL) of serum tested.

L. Serum Carotenes (See Nutritional Evaluation of Foods, p. 231)

M. Detection of Adult Carriers of a Hereditary Fat Metabolism Disease

REFERENCES: O'Brien, J.S., Okada, S., Chen, A., and Fillerup, D. Tays-Sachs disease: Detection of heterozygotes and homozygotes by serum hexoseaminidase assay. *N. Engl. J. Med.* 283:15–20 (1970).

Brady, R. Hereditary fat-metabolism diseases. *Sci. Am.* 229(2):88 -97 (1973).

Tay-Sachs is one of about 10 known fat metabolism diseases that are hereditary and caused by the excessive accumulation of lipids in nerve cells. This disease occurs predominantly in children of Ashkenazic Jewish descent, although it has been reported among other groups at much lower frequencies. The autosomal recessive inheritance is manifested by the absence of an enzyme, hexosaminidase A (Hex A), in Tay-Sachs patients, who die in early childhood. Carriers who have only one functional gene for Hex A have one-half the enzyme activities of noncarriers. The method of detection of carriers is therefore based on assaying blood serum for Hex A activity.

Hex A is responsible for splitting off a molecule of *N*-acetylgalactosamine from a cerebroside known as ganglioside GM_2. This hydrolysis is the first step in lipid breakdown, and in its absence accumulation of lipid in nerve cells results. Although individuals who are heterozygous for this trait produce less enzyme, the amount is sufficient for normal metabolic function, and no symptoms appear. The ganglioside GM_2 molecule can be visualized as a ceramide having a terminal molecule of *N*-acetylgalactosamine:

Hex A

Ceramide-*O*-glucose ——— galactose ——— *N*-acetylgalactosamine

N-acetylneuraminic acid

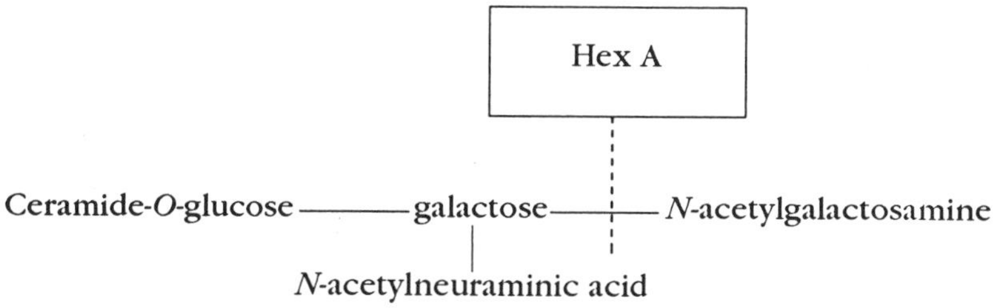

The Tay-Sachs assay is carried out by testing for Hex A activity on a synthetic sugar, 4-methylumbelliferyl-*N*-acetyl-β-D-glucosaminide, which on hydrolysis is converted to *N*-acetylglucosaminide and 4-methylumbelliferone. The latter fluoresces upon exposure to UV light. Therefore, the activity of the enzyme in a given aliquot of serum can be determined by measuring the amount of fluorescence produced against a standard curve generated for 4-methylumbelliferone. The reaction is shown below:

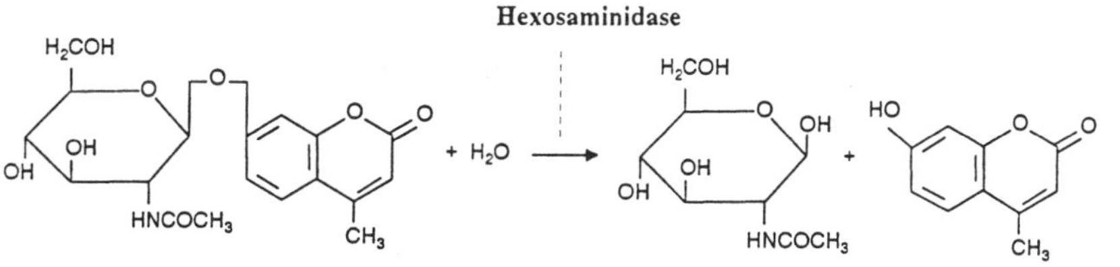

Hexosaminidase

4-Methylumbelliferyl *N*-acetylglucosaminide + H_2O → *N*-Acetylglucosamine + 4-Methylumbelliferone

A Turner fluorimeter, Model 110, will be used for these studies. It consists of two sets of filters, primary and secondary, through which a light source of a given wavelength is projected onto the sample. The compound in the sample absorbs photons, causing the molecule to become excited. The excitation is rapid and is followed by either of two events, depending on the atomic structure of the molecule: (1) a return to ground state with emission of energy as heat or light (which can be measured as absorption) or (2) emission of light of a longer wavelength (fluorescence). Whereas absorbance (A) is a logarithmic function, fluorescence is a linear function and is read directly as %F.

Procedure for Using the Turner Fluorimeter

In this instrument, zero is set with a black "dummy" instead of by adjusting the zero point with the "blank" solution as is done in a spectrophotometer. The fluorescence readings for the "blank" tubes of the assays are then subtracted from every value.

1. To turn on the instrument, rotate the "zero" dial clockwise past a slight resistance. Open the door and check for a bluish light. If the UV bulb is not yet on, turn the "zero" dial further to the right. Allow a warm up time of 5 min.
2. Insert the appropriate filters in the instrument. Primary filters go on the right side of the instrument, secondary filters on the left. Set the sensitivity knob to the desired setting (1×, 3×, 10×, 30×).
3. With the door open, rotate the "zero" knob to bring the needle to "0."
4. Place the black dummy in the holder on the inside of the door, and read the samples. Close the door securely. Turn the "fluorescence" dial to "0," and then turn the "blank" knob to bring the needle to "0."
5. Use 13 × 100 mm borosilicate tubes to read samples.
6. Read the reagent blank first. Readings are taken in %F by rotating the "fluorescence" dial until the needle on the meter returns to "0."
7. Turn the instrument off by turning the "zero" knob counterclockwise until the light goes out.
8. Return the filters to the appropriate slats in the storage box.

Procedure for Tay-Sachs Assay

1. Dilute 0.1 mL of serum in 0.9 mL of 0.04 M citrate-phosphate buffer, pH 4.4. Put on ice for 10 min.
2. Transfer 0.10 mL of the above to
   a. Three screw-cap tubes and
   b. Four (13 × 100 mm) borosilicate glass tubes. Cover these with parafilm and put on ice.
3. Heat the three screw-capped tubes at 50°C for 3 hr, and then put on ice to cool.
4. Store all tubes in the refrigerator until the next laboratory period or proceed with Step 5.
5. Add 0.1 mL of 1 mM methylumbelliferyl-*N*-acetyl-β-D-glucosaminide to the three heat-treated tubes and to three of the unheated tubes.
6. Incubate the seven tubes at 37°C for 1 hr.
7. Remove to ice and add 5 mL of 0.17 M glycine-carbonate buffer, pH 9.9, to all tubes to stop the reaction. To the fourth untreated tube, add 0.1 mL of substrate after the buffer. (This tube will serve as the blank.)
8. Use the black "dummy" to set the instrument at 0%F and read all the tubes in the Turner Fluorimeter using filters: primary, 7-60 (365 nm); secondary, 48 + 2A (450 nm), and a 1% gray filter. The sensitivity is set at 10×, or it is adjusted to give about 80%F with the tube containing 6 nmol of standard solution (see Step 9 below). Subtract the value for the blank (the tube containing 0 nmol of standard) from all the experimental tubes before reading the %F values against the standard curve.
9. Construct a standard curve by diluting 0.1 mL of stock 4-methylumbelliferone (0.2 mg/mL) in 9.9 mL of glycine-carbonate buffer, pH 9.9. Deliver 0, 0.1, 0.2, 0.4, 0.5, and 0.6 mL to six borosilicate test tubes, and bring the volume to 5 mL with the same buffer. (These volumes correspond to 0, 1, 2, 4, 5, and 6 nmol of 4-methylumbelliferone, respectively.)
10. Convert the fluorescence readings obtained above to nmol of umbelliferone, and calculate (a) the total hexosaminidase and (b) the heat-stable fraction in nmol/mL·hr.

(a) = average nmol/aliquot × 100 (dilution factor) in untreated serum.

(b) = average nmol/aliquot × 100 in serum heated at 50°C for 3 hr.

11. Calculate the % Hex A in serum and record the results in Table 9.12.

$$\% \text{ Hex A} = \frac{(a) - (b)}{(a)} \times 100$$

Compare your results with the values shown below:

	Controls	*Carriers*	*Tay-Sachs Patients*
Total hexosaminidase (nmol/mL · hr)	333-775	288-644	284–1232
Percent Hex A	49-68	26-45	0–4

N. Blood Typing

Blood type is determined by the antigens on the surface of the blood's red cells. The major antigens are the A, B, and Rhesus antigens. There are also many minor antigens, such as Lewis, Kell, and Duffy. The Rhesus antigens are also subdivided into different groups, of which D is the most important. In this experiment, we will be testing for the presence (or absence) of A, B, and RhD antigens. In O-negative blood, the surfaces of the red cells have none of these antigens. In AB-positive blood, all these antigens are present. It is important to note that in the following experiments we are not going to do complete blood typing, as might be done in a professional blood bank. The blood bank also normally cross-matches the donor's blood with the recipient's to make sure that there are no other incompatibilities.

Procedure for Blood Typing: Rh Typing

1. To two drops of whole blood (green-capped tube may be used or a finger-prick incision may be made) on a glass slide add one drop of RhD serum. Mix the blood and antiserum by gently rotating the slide back and forth.

2. Read for the presence of agglutination after 3 to 4 min.

The test is temperature dependent but will work well at room temperature if sufficient reaction time is allowed. A positive agglutination test results when RhD antigens are present on the surface of the blood cells.

ABO Typing

1. Place one drop of anti-A serum on one side and one drop of anti-B serum on the other of a clean glass slide.
2. Add to each drop of serum a volume of fresh blood equal to approximately one-half the volume of antiserum used or an amount sufficient to produce a final cell concentration of approximately 10% to 15%. Mix the blood and serum on each side.
3. Rotate the slide and examine for agglutination within a 2 min period.
4. Determine the blood type according to the following table and record the results in Table 9.12.

Blood Type	*Agglutination with* *Anti-A serum*	*Anti-B serum*
A	+	−
B	−	+
O	−	−
AB	+	+

URINE ANALYSIS

A. Urine Collection

A 2-L bottle, containing some thymol as preservative, will be provided. Collect all urine for a period of 24 hr starting after the first morning void, and concluding with the first sample of the next day. Metabolites are excreted in the urine at varying concentrations during the day, thus necessitating the 24-hr collection. The results are expressed on a per-day basis.

1. Measure and record the volume of the urine excreted to ± 10 mL.
2. Remove a 100-mL representative sample, and add 1.5 mL of 6 N HCl. Save at 4°C for the next laboratory period for the catecholamine extraction.

B. Urine Analysis by Use of a Commercial Reagent Strip

Eight separate pads are located on the plastic strip, *Ames N-Multistix* (Miles Inc., Diagnostic Division, Elkhart, IN). Place the strip in a horizontal position on a paper towel, and proceed as follows:

1. Using a sample of well-mixed, unfiltered urine, deliver a sufficient amount to wet each pad. Avoid wetting the adjacent areas.
2. Do one test at a time, and record your results in the unit corresponding to the color development obtained.
3. Compare test areas to corresponding colors on the reagent bottle label at the *specified times*. Consult Table 9.4 for more information.

TABLE 9.4. Ames Multistix (Urine Analysis)

Test	Reading time (sec)	Color development (Normal or negative→high)	Unit equivalents
Urobilinogen	45	Chartreuse → copper	0.1–12 Ehrlich units/dL
Nitrite	40	Buff → pink	– to +++
Blood	40	Gold → dark blue-green	– to +++
Bilirubin	20	Ivory → blush	– to +++
Ketone	15	Tan → burgundy	5–160 mg/dL
Glucose	30	Light blue → green-brown	0–2 g/dL
Protein	Not critical	Pale yellow → light aqua	Trace to 2 g/dL
pH	Not critical	Orange → aqua	5–8.5

4. For an explanation of the procedure and the normal values, consult Table 9.5.

C. Total Solids in Urine

Optical urinometry is an excellent diagnostic tool for renal dysfunction. Measurement of specific gravity is made with a hand refractometer, which allows direct measurement of specific gravity.

Specific gravity is the ratio of the weight of a substance to the weight of an equal volume of water at a specified temperature (usually 20°C). Its measurement is an important part of routine urinalysis and helps to determine the concentrating ability of the renal tubules. In cases of renal tubular damage, this renal function is generally lost first. Elevated values of specific gravity are commonly found in patients with uncontrolled diabetes.

Normal values for a 24-hr specimen range from 1.015 to 1.025. Excretion values range from 40 to 70 g/day (expressed as total solids.)

Procedure for Determining Total Solids

1. Place two to three drops of sample on the glass surface of the AO Goldberg refractometer, and bring the upper hinged surface down firmly. Blot any excess fluid outside the glass area with a tissue; be careful not to disturb sample.
2. Hold the meter toward a source of light so that the beam passes through the sample and the prism.
3. Read the specific gravity from the proper scale at the sharp line of contrasting light and dark areas that falls across the scales. (See Fig. 9.5 for an illustration of the field seen through the refractometer optical stem.) The scale on the left side will read 1.000 specific gravity when water is sampled.
4. Convert to total solids per 100 mL using the values in Table 9.6, and calculate the amount excreted in 24 hr. Record the results in Table 9.12.

D. Determination of Protein in Urine

Proteinuria is the condition in which large amounts of proteins are found in the urine. It is most commonly a result of the presence of serum albumin, since albumin is the most abundant of the serum proteins and has the smallest molecular weight,

TABLE 9.5. Analysis by the *Ames Multistix* Method[a]

Test	Chemical principle	Performance range of test strip	Normal values expected	Clinical significance
pH	Double pH indicators for broad range assay	pH 5–9	*ca* 6	Conditions leading to ketosis cause acidic urine. Bacterial infections lead to either acid or alkaline pH depending on the end products of bacteria causing infection
Protein	Presence of protein results in color change of yellow bromphenol blue to blue-green	5–20 mg%	Under 10 mg%	Values greater than 20 mg% are indicative of proteinuria. Test serves as indicator of kidney or urinary tract disease
Glucose	(a) Glucose→ gluconic acid + H_2O_2 (b) H_2O_2 + reduced dye → H_2O + oxidized dye	Up to 100 mg%	None	Improper metabolism of glucose as in diabetes
Ketones	Reaction of acetoacetic acid or acetone with nitroprusside	10 mg% (+) 30 mg% (+ +) 80 mg% (+ + +)	None	Decreased intake or utilization of carbohydrates due to dietary imbalance, or diabetes mellitus

Bilirubin	Bilirubin is coupled with diazotized dichloroaniline	0.2–0.4 mg% but generally a qualitative test (0–4 +)	None	Liver malfunction as in obstructive jaundice
Blood	Hemoglobin causes the reductions of cumene hydroperoxide resulting in the oxidation of the dye ortho-tolidine, which changes color	5–15 RBCs/μL	None	Kidney malfunction
Nitrite	Nitrate is reduced to nitrite by bacterial reductase action—test is based on a color change on test strip when nitrite is present	0.03–0.06 mg%	None	Bacterial infection
Urobilinogen	Urobilinogen reacts with *p*-dimethylaminobenzaldehyde to produce a reddish color	As low as 0.1 Ehrlich units %	None	Various forms of hepatitis involving impairment or destruction of liver cells

[a]N-Multistix reagent strips, Ames Division, Miles Laboratories, Inc., Elkhart, IN.

TABLE 9.6. Total Solids in Urine (Urine Solids, 20°C)

Specific gravity	Total solids (g/100 mL)
1.000	0.0
1.001	0.1
1.002	0.4
1.003	0.7
1.005	1.0
1.006	1.2
1.007	1.5
1.008	1.6
1.009	1.9
1.010	2.2
1.011	2.4
1.012	2.5
1.013	2.8
1.014	3.1
1.015	3.3
1.016	3.6
1.017	3.8
1.018	4.0
1.019	4.3
1.020	4.6
1.021	4.8
1.022	5.1
1.023	5.4
1.024	5.7
1.025	5.9
1.026	6.2
1.027	6.5
1.028	6.8
1.029	7.0
1.030	7.3
1.031	7.6
1.032	7.9
1.033	8.3
1.034	8.6
1.035	8.8
1.036	9.1
1.037	9.4
1.038	9.8
1.039	10.1
1.040	10.3

thus permitting the greatest diffusion through damaged membranes. However, significant amounts of globulins can also be found in urine. The normal excretion of total protein in a 24-hr period ranges from 0.020 to 0.075 g. Albumin constitutes 90% to 100%, and globulin, 0% to 10%. The determination of protein (or albumin) may be of assistance in following the course of kidney disturbances, but the results can only be interpreted in the light of other clinical findings.

Total protein will be determined by performing a biuret reaction. Before adding the biuret reagent, the proteins in the urine will be precipitated by the addition of trichloroacetic acid (TCA) to a sample of filtered urine, and the pellet obtained after centrifugation will be resuspended in alkali (3% NaOH). Follow the protocol shown in Table 9.7 and record the results in Table 9.12.

E. Catecholamines in Urine

REFERENCES: *Manual of Fluorimetric Clinical Procedure,* G.K. Turner Assoc., Palo Alto, CA (1977).
Tietz, pp. 803–817.
Zubay, pp. 1051–1053.

TABLE 9.7. Procedure for Protein Determination[a]

Additions (mL)	Tube no. 1 (Blank)	2	3	4	5	6	7	8
BSA (2 mg/mL)	0	0.1	0.2	0.3	0.4	–	–	–
Filtered urine						20	20	20
H_2O	20	19.9	19.8	19.7	19.6	–	–	–
50% TCA	2.0 →							
	Mix and put in ice for 3 min; centrifuge in clinical centrifuge for 5 min. Discard supernatant and resuspend each pellet in 1.5 mL of 3% NaOH. Mix.							
Biuret reagent	1.5 →							
	Incubate at 37°C for 15 min and read against the blank.							
% $T_{540\ nm}$								
$A_{540\ nm}$								

[a]Calculate the amount of protein per mL of urine by comparing the $A_{540\ nm}$ of the urine tubes with the standard curve generated from tubes 1 to 5. Express the results in g/day.

The catecholamines are a group of three compounds synthesized from L-tyrosine in the adrenal medulla. The first of these, 3,4-dihydroxyphenethylamine (dopamine), is a precursor to norepinephrine (noradrenaline), which in turn becomes converted to epinephrine (adrenaline) upon methylation (see pathway of biosynthesis in Fig. 9.6). These hormones are released into the bloodstream in response to outside stimuli such as cold, shock, or pain. A series of physiological reactions results, including a rise in blood pressure, increased heart and metabolic rates, and altered smooth muscle activity. The biochemical manifestation of increased hormone output is a rapid breakdown of glycogen, resulting in raised levels of lactic acid in muscle tissue and of glucose in the blood. Elevated values of catecholamines are associated with tumors of neural origin, such as blastomas and gangliomas. Normal values in urine per 24 hr are 10 to 15 μg for epinephrine and 30 to 50 μg for norepinephrine. Total catecholamine values range from 40 to 100 μg/day.

In the method used here, catecholamines are isolated by chromatographic separation on alumina at pH 8.5. They are eluted with dilute acetic acid and are oxidized to lutin derivatives by treatment with ferricyanide and Zn^{++}. Ascorbic acid is then added to inactivate excess oxidizing agent. Measurements are made by fluorimetry against internal standards of known concentration. To determine norepinephrine and epinephrine concentrations separately, the difference in excitation and emission spectra of the unknowns are compared with the corresponding values for standards of known concentration.

For description and use of the fluorimeter, see Section M.

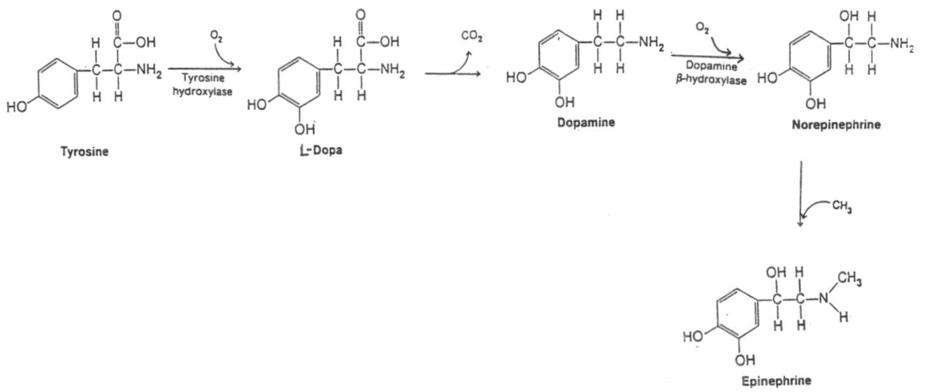

Fig. 9.6. The biosynthetic pathway of epinephrine.

Procedure for Extraction and Quantitation

Isolation

1. Filter an aliquot of the urine sample, and transfer 50 mL into a 100-mL beaker. Place on a magnetic stirrer, and adjust to pH 8.5 by dropwise addition of 1 N NaOH. Continue stirring until the pH *has stabilized.*
2. Add 1 g of alumina (prewashed) and 0.5 g of EDTA. Continue stirring for 5 min to mix thoroughly.
3. Distribute the suspension equally into two 50-mL centrifuge tubes, and centrifuge in a clinical centrifuge for about 5 min. Discard the clear supernatants by decanting.
4. Wash the alumina twice with 10-mL portions of distilled water by resuspending, centrifuging, and discarding the washes. Avoid loss of alumina during these washings. Repeat the centrifugation step if necessary.
5. During the second wash, combine the alumina from the two tubes by transferring the suspension from one tube to the other before the centrifugation.
6. Extract the catecholamines by adding 10 mL of 0.2 N acetic acid, cover the tube with parafilm, and mix on a Vortex mixer for 5 min. Centrifuge and collect the supernatant in a clean 50-mL centrifuge tube.
7. Repeat the extraction with an additional 10 mL of 0.2 N acetic acid, and add this supernatant to the first extract in the centrifuge tube.
8. Centrifuge the extract for 5 min to remove any alumina that might have been carried over with the extracts, and carefully remove *ca* 12 mL of the supernatant to a clean test tube without disturbing the pellet. This extract is now ready for the catecholamine determinations.

Catecholamine Determinations

1. Set up a series of six tubes as shown in Table 9.8 (use 13 × 100 mm borosilicate tubes).
2. Add two drops of bromthymol blue, and titrate to approximately pH 6 by dropwise addition of 2 N K_2CO_3. Stir vigorously, and add 0.5 mL of 0.2 M phosphate buffer to each tube.

TABLE 9.8. Procedure for the Determination of Catecholamines

Additions (mL)	Tube no. 1 (Blank)	2 (U_1)[a]	3 (U_2)[a]	4 (U_3)[a]	5 (N)	6 (A)
Acetic acid extract	2.0	→	→	→	→	→
Standards:						
Norepinephrine (0.25 μg/mL)	–	–	–	–	1.0	–
Epinephrine (0.25 μ/mL)	–	–	–	–	–	1.0
H_2O	1.0	1.0	1.0	1.0	–	–

[a]U_1 through U_3 are replicate samples of the unknown extract. Use an average value.

3. a. Prepare about 10 mL of a mixture of 5 N NaOH:2% ascorbate (9:1). This reagent deteriorates rapidly and should be prepared just prior to use.
   b. Add 0.1 mL of 0.25% $ZnSO_4$ and 0.1 mL of 0.25% K ferricyanide to tubes 2 through 6, and mix thoroughly.
   c. After 2 min, add 1 mL of the ascorbate reagent to the six tubes, and mix again.
   d. Now add the same amounts of the oxidizing agents ($ZnSO_4$ and K ferricyanide) to tube 1, the blank.
4. After 10 to 20 min read the tubes in a Turner fluorimeter to determine the total catecholamines with filter set I: primary, 405 (420 nm); secondary, 65A (485 nm); and sensitivity setting, 30×.
5. Now read the same tubes for determining the individual catecholamines using filter set II: primary, 2A + 47B (438 nm); secondary, 2A–15 (540 nm); and sensitivity setting, 10×.

Calculations

Total catecholamines

$$\frac{U - B}{N - U} \times 0.25 \times \frac{24 \text{ hr volume (mL)}}{5.0} = \mu/\text{day}$$

where U = percent fluorescence of the average of the three values for the urine sample; B = percent fluorescence of the blank; N = percent fluorescence of the sample + norepinephrine (0.25μg/mL).

Individual Catecholamines

Because $N_I - U_I$ = fluorescence due to 0.25 μg norepinephrine standard with filter set I, $(N_I - U_I) \times 4$ = fluorescence per μg of norepinephrine added. Calculate the following values for both standards using the two filter sets.

$$n_I = (N_I - U_I) \times 4$$

$$n_{II} = (N_{II} - U_{II}) \times 4$$

$$a_I = (A_I - U_I) \times 4$$

$$a_{II} = (A_{II} - U_{II}) \times 4$$

Also calculate the fluorescence resulting from the catecholamines present in the urine sample (use the average of the three values).

$$u_I = U_I - B_I$$

$$u_{II} = U_{II} - B_{II}$$

Using these values calculate the amount of norepinephrine (y) in the urine sample (equivalent to 5.0 mL of urine analyzed):

$$y = \frac{(u_I a_{II}/a_I) - u_{II}}{(n_I a_{II}/a_I) - n_{II}}$$

and for epinephrine (x):

$$x = \frac{u_{II} - y\, n_{II}}{a_{II}}$$

and for 24 hr samples:

$$\frac{y \text{ or } x}{5.0} \times \text{total urine volume (mL)}$$

Record these values in the summary table (Table 9.12).

F. Coproporphyrin in Urine

The class of compounds that are precursors of the heme proteins contain porphyrin rings, which may complex with metal ions. In the coproporphyrins, the copper ion complexes with neighboring nitrogens as shown below:

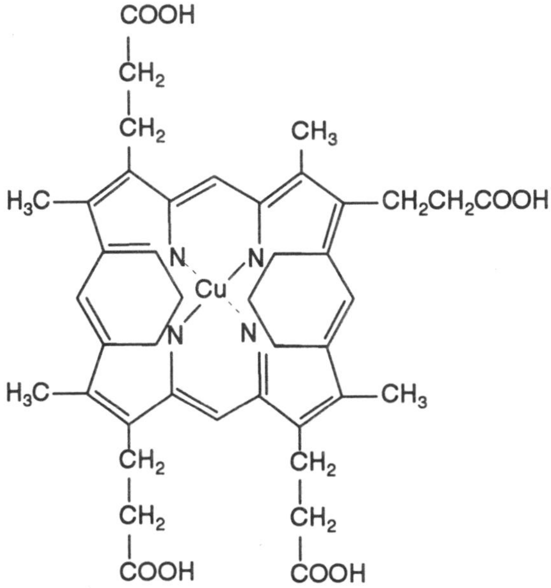

The uroporphyrins, which are also excreted normally in urine in small amounts, are similar in structure except that the methyl groups are replaced by acetic acid molecules.

Normal excretion of coproporphyrin in a 24 hr period ranges from 25μg to 300 μg. Increased urinary output is associated with a number of disorders such as lead poisoning, poliomyelitis, rheumatic fever, infectious hepatitis, and certain anemias. Examples of these values are tabulated below:

Lead poisoning	500 to 3000 μg/day
Poliomyelitis	300 to 900
Pernicious anemia	150 to 400
Hodgkin's disease	200 to 1000
Acute alcoholism	250 to 500

In the method used here, coproporphyrin is extracted into ethyl acetate at pH 4.8. If the urine is fresh, much of the porphyrin will be in the precursor form and must be treated with a weak iodine solution, but if the urine is allowed to stand for at least 24 hr this treatment can be omitted. Then the fluorescent coproporphyrin is reextracted from ethyl acetate into 1.5 N HCl and measured directly in the fluorimeter. The fluorescence is compared to a standard of known concentration.

Procedure for Extraction and Quantitation

1. Transfer 10 mL of unfiltered urine to a 250-mL separatory funnel. Add 5 mL of 5 M acetate buffer, pH 4.8, 10 mL of water, and 100 mL of ethyl acetate. Shake the funnel for 2 min, allow the phases to separate, and discard the aqueous phase.
2. Wash the ethyl acetate with two 10-mL portions of 1% sodium acetate, and again discard the aqueous phase.
3. If the urine is fresh, add 10 mL of 0.005% iodine, shake gently for 1 min, and discard the aqueous phase.
4. Extract the coproporphyrin from the ethyl acetate with three 5-mL portions of 1.5 N HCl, collecting the extracts in a 25-mL graduated cylinder. Bring the volume to 25 mL with 1.5 N HCl and read this sample in a fluorimeter having the following filters inserted: primary, 405 (420 nm); secondary, 25 (610 nm).

The fluorimeter is set to 0%F with a dummy cuvette (black tube). Then the fluorescence values of the sample (0%F), of a standard containing 0.05μg/mL, and of a blank (1.5 N HCl) are read at a sensitivity setting of 30×. (Use 13 × 100 mm borosilicate tubes as cuvettes).

Calculation

The fluorescence of coproporphyrin in 1.5 N HCl saturated with ethyl acetate is about 8% higher than in pure 1.5 N HCl. A correction factor of 0.925 is used to take account of this factor.

The standard contains 0.05μg/mL; therefore, the coproporphyrin content per mL in the original urine sample is

$$\frac{F_{\text{unknown}} - B}{F_{\text{standard}} - B} \times 0.925 \times 0.05 \times 2.5 = \mu g/mL$$

where F = percent fluorescence of unknown and standard, respectively; B = percent fluorescence of the blank; and 2.5 = factor by which the urine was diluted during the procedure. The daily excretion is obtained by multiplying this value by the volume of urine excreted in 24 hr. Record this value in the summary table (Table 9.12).

NUTRITIONAL STUDIES

REFERENCES: Association of Vitamin Chemists. *Methods of Vitamin Assay*, 3rd ed. Wiley Interscience, New York (1966).

Bieri, J.G., and McKenna, M.C. Expressing dietary values for fat-soluble vitamins: Changes in concepts and terminology. *Am. J. Clin. Nutr.* 34:289–295 (1981).

Booth, V.H. *Carotene: Its Determination in Biological Materials*. W. Heffer & Sons, Cambridge, MA, (1957).

Booth, V.H. The extraction of pigments from plant material. *Analyst* 84:464–465 (1959).

Davidson, I., and Henry, J.B. *Clinical Diagnosis by Laboratory Methods*, 16th ed. W.B. Saunders, Philadelphia (1979).

Gebhardt and Matthews (1985).

Oser, pp. 578–591.

Sebrell, W.H., Harris, J.R., and Harris, R.S. *The Vitamins*, vol. 1. Academic Press, New York (1967).

A. Introduction

The United States Department of Agriculture and the Department of Health and Human Services are charged with assessing and overseeing the dietary needs of the population. The dietary guidelines that have been issued consist of about 40 different nutrients, including:

1. Vitamins and minerals (fruits and vegetables).
2. Amino acids (protein sources).
3. Essential fatty acids (fats and oils).
4. Carbohydrates (energy sources).

Since no single food supplies all of these required nutrients, a variety of foods is required for a balanced diet. Two of these categories are considered to demonstrate some of the methods available to analyze a given food for its nutritional value.

Amino Acids Derived From Protein Sources

As more people rely on plant instead of animal protein, amino acid analysis of protein sources becomes increasingly important. Because most plants do not contain all of the essential amino acids, a combination of these foods is required to fulfill the complete requirement. It is therefore useful to analyze common plant materials for their amino acid composition.

Of the 22 amino acids in proteins, 10 are not synthesized by the human body. The 10 are: tryptophan, leucine, isoleucine, lysine, valine, threonine, methionine, phenylalanine, arginine, and tyrosine. These are referred to as *essential amino acids*, and they must be present in the diet at the same time and in sufficient amounts in order to be effective.

Protein quality is determined by the amount of amino acids available as nutrients. This quantity varies from food to food. The Food and Nutrition Board of the U.S. Academy of Sciences recommends 0.213 g of protein per pound of body weight. If one uses meat, which has the highest quality protein (*ca* 75%), then the daily requirement rises to 0.370 g/lb. In plants, the portion available as a protein source tends to be lower (*ca* 50%). Then it becomes necessary to raise the daily consumption of the products accordingly. It is also necessary to mix the various plant sources so that a mixture of grains, nuts, and legumes provides all the essential amino acids in quantities sufficient to meet the daily requirement.

One of the methods used here will determine tryptophan microbiologically [see Section B, p. 223] A second more sensitive method [see Section C, p. 226] determines the concentration of phenylalanine, another essential amino acid, fluorimetrically. High-performance liquid chromatography (HPLC) can also be used to determine the amino acid content of foods. It is a method with greater sensitivity (to nanogram levels), but it is expensive and time consuming because it requires samples of higher purity.

For a vitamin assay, an extraction, purification, and quantitation method for carotenes (α and β found in spinach or other green leafy vegetable) is described. Serum levels of carotenes will also be determined.

Evaluation of Vitamin A Source Materials

Included in the class of yellow, naturally occurring lipid pigments are the carotenes, certain hydroxylated carotenoids, and some epoxides. Although a number of carotenoids occur naturally, only the α and the β forms have provitamin A activity, and the conversion to vitamin A is only 60% efficient:

Beta carotene

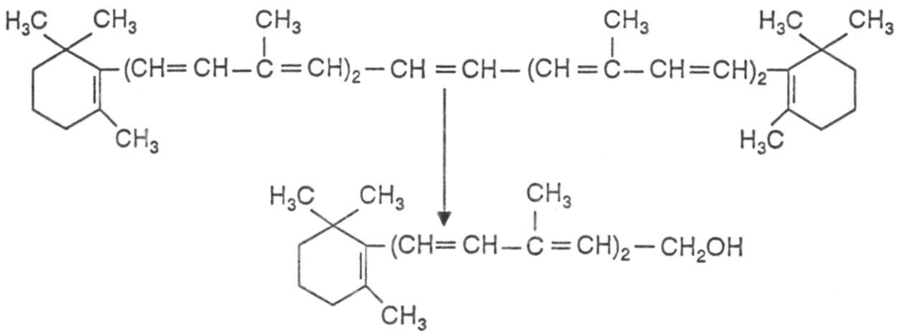

Retinol (Vitamin A)

Since the carotenes are involved in photosynthesis, they are always found in conjunction with chlorophyll in green plants. The most abundant of the carotenes is the β form, which is the precursor to vitamin A. Carrots contain a considerable amount of α-carotene as well as β-carotene. Carotenes and carotenoids are important in animal nutrition because of the conversion of some of them to vitamin A. Hypovitaminosis A is a major nutritional problem throughout the world, even in North America, especially among Mexican-Americans, and is a cause of blindness in young children. Analysis of serum levels of carotenes is important in screening for vitamin A deficiencies. Normal values of serum carotenoids range from 25 to 250 μg/dL.

The destruction of vitamin A and carotenes in foods, especially in storage grains, is a serious problem. Analytical methods for quantitative determinations of carotenoids in foods are therefore important to develop, as are procedures aimed at preventing this destruction. Such methods are also valuable for the selection and development of varieties of plants with higher provitamin A content.

B. Microbiological Determination of Tryptophan

Experimental Procedures

1. Extraction and alkaline hydrolysis.
   a. Weigh out 1.5 g of kidney beans or cheddar cheese and 2.0 g of egg white (mainly albumin). These foods contain approximately 8% and 10% of protein, respectively.
   b. Transfer the egg white to a glass homogenizer, and gradually add 10 mL of 0.1 M phosphate buffer, pH 7.0. Use some of the buffer to transfer the sample quantitatively from the weighing dish to a centrifuge tube. Grind until the sample is evenly dispersed. For the beans or the cheese, use a motar and pestle for grinding; then transfer the paste with the buffer into a centrifuge tube.
   c. Centrifuge the suspensions, and discard the insoluble debris.
   d. Measure 1 mL of each supernatant and transfer to 20 mL ampules with a Pasteur pipette. Add 9 mL of 6 N NaOH, mix well, and seal the ampule by pulling the glass tip over an intense flame.
   e. Incubate at 110°C for 24 hr.
2. Open the ampules after they have cooled, and transfer the extracts to 50-mL beakers. Titrate each with 6 N HCl to pH 7.0 while mixing on a magnetic stirrer. **The pH will drop rapidly.** Change to 1 N HCI at a pH of *ca* 8.5, and finish the titrations with the more dilute acid.
3. Transfer the neutralized hyrolysates to 40-mL centrifuge tubes, and centrifuge at 15,000 rpm for 15 min. Decant the supernatants into 25-mL graduated cylinders, and bring the volumes to 20 mL with 0.1 M phosphate buffer, pH 7.0. Mix well. The hydrolysates could be frozen at this point if the experiment is to be continued on another day. The two tubes now contain approximately 7.5 mg of kidney beans or cheddar cheese and 10 mg of egg white per mL.
4. Set up a series of 16 culture tubes (18 × 150 mm). Deliver the varying amounts of tryptophan solution (200 μg/mL) to tubes 1 to 6 for the standard curve, and of water, as indicated in Table 9.9. To tubes 7 to 16 deliver only the

TABLE 9.9. Microbiological Determination of Tryptophan

Tube no.	L-tryptophan standard (200 μg/mL) mL	L-tryptophan standard (200 μg/mL) μg	Extracts: Cheddar cheese[a] (mL)	Extracts: Egg white (mL)	H_2O (mL)	Growth medium (mL)	Autoclave[b]	Sterile glucose (4%) **(ADD ASEPTICALLY)** (mL)
1	0	0			1.00	5.0	15 min	0.30
2	0.05	10			0.95	5.0	15 min	0.30
3	0.10	20			0.90	5.0	15 min	0.30
4	0.15	30			0.85	5.0	15 min	0.30
5	0.20	40			0.80	5.0	15 min	0.30
6	0.25	50			0.75	5.0	15 min	0.30
7			0.2		0.80	5.0	15 min	0.30
8			0.4		0.60	5.0	15 min	0.30
9			0.6		0.40	5.0	15 min	0.30
10			0.8		0.20	5.0	15 min	0.30
11			1.0		0	5.0	15 min	0.30
12				0.2	0.80	5.0	15 min	0.30
13				0.4	0.60	5.0	15 min	0.30
14				0.6	0.40	5.0	15 min	0.30
15				0.8	0.20	5.0	15 min	0.30
16				1.0	0	5.0	15 min	0.30

[a]Kidney bean extract can also be used.
[b]Autoclave at 120°C and 15-lb pressure.

amounts of water indicated. Add 5 mL of tryptophan-free growth medium to all tubes. (The minimal growth medium contains, per L: K_2HPO_4, 10.5 g; KH_2PO_4, 4.5 g; $[NH_4]_2SO_4$, 1.0 g; Na citrate · $2H_2O$, 1.0 g; Mg SO_4, 0.05 g; glucose, 2 g). To three other culture tubes (tubes 17, 18, and 19), transfer *ca* 5 mL of the two food extracts and 5 mL of a 4% glucose solution. Plug the tubes and sterilize by autoclaving for 15 min (120°C, 15-lb pressure).

5. Allow the tubes to cool to room temperature, and transfer aseptically 0.3 mL of sterile glucose, and the amounts of the extracts indicated in the protocol (see Table 9.9) to the appropriate tubes.
6. A culture of *Escherichia coli* 514 (a mutant with a *try* deletion) will be provided. Harvest the cells by spinning at top speed in a clinical centrifuge for 10 min, discarding the supernatant, and resuspending the cells in 5 mL of sterile saline. Transfer 1.0 mL to a second tube containing 5 mL of sterile saline solution to effect a cell dilution as well as washing out nutrients left from the original growth medium. **Use aseptic techniques.**
7. Using a sterile pipette, inoculate each tube with two drops of the cell suspension.
8. Incubate all tubes in a shaking water bath at 37°C for 48 hr.
9. Mix the cultures vigorously in a Vortex mixer and read the absorbance in a spectrophotometer against the blank at 450 nm.
10. Construct a standard curve from the $A_{450\ nm}$ readings of tubes 1 to 6, plotting absorbance as a function of L-tryptophan (in μg).
11. Determine the amount of L-tryptophan in the tubes having absorbance values that fall within the linear range of the standard curve. It is necessary to use only the average of those values that fall within the linear portion of the standard curve, because the high concentrations of salt that result from base and acid additions are inhibitory to bacterial growth. Multiply this value by 2, because the amino acid forms a racemic mixture during the hydrolysis and heat treatment, and the organism can only utilize L-tryptophan for growth. Calculate an average tryptophan content in μg per g of beans or cheese and

of egg white and the percent of tryptophan by weight for each sample assayed.

C. Fluorimetric Determination of Phenylalanine

REFERENCES: *Manual of Fluorimetric Clinical Procedures*. G.K. Turner Assoc., Palo Alto, CA (1977).

McCaman, M.W., and Robins, E. Fluorimetric method for the determination of phenylalaline in serum. *J. Lab. Clin. Med* 59:885–890 (1962).

As an alternative method for assessing the nutritional value of a protein source, the following technique for the determination of one of the essential amino acids, phenylalanine, will be used. The food to be analyzed is first ground or homogenized to facilitate the extraction of proteins, which are then transferred to an ampule. A strong alkali is added. The ampule is sealed and incubated at 110°C for 24 hr to hydrolyze the proteins. The hydrolysates are then made slightly acidic, and the phenylalanine concentration is determined by a fluorimetric assay.

The phenylalanine assay that is used is a modification of the procedure that McCaman and Robins developed for the detection of phenylalanine in the serum to screen newborn infants for phenylketonuria. The method is dependent on the formation of a fluorescent complex of phenylalanine in the presence of ninhydrin and a small peptide, L-leucyl-L-alanine.

The hydrolyzed food extracts that were previously prepared are first neutralized and then made slightly acidic by addition of 0.6 N TCA. Ninhydrin, succinate buffer, and the peptide are added to varying amounts of extract and the mixture incubated for 2 hr at 60°C. After cooling, an alkaline copper tartrate solution is added, and the fluorescence produced by excitation at 365 nm is read at 500 nm. The samples are read against a standard curve constructed for known amounts of phenylalanine.

1. Extraction and alkaline hydrolysis.

   (See section B, p. 223—this procedure is the same as for the microbiological determination of tryptophan.)

2. Transfer 0.50 mL of hydrolysate from the ampule to a 1.5-mL microfuge tube, and add 0.55 mL of 6 N HCl. Mix and incubate for 5 min at room temperature.

TABLE 9.10. Fluorometric Determination of Phenylalanine[a,b,c]

Additions (μL)	Tube no. 1	2	3	4	5	6	7	8	9	10	11	12
Phenylalanine (0.05 mg/mL)	0	10	20	30	40	50						
0.3 N TCA	50	40	30	20	10	0						
Hydrolysates:												
Cheese							20	30	40			
Egg white										20	30	40
0.6 N TCA							30	20	10	30	20	10
Reagent Mix (mL)	0.8 →											
	Mix. Incubate at 60°C for 2 hr. Cool to room temperature.											
Copper Reagent (mL)	5.0 →											
	Mix. Incubate at room temperature for 15 min.											
%F												
μg phe/aliquot	0	0.5	1.0	1.5	2.0	2.5						
μg/phe/mL of extract												

[a]Use 13 × 100 mm borosilicate tubes.
[b]Use the following settings on the fluorimeter:
Sensitivity: 10×.
Filters:
1° 7-60.
2° 2A + 65A.
[c]Reagents used:
1. Succinate buffer, 0.3 M, pH 5.8.
2. Ninhydrin, 30 mM.
3. L-leucyl-L-alanine, 5 mM.
4. Copper reagent (per L):

Na_2CO_3	1.6 g
K Na $C_4H_4O_6 \cdot 4H_2O$	0.1 g
$CuSO_4 \cdot 5H_2O$	0.06 g

5. TCA: 0.3 N, 0.6 N.
6. Phenylalanine standard: 0.05 mg/mL.

3. Centrifuge for 10 min in a microfuge, and decant the supernatants into new tubes. Transfer the amounts of supernatant indicated in Table 9.10 to a series of borosilicate tubes and proceed with the assay. (Note that each

hydrolysate now represents 7.5 and 10 mg/mL of bean or cheese and egg white extracts, respectively).

4. Prepare 16 mL of the ***Reagent Mix*** as follows:

10 mL	0.3 M succinate buffer, pH 5.8
4 mL	30 mM ninhydrin
2 mL	5 mM L-leucyl-L-alanine

   Add 0.8 mL of this solution to each tube as indicated in Table 9.10.

5. Incubate all tubes for 2 hr at 60°C; then allow them to cool to room temperature.
6. Add 5 mL of copper reagent, mix, and allow the tubes to incubate at room temperature for 15 min.
7. Read %F at 10× sensitivity (or adjust to bring readings of standard tubes within range) with primary filter 7-60 (365 nm) and secondary filter 2A + 65A (500 nm).
8. Construct a standard curve from the readings of tubes 1 to 6. Calculate the average phenylalanine content per mL of extract assayed and the amount (in μg) of this amino acid per g of cheese or kidney bean, and egg white. Also calculate the phenylalanine content (in %) of the protein for each food stuff. (Assume that cheese and beans contain *ca* 8% and egg whites contain *ca* 10% of protein). Record the results in the summary table (Table 9.12).

D. Determination of Carotenes in Vegetables

Care must be taken to prevent destruction by oxidation or isomerization of the carotenes during extraction. During the extraction, the material should be protected from bright light. After cutting the leaves, immediately proceed with the extraction procedure to minimize the enzymatic destruction of carotenes in damaged green leaves.

Extraction

1. Weigh 1.0 g of a dark green leafy vegetable, taking care to obtain a representative sample.
2. Place sample in a mortar. Add 1 g of sand and 10 mL of a 1:1 mixture of acetone and hexane containing 0.1% hydroquinone.

3. Cut the vegetable into very small pieces, grind, and allow tissue to settle.
4. Transfer the extract with a Pasteur pipette to a 125-mL Erlenmeyer flask, and grind the residue thoroughly. Add another 10 mL of the acetone:hexane mixture. Again allow settling and remove the extract. Repeat the grinding and extraction steps until the extract is practically colorless (five or more extractions are usually necessary).
5. Tranfer about 20 mL of a 50% ammonium sulfate solution into a 250-mL separatory funnel. Add a small layer of hexane on top, and pour the combined extracts onto this layer.
6. Wash the extract by adding about 100 mL of H_2O, and shake. Remove and discard the lower aqueous layer, and repeat the H_2O washes three more times. After each wash add *ca* 10mL of 50% ammonium sulfate to break any emulsion that may have formed.
7. Transfer the washed extract to a graduated cylinder, and bring the volume of this extract to 20 mL with hexane.

Separation of Vegetable Carotenes on a Silica Sep-Pak Cartridge

1. The "long body plus" Sep-Pak Column is a Waters Co. (Milford, MA) product with a black color-coded ring. It contains silica gel, which is suitable for the separation of nonpolar materials. The black ring should be at the bottom of the cartridge, i.e., the sample is loaded onto the end of the cartridge distal to the black ring (see Fig. 9.7).
2. Attach the cartridge to a syringe with Luer fittings, and mount the syringe on a ring stand. Pass through 10 mL of hexane, and discard the effluent.
3. Load 2 mL of the sample onto the cartridge and allow it to enter the gel completely. (If necessary, gentle pressure can be applied to facilitate elution after the sample has been loaded).
4. Start collecting in a graduated tube or cylinder when the yellow carotene band has reached the bottom of the cartridge and elute with enough solvent (10% acetone in hexane) to remove the entire band (or until there is no more color in the eluant: 3.0 to 3.5 mL will usually be required).

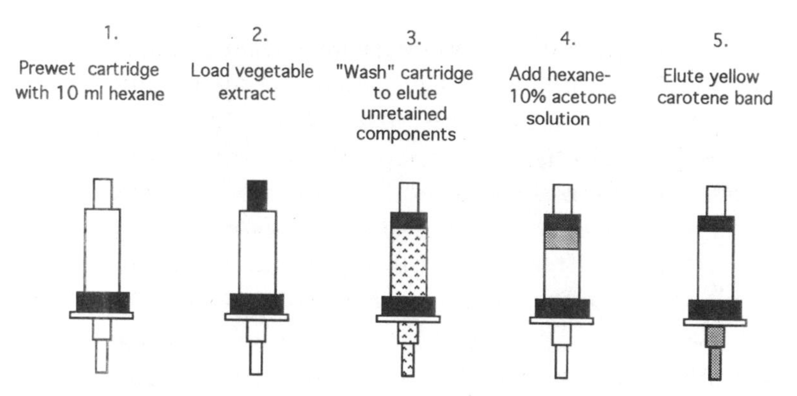

Fig. 9.7. Elution of carotenes from a silica Sep-Pak cartridge.

5. Bring the volume to 4 mL with the same solvent, and read directly at 450 nm in the spectrophotometer against a hexane blank. When making the calculation for the amount of carotenes in the sample, be sure to take into account the total volume of extract obtained from the 1-g sample of starting material.
6. Construct a standard curve as shown in Table 9.11.

Carotenes are split in half by action of an oxidase. Assuming an efficiency for this oxidation reaction of 60%, the number of International Units of vegetable can be calculated (1 g of vitamin A is equivalent to 3×10^6 IU). The vitamin A content of foods can also be expressed in *retinol equivalents*. This term is used to take into account both the loss resulting from the conversion of β-carotene to retinol and the variability in human absorption. An estimate of 33% availability from diets, coupled with the conversion of about 50%, results in an estimation of overall utilization of only one-sixth of the carotenes ingested. By this convention, 1 μg of β-carotene in food becomes 0.167 μg of retinol equivalent.

Spinach is a good source of β-carotene, but other fruits and vegetables, such as sweet potatoes, carrots, yellow squash, and dried apricots, also have similarly high β-carotene contents. From the results obtained, it is simple to calculate an amount of these foods that needs to be consumed in order to meet the recommended daily allowance (RDA) of 5000 IU of vitamin A. *For example*: If there is 3×10^6 IU/g of vitamin A and one needs to obtain 5000 IU,

TABLE 9.11. Standard Curve for Carotene Determinations

Additions (mL)	Tube no. 1	2	3	4	5	6	7	8
Carotene standard (8 μg/mL)	0	0.1	0.2	0.4	0.8	1.0	1.2	1.4
Hexane	4.0	3.9	3.8	3.6	3.2	3.0	2.8	2.6
Carotene concentration (μg/mL)	0	0.2	0.4	0.8	1.6	2.0	2.4	2.8
$\%T_{540\ nm}$								
$A_{450\ nm}$								

then 1.72 mg of vitamin A needs to be consumed per day. Since 60% of β-carotene is converted to vitamin A, 2.87 mg of β-carotene is required per day. In the case of spinach, which provides about 75μg of β-carotene/g of green leaves, a portion of *ca* 38.5 g of this vegetable will meet the daily requirement.

7. For the extract you have prepared, calculate the following:
   a. The amount of β-carotene in the vegetable (μg/g).
   b. Number of International Units of vitamin A provided/g of vegetable.
   c. "Retinol equivalents" (μg/g).
   d. The portion of this vegetable that would meet the RDA guidelines for vitamin A (in g/day).

E. Determination of Carotenes in Serum

1. Place 2.0 mL (or whatever amount is left from the previous experiments) of serum into a screw-capped test tube, and add 2.0 mL of 95% EtOH and 4.0 mL of petroleum ether.
2. Stopper and shake vigorously for 5 min while releasing the pressure occasionally.
3. Centrifuge for 5 min at top speed in the clinical centrifuge.
4. Transfer the upper layer of petroleum ether extract to a cuvette.

TABLE 9.12. Summary of Data

	Normal value	Unknown
Serum determinations		
1. Blood glucose	90–120 mg/dL	
2. BUN	6–22 mg/dL	
3. Hematocrit	35–50%	
5. Cholesterol		
Total	100–200 mg/dL	
HDL		
Men	>55 mg/dL	
Women	>65 mg/dL	
LDL	<130 mg/dL	
6. Triglycerides	40–200 mg/dL	
7. Serum proteins (g/dL)		
Total	6.2–8.1	
Albumin	3.5–4.2	
α_1	0.2–0.4	
α_2	0.9–1.3	
β	0.9–1.3	
γ	1.4–2.4	
8. ALT	5–35 IU/L	
9. Carotenes	50–250 μg/dL	
10. Tay-Sachs disease		
Hexosaminidase:		
Total	330–780 nmol/mL·hr	
Hex A	49–68%	
11. Blood type	A,B,AB,O/Rh^+, Rh^-	
Urine analyses		
	(per day)	
12. Volume	500–2000 mL	
13. Albumin	0.020–0.075 g	
14. Total solids	40–70 g	
15. Catecholamines:		
Total	40–100 μg	
Norepinephrine	30–50 μg	
Epinephrine	10–15 μg	
16. Coproporphyrin	25–300 μg	
Nutritional studies		
17. Amino acid	% Tryptophan	% Phenylalanine
Cheddar cheese	0.10–0.50	0.25–1.25
Egg white	0.25	0.10–0.50
18. β-Carotenes		
Spinach	50–100 μg/g	
Other		

5. Use 4 mL of petroleum ether as a blank, and read at 450 nm. Determine the concentration of carotenes in the serum from the standard curve (Section D.6, p. 230). Be sure to take the value for the amount of serum used in the extraction into account. Express the values in μg/dL, and record these data in the summary table (Table 9.12).

CHAPTER 10

Cell Components

A. Isolation of Cell Fractions and Studies of Electron-Transport Chain in Mitochondria and Microsomes

Day 1

1. Isolation of rat liver cell fractions.
2. Determination of P/O ratios and RCRs.

Day 2 Study of "mixed-function oxygenase" system in microsomes.

1. DCPIP assay.
2. Assay for Cyt-b_5 reductase.
3. Function of Cyt-P_{450}.

B. Determination of Fractionation Effectiveness

Day 3

1. Study of electron transport. Reconstruction of oxidative function of the chain.
2. Protein determinations (Biuret method).
3. Assay of fractions for succinoxidase activity.

Day 4 Assay of fractions for the following enzyme activities.

1. Glucose-6-phosphatase.
2. Acid phosphatase.
3. Lactate dehydrogenase.

C. *Studies of Nuclear Fraction*

Day 5

1. Extraction of RNA and DNA from fractions.
2. Isolation of double-stranded DNA from nuclear fraction.

Day 6 (Omit if done in conjunction with the Nucleic Acids module.)

1. Diphenylamine and orcinol assays.
2. Melting curve for duplex DNA.

REFERENCES:

Alberts, B., Bray, D., Lewis, J., Raff, M., Roberts, K., and Watson, J.D. *Molecular Biology of the Cell*, 2nd ed. Garland Publishing Inc., New York (1989).

Birnie, G.D. *Subcellular Components—Preparation and Fractionation*, 2nd ed. University Park Press, Baltimore, pp. 36, 79–81, 190–195 (1972).

Britten, R.J., Graham, D.E., and Neufeld, B.R. Analysis of repeating DNA sequences by reassociation. *Methods Enzymol*. 29:363–418 (1974).

Cooper, Chs. 1, 2.

deDuve, C., and Baudhuin, P. Peroxisomes (microbodies and related particles). *Physiol. Rev*. 46:323–357 (1966).

deDuve, C., Wattlaux, R., and Baudhuin, P. Distribution of enzymes between subcellular fractions in animal tissues. *Adv. Enzymol*. 24:291–358 (1962).

Estabrook, R.W. Mitochondrial respiratory control and the polargraphic measurement of ADP:O ratios. *Methods Enzymol*. 10:41–47 (1967).

Hinkle, P.C., and McCarty, R.E. How cells make ATP. *Sci. Am*. 238(3):104–123 (1978).

Hinkle, P.C., Kumar, M.A., Reseter, A., and Harris, D.L. Mechanistic stoichiometry of mitochondrial oxidative phosphorylation. *Biochemistry* 30:3576–3582 (1991).

Hrycay, E.G., and Prough, R.A. Reduced nicotinamide adenine dinucleotide-cytochrome b_5 reductase and cytochrome b_5 as electron carriers in NADH-supported cytochrome P_{450}-dependent enzyme activities in liver microsomes. *Arch. Biochem. Biophys* 165:331–339 (1974).

Kappas, A., and Alvares, A. How the liver metabolizes foreign substances. *Sci. Am.* 233:22-31 (1975).

Lehninger, Ch. 5.

Nicholls, D.G. *Bioenergetics—An Introduction to the Chemiosmotic Theory*. Academic Press, New York (1982).

Rawn, Chs. 1 and 14.

Rendina, G. *Experimental Methods in Biochemistry*. W.B. Saunders, Philadelphia (1971).

Robyt and White, pp. 256-266.

Stryer, pp. 397-423, 566, 770-786.

A. The Cell

The cell is the basic unit of all living organisms. Cells produce like cells that carry all the genetic information that will again be passed on to the next generation. Some organisms, the *prokaryotes*, consist of single cells that are capable of carrying out all metabolic functions but are not highly organized. Organisms of greater complexity consist of *eukaryotic* cells, which are highly organized. Such a eukaryotic cell is schematically represented in Figure 10.1, which illustrates the typical shapes of various

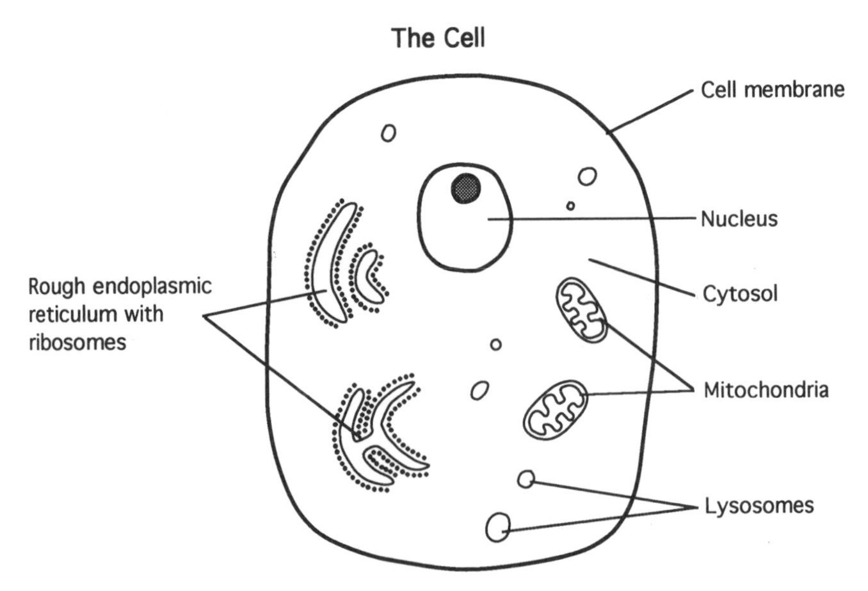

Fig. 10.1. Schematic representation of an animal cell. Not shown in the diagram are vacuoles, Golgi bodies, and smooth endoplasmic reticulum. The lysosomes vary in size and shape and contain degradative enzymes.

organelles. A great deal of information about the inner structure of the individual cell components has been obtained by electron microscopy.

The specialized organelles are surrounded by membranes that not only envelop their contents but, in some instances, also take part in metabolic functions. The earliest studies of cells were done using a light microscope, and many staining techniques were developed to study the inner structures. The introduction of the electron microscope made it possible to observe the internal organization of organelles in greater detail. From these photographs, it has been possible to localize certain cell functions. For biochemical analysis, a more useful method of studying cell structure involves the disruption of the cell's membrane and the isolation of the organelles through the technique of *fractionation*. By isolating and concentrating suspensions of a given organelle, it is possible to study the chemical reactions that these structures carry out in the intact cell. It is this approach that will be pursued in the following experiments.

The cell fractions that are to be studied are the following: (1) the nuclear fraction, (2) the mitochondria, (3) the microsomes, and (4) the cytosol. The *nucleus* is the largest and the most dense body in the cell, and therefore this fraction can be recovered at relatively low centrifugal force. It contains most of the cell's DNA and is the site of DNA replication and of RNA synthesis and transcription. The *mitochondria* (Fig. 10.2) will be studied in some detail. They are responsible for

A Mitochondrion

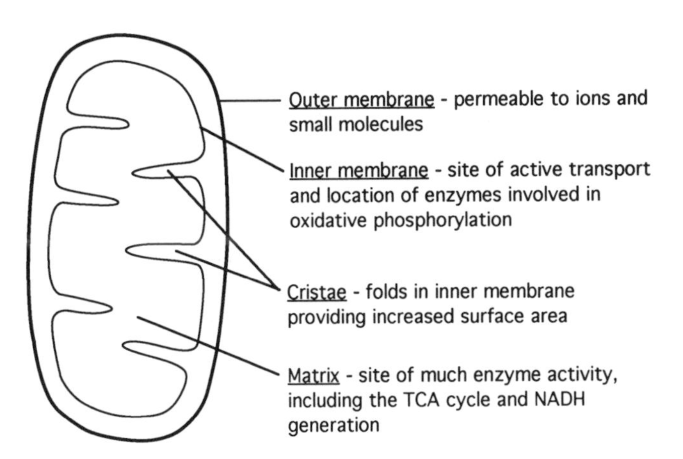

Fig. 10.2. The *mitochondrion*—site of bioenergetic reactions.

energy production and storage. The *microsomes* are vesicles formed when cells are disrupted by homogenization. They consist of "endoplasmic reticulum" fragments that have circularized, and they may be smooth or rough depending on whether they are coated with ribosomes. Microsomes are involved in protein synthesis and transport. They also have a role in the breakdown of nonpolar substances such as lipids and some hydrophobic foreign substances by rendering them more water soluble and thus more readily excreted. By differential centrifugation, microsomes can be subdivided into rough and smooth fractions, because those coated with ribosomes are of higher density.

Another common organelle in the cell is a group known as *lysosomes*. Lysosomes are difficult to isolate by differential centrifugation, because they are not of uniform size and they rupture easily during handling. They contain a variety of degradative enzymes that are involved in the breakdown of cellular material. The *cytosol* is the fluid in which all the other microbodies are suspended. It represents about half the cell volume and is the site of many biochemical reactions. In the cytosol are located soluble enzymes involved with glycolysis and biosynthesis of molecules as diverse as sugars, fatty acids, amino acids, and nucleotides.

The arrangement of macromolecules within the cell is as important to cellular function as is the catalytic reactivity of these macromolecules. Compartmentalization provides efficiency by bringing related components together (*e.g.*, mitochondrial electron transport chain) or by separating molecules that would interfere with each other (*e.g.*, hydrolytic enzymes localized in lysosomes). In this module, several subcellular fractions from rat liver will be prepared, and various properties of these organelles will be examined (see Fig. 10.3).

To study specific biochemical reactions, it is necessary to disrupt the cell to obtain a cell-free system. Several mechanical devices are available to cause membrane breakage, but the procedure must be selected with care to avoid the loss of the organelle's integrity as well. For the isolation of intact organelles, it is important to provide an environment in which the osmotic pressure is similar to that of the intact cell. For this purpose the cells are suspended in a 0.34 M sucrose solution, or 5% Ficoll-400 is used as suspension medium. Ficoll is a commercial high molecular weight (MW *ca* 4×10^5) polymer

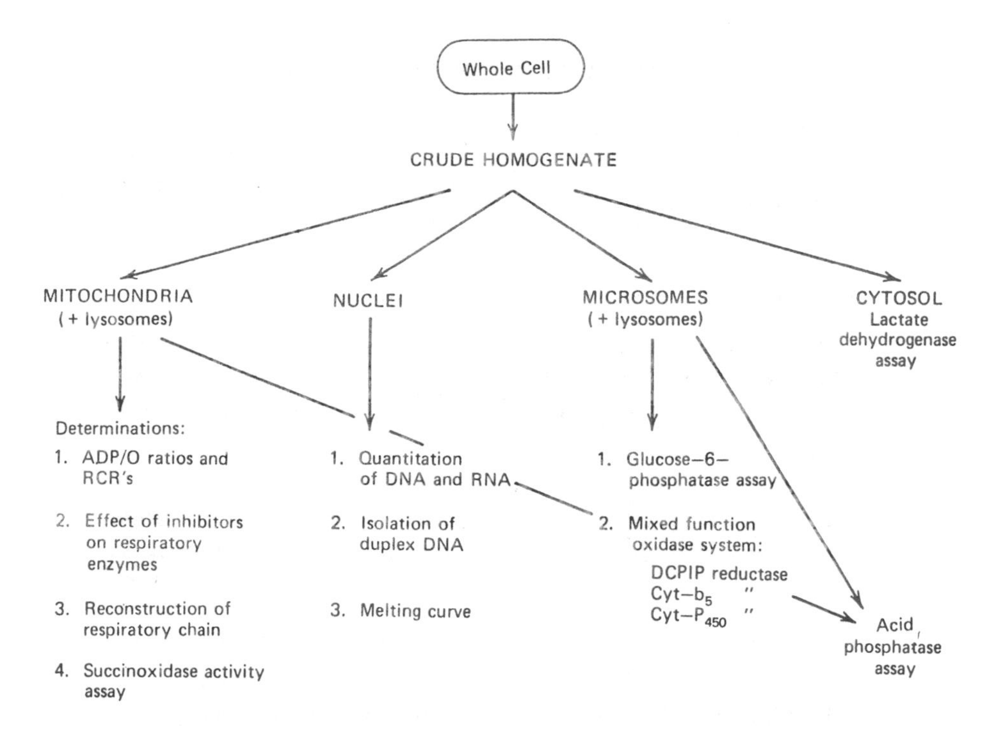

Fig. 10.3. Liver fractionation. Outline of assays performed in this study.

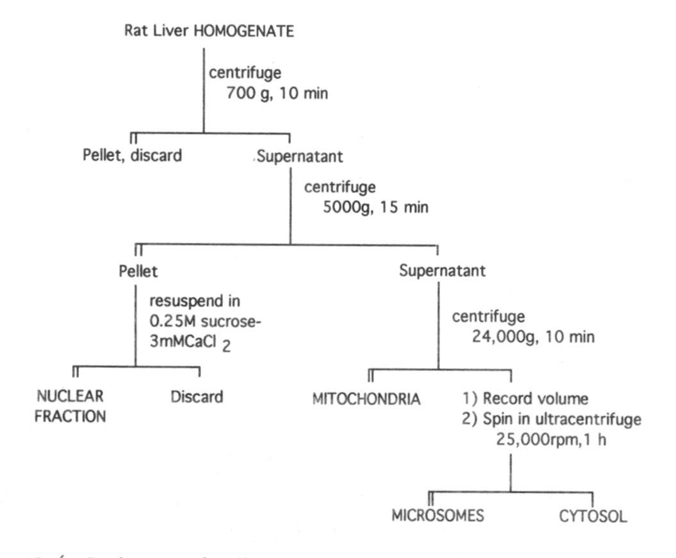

Fig. 10.4. Isolation of cell components by differential centrifugation.

of sucrose that has the advantage of being a high-density product that is inert and protects the sensitive membranes of organelles from damage.

A commonly used method to rupture cells is to grind them in a glass homogenizer. First the tissue is chopped or mashed, usually with scissors, and then the cell membranes are disrupted in a sucrose or Ficoll solution. This *crude homogenate* is then centrifuged. Figure 10.4 shows an outline of a series of centrifugations that result in the separation of fractions containing cell components of varying densities. Such a schedule of differential centrifugations permits the recovery of nuclei, mitochondria, and microsomes.

A rat will be sacrificed by decapitation and the liver removed and weighed. The liver is then immersed in a cold sucrose solution, and the tissue is ground in 2-g portions in a Potter-Elvehjem homogenizer. This tissue grinder consists of a round-bottomed, heavy-walled glass tube and a close-fitting Teflon pestle. The pestle is mounted on a steel rod that is attached to the drive shaft of an electric motor with adjustable speeds to about 1000 rpm. The tissue is ground by moving the glass tube up and down manually, causing the homogenate to pass repeatedly between mortar and pestle. Care should be taken to keep the homogenate cold.

The method used to separate the cell organelles relies on differential centrifugation to resolve components of different densities. In designing this procedure, yield and resolution had to be sacrificed somewhat to obtain intact mitochondria, which requires gentle treatment. The activities of the systems studied may also be affected by the rat's diet, age, sex, and by whether the animal was fasting when sacrificed.

The effectiveness of the separations achieved will be evaluated by analyzing the fractions for various marker enzyme activities. The distribution of total protein, DNA, and RNA among the fractions will be determined.

Oxidative Phosphorylation

The mitochondria will be studied on the day of their isolation, as they are easily disrupted and phosphorylation is difficult to demonstrate. The oxygraph will be used to determine oxygen consumption, and ADP/O and RCR ratios will be calculated. Also, the sequence of respiratory enzymes will be demonstrated by studying the effects of various inhibitors on oxygen consumption.

Microsomal Transformations of Foreign Substances

The mixed-function oxygenase system of microsomes will be demonstrated. One of the functions of the liver is to permit the conversion of lipophilic to hydrophilic substances, thereby enabling the more water-soluble products to be excreted. Although this mechanism most probably evolved in animals to remove metabolic products from the body, it is also useful to remove foreign substances such as drugs, food additives, and environmental pollutants from the body.

Isolation of Duplex DNA From Nuclei

DNA will be isolated from the nuclear fraction, and the melting temperature for rat liver DNA will be determined. From this study, the G-C content of the DNA can be calculated.

B. Isolation of Rat Liver Cell Fractions (see Fig. 10.4)

These experiments are carried out by students working in groups of two. **CARRY OUT ALL PROCEDURES ON ICE.**

1. Kill a rat by decapitation, and exsanguinate it under running water. Rapidly remove the liver, and place it in a beaker containing 0.25 M sucrose–1 mM EDTA (sucrose-EDTA). Blot the liver carefully with tissues and place in a weighed beaker. Record the weight, and cover the liver with cold sucrose-EDTA solution.
2. Mince the liver with scissors, decant, and discard the pink sucrose solution, including the floating debris. Add fresh sucrose-EDTA solution. Continue to mince and rinse until the sucrose solution is nearly colorless (three to four times).
3. Transfer the minced liver to a Potter-Elvehjem homogenizer in about 2-g portions with *ca* 18 mL of sucrose-EDTA solution. Homogenize the extract with a Teflon pestle (two to three strokes only), keeping the homogenizer in an ice bath. (Excessive grinding or heating damages and inactivates mitochondria.)
4. Repeat the procedure until the entire liver has been homogenized. Combine all homogenates, record the volume, and save three 3-mL portions and freeze. Label the tubes "HOMOGENATE."

5. Carefully layer 20 mL of homogenate onto an equal volume of 0.34 M sucrose–1 mM EDTA in a 50-mL Sorvall centrifuge tube. Repeat this procedure using enough tubes to accommodate all of the homogenate.
6. Centrifuge for 10 min at 700 *g* in a refrigerated Sorvall centrifuge using the SS-34 rotor.

Isolation of Mitochondria

7. For each tube, transfer the supernatant from step 6 to another 50-mL centrifuge tube, and centrifuge for 15 min at 5,000 *g*. With the pellet, proceed to step 12.
8. Pour off the supernatant and the pink, partially sedimented layer, and save this suspension for the isolation of microsomes. Transfer the pellets quantitatively with about 10 mL of sucrose-EDTA to a clean 50-mL centrifuge tube, and disperse the pellet by homogenizing manually with a Teflon pestle. Bring the volume to 40 mL, and centrifuge for 10 min at 24,000 *g*.
9. Repeat step 8 to wash the mitochondria. Discard this supernatant.
10. Immediately before conducting the respiratory control experiments, resuspend the combined mitochondrial pellets in 10 mL of sucrose-EDTA solution. After the oxidative-phosphorylation study, divide the remaining suspension into three equal portions and freeze. Label the tubes "MITOCHONDRIAL FRACTION."

    PROCEED WITH THE OXIDATIVE PHOSPHORYLATION EXPERIMENTS USING THESE FRESH MITOCHONDRIA.

11. At this point, one student could continue with the isolation of the nuclear fraction, outlined in steps 12 and 13, while the other student proceeds with the isolation of the microsomes (beginning with step 14).

Isolation of Nuclei

12. Resuspend each tube of the nuclear-fraction pellet from step 6 in 10 mL of 0.25 M sucrose–3 mM $CaCl_2$. Filter the suspension through two layers of cheese cloth; wash cellular particles with 10 mL of 0.25 M sucrose–3 mM $CaCl_2$.

Discard the connective tissue and debris. Centrifuge the filtrate for 10 min at 1500 *g*.

13. Discard the supernatant, and resuspend the combined pellets in 10 mL of 0.25 M sucrose–3 mM $CaCl_2$. Freeze in four equal portions. Label the tubes "NUCLEAR FRACTION."

Isolation of Microsomes

14. Measure and record the volume of the supernatant from the mitochondrial isolation (step 8). Centrifuge a portion of this fraction for 1 hr in the Beckman ultracentrifuge at 25,000 rpm.
15. Pour off the supernatant, measure its volume, and save four 3-mL aliquots. Discard the rest. This is the soluble portion of the cell. (This fraction will be contaminated with material released by disruption of the nuclei, mitochondria, and cell membranes.) Label the tubes "CYTOPLASM."
16. Combine the pellets in about 5 mL of sucrose-EDTA solution, and use a hand-held glass homogenizer to resuspend the pellet. Freeze in four aliquots. Label the tubes "MICROSOMAL FRACTION."
17. To accomplish rapid freezing, immerse the tubes in liquid N_2 or a dry-ice EtOH bath.

Use of the Ultracentrifuge

The preparative ultracentrifuge is used to sediment the smallest particulate matter from tissue homogenates. Heavier organelles are resolved by conventional high-speed centrifuges that achieve speeds only up to about 20,000 rpm. The major differences between these two centrifuges is the vacuum system contained in the ultracentrifuge. At speeds greater than 20,000 rpm, the heat of friction between the rotor and air becomes important. This source of heat can be removed when the rotor chamber is evacuated (see Fig. 10.5 for illustration of a typical control panel).

Directions for operation of the Beckman Model L-2 Ultracentrifuge (other instruments are slightly different, but the basic controls are the same) follow:

1. Turn the main power switch on.
2. Put the brake switch on.

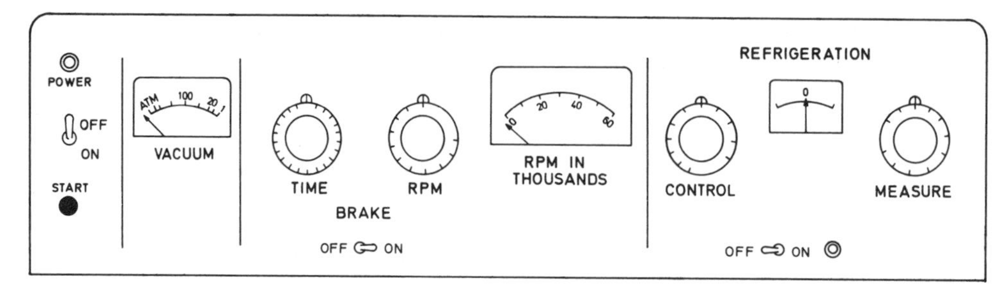

Fig. 10.5. Control panel for the Beckman Model L-2 Ultracentrifuge. (Adapted from *Operating Manual or the Model L-2 Preparative Ultracentrifuge*, Beckman Instruments, Inc., Palo Alto, CA; courtesy Spinco Division, Beckman Instruments, Inc.)

3. Turn on the refrigeration, and set the temperature control at 4°C. The dial on the right allows the temperature to be read by means of the null meter (between the dials on the panel).
4. Put the vacuum switch in the ON position.
5. Set the timer for the length of time of the centrifugation.
6. Set the speed (in rpm).
7. When the vacuum reaches 100 microns and the temperature has reached 4°C, press the start button on the left side of the panel under the main power switch.

Use and Calibration of the Oxygraph

Principle of Operation of the Polarograph

The "oxygen electrode" consists of a silver reference anode in contact with a KCl electrolyte solution and a platinum cathode. A constant voltage of 0.62 V is used to polarize the electrode. A Teflon membrane covers the electrode surface, which is in contact with the solution, thereby preventing contamination of the platinum cathode and the leakage of KCl into the solution. Oxygen, however, readily diffuses across the Teflon membrane and is electrolytically reduced to water, thereby depolarizing the electrode and allowing current to flow. The current flow, which is directly proportional to oxygen tension over a very wide range, can readily be measured and recorded after suitable amplification.

Operating Instructions for the Oxygraph

The oxygraph consists essentially of a Clark-type oxygen electrode, a polarizing circuit, a measuring circuit, and a potentiometric recorder that records continuously the oxygen content of the solution as a function of time (see Fig. 10.6).

1. The jacket of the oxygraph cell must contain water to prevent overheating by the magnetic stirring apparatus. Record the room temperature.
2. Turn on the instrument (Servo and Chart switches should be OFF), and open the pen. Make sure that the pen tip is making contact with the chart paper. Any one of a number of recording instruments can be used, and each one will come with its own set of instructions. However, the chart paper should have a scale from 0 to 100, and the instructions for the Servo instrument are generally applicable. In particular, the Linear Model 1201 recorder is suitable for this application.
3. a. Fill the cell with fully oxygenated RC buffer solution (0.25 M sucrose, 2 mM EDTA, 20 mM K phosphate buffer, 4 mM $MgCl_2$, pH 7.4) and replace the stopper, ***making sure there are no air bubbles***. Turn the Servo and the pen switches on. Adjust the potential (sensitivity) knob on the control box until the pen gives a stable tracing at *ca* 95 on the chart paper.

   b. Measure and record the volume of buffer used to fill the cell.
4. *Calibration*. Once the initial adjustments have been made, the difference between the upper and lower pen positions should represent the output due only to oxygen. To check this, dithionite is added to reduce the oxygen in the cell. Remove the stopper, add a few crystals of $Na_2S_2O_4$, replace the stopper, and record the change in potential. Once the pen has stabilized, wash the cell thoroughly with H_2O. Since the volume of the cell is known, the content of oxygen in the cell can be calculated as described below.

   ONCE ADJUSTED, DO NOT CHANGE THE SENSITIVITY SETTING.

   a. The solubility of oxygen in water is temperature dependent. At 23°C it is 0.272 μmol O_2/mL H_2O (Lange's *Handbook of Chemistry*, 9th ed). Since in oxidative phosphorylation studies, the ADP/O ratios are ex-

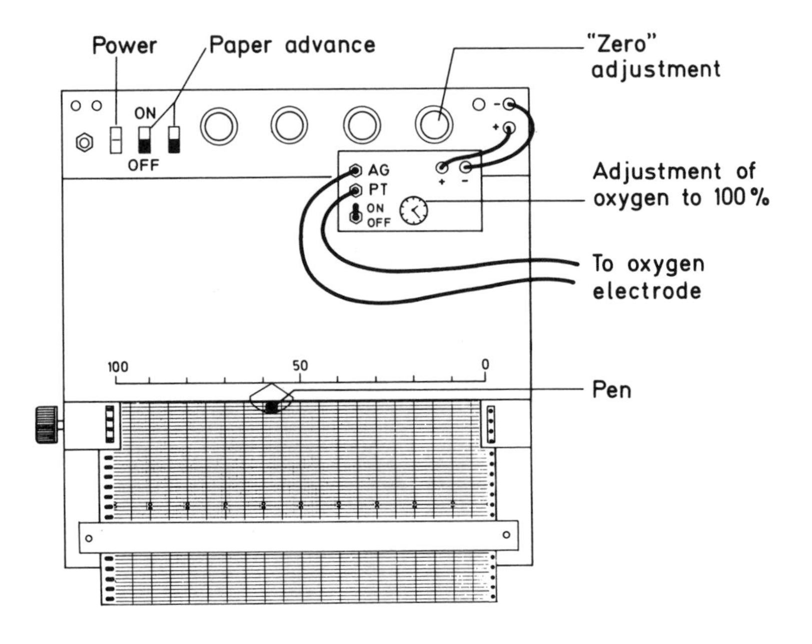

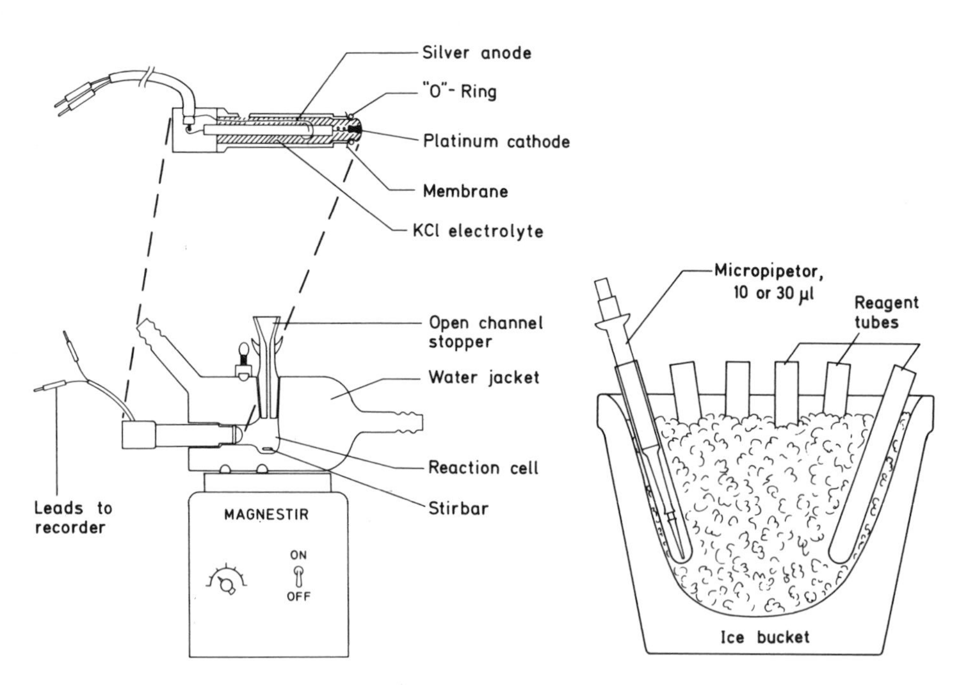

Fig. 10.6. The oxygraph consists of a Clark-type oxygen electrode inserted into a jacketed chamber containing a reaction vessel. The electrode is attached to a recorder, which measures oxygen consumption as a function of time. The reagents and inhibitors that are used in the study are readily available in test tubes on ice, and automatic pipetting devices are inserted into each of the tubes. (Diagram of oxygen electode adapted from literature provided by Yellow Springs Instrument Co., Yellow Springs, OH; courtesy of Yellow Springs Instrument Co., Inc.)

pressed in atoms of O, the oxygen content of the air saturated buffer becomes 0.544 μmol or 554 nmol O/mL.

b. The oxygen content of the oxygraph cell can be determined from the graph shown in Figure 10.7. For the appropriate temperature, record the number of μmol O_2/L, multiply this value by the volume of the cell, divide by 1000, and multiply by 2. This is the number of μmol of O atoms present at full saturation. This value represents the number of squares on the chart paper traveled by the pen during the calibration. For example, if this number is 95 and the cell volume is 1.3 mL, then each square represents (544 nmol O/mL × 1.3 mL)/95 = 7.44 nmol of O atoms at 23°C.

Determination of P/O and RCR Ratios

When substrate is added to tightly coupled mitochondria suspended in buffer, a slow rate of oxygen uptake is observed (see, for example, lines B–C for malate + glutamate and D–E for succinate in Fig. 10.8). On addition of ADP to the reaction mixture, an increase in the rate of oxygen utilization is observed. The length of time that the new rate will hold depends on the amount of ADP added, and the concentration of oxygen utilized is proportional to the amount of ADP phosphorylated to ATP. Therefore, the oxygraph tracings can be used to calculate an ADP:O ratio.

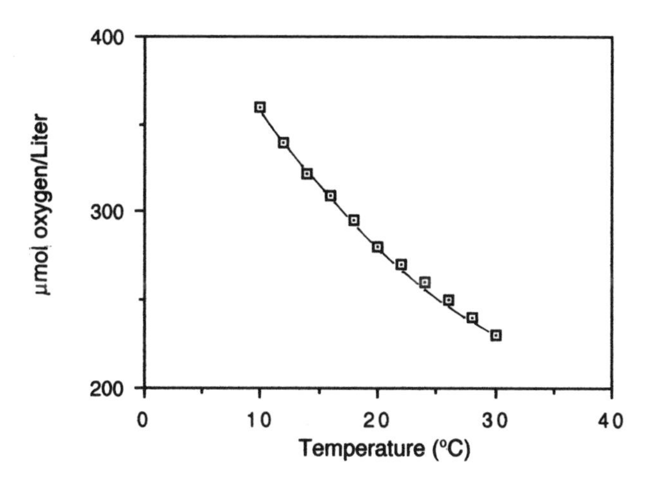

Fig. 10.7. Relationship between oxygen content of air-saturated water and temperature. (Estabrook, p. 45.)

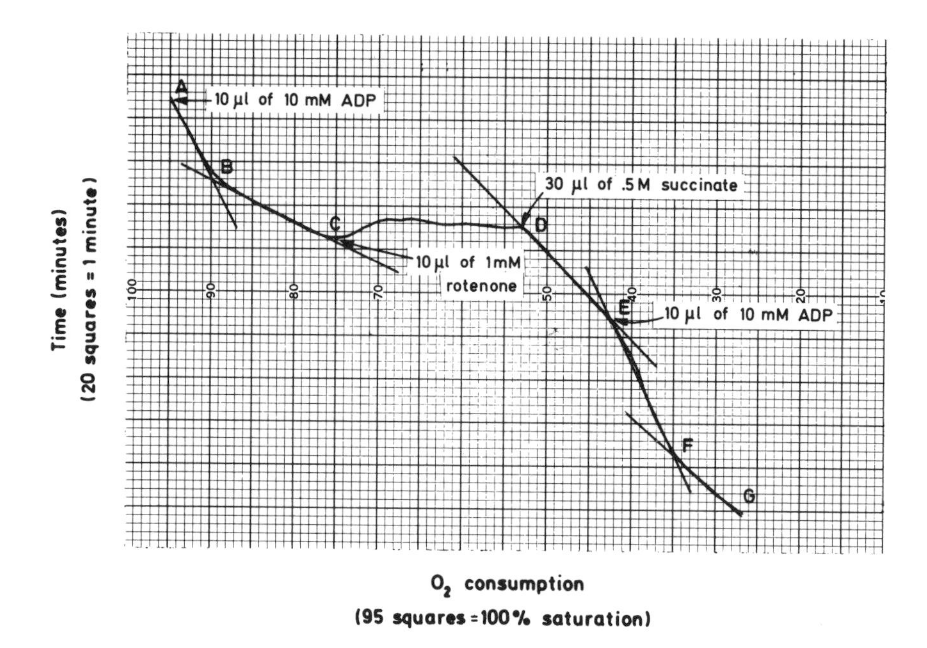

Fig. 10.8. Mitochondrial oxidative phosphorylation. Oxygraph tracing following addition of 30 μL of 0.5 M malate + 30 μL of 0.5 M glutamate to 0.1 mL of mitochondrial suspension in "RC mix." Cell volume = 1.3 mL; temperature = 23°C.

1. Extend the slopes of the lines as shown, and count the number of squares resulting from ADP-stimulated oxygen utilization (*e.g.*, A–B, 9 squares; E–F, 16 squares).
2. Determine the amount of oxygen consumed that is represented by each square. In the calibration procedure used earlier, it was established that for the tracing shown in the diagram, 7.44 nmol O/square is utilized (for the malate + glutamate + ADP reaction, 7.44 × 9 or 67 nmol O; for the succinate + ADP reaction, 7.44 × 16 or 119 nmol of oxygen atoms).
3. *Calculation of ADP/O.* The number of nmol of ADP added (0.01 mL of 10 mM or 100 nmol) is divided by the oxygen consumed. (Thus, we obtain P/O ratios of 1.49 and 0.84, respectively.)
4. The respiratory control ratio (RCR) is a measure of how "tightly coupled" are the mitochondria in suspension. RCR is defined as the ratio of the rate of oxygen utilization in the presence of ADP to the rate of oxygen utilization to

which the mitochondria return after all the ADP has been phosphorylated (For example, in Fig. 10.8, the RCR ratios are AB/BC = 4, and EF/FG = 1.9. Note that although these two values should be equal, the RCR ratio decreases as the mitochondria rapidly lose their ability to carry out the reaction outside the cell).

C. Mitochondrial Respiration and Oxidative Phosphorylation

The mitochondrion (Fig. 10.2) is often referred to as "the powerhouse of the cell" because it is the site of most of the cell's energy production. In the process of oxidative phosphorylation, this energy is captured in the high-energy compound adenosine triphosphate (ATP), which in turn can be readily utilized for the energy-requiring processes of the cell. The substrates for energy production are derived from the metabolism of carbohydrates, fats, and amino acids. In the final stages, electrons are fed into the mitochondrial electron-transport system by the appropriate substrate, where they are shuttled through a series of electron carriers to oxygen. For example, isocitrate may be oxidized to α-ketoglutarate and carbon dioxide in the citric acid cycle, thereby giving up electrons. NAD^+ accepts these electrons and undergoes a reduction to NADH. NADH in turn is oxidized back to NAD^+, and the electrons removed from it are transferred to another compound. The electrons originally removed from isocitrate are then passed through the electron-transport chain, which is a system of enzymes and coenzymes, until they finally react with oxygen. The series of oxidoreductions of the electron-transport system is accompanied by the release of energy, some of which is lost as heat, but in a few specific oxidoreductions this energy is captured and preserved as the bond energy of ATP by the esterification of inorganic phosphate with ADP.

In the transfer of electrons from 1 mol of NADH to oxygen, about 52,000 calories are released, whereas 1 mol of succinate produces about 37,000 calories. The formation of 1 mol of ATP from ADP and inorganic phosphate takes up about 7,700 calories. If 3 mol of ATP are formed for each mol of NADH oxidized, the energy of electron transport converted to the chemical energy of ATP is about 23,100 calories, and therefore the efficiency of energy conservation is about 45%.

Electron Transport Chain

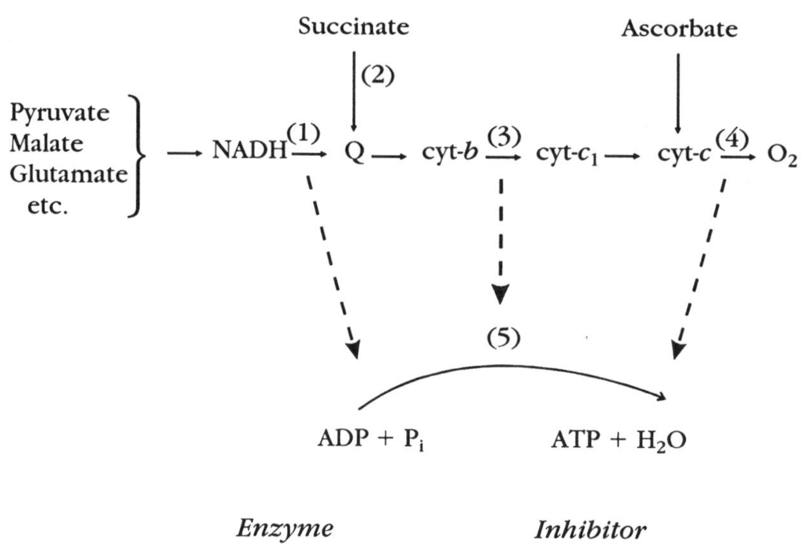

Enzyme	*Inhibitor*
(1) NADH dehydrogenase	Rotenone
(2) Succinoxidase	Malonate
(3) Cytochrome *b* oxidase	Antimycin A
(4) Cytochrome *c* oxidase	KCN
(5) ?	CCP

The components of the mitochondrial transport system as illustrated in the above diagram will be studied in this section, and the relative positions of the oxidative enzymes in the chain will be demonstrated.

The phosphorylation system exerts a regulatory effect on the rate of respiration, *i.e.,* the rate of electron flow. Thus *the amounts of ADP and inorganic phosphate limit the respiration rate. Since in most cells inorganic phosphate is present in excess, it is the ADP that controls the rate of respiration*. In intact organisms, there is a high degree of respiratory control, and the mitochondria are said to be "tightly coupled." Indeed, the phenomenon of respiratory control can be utilized experimentally to determine whether intact mitochondria have been isolated from the original tissue, since structural damage to the mitochondrion results in loss of this control.

Oxidative Phosphorylation

Under the best available experimental condition, ADP/O ratios are difficult to measure accurately *in vitro*. Researchers working in this field have consistently reported obtaining P/O ratios of 2.5 or less in the oxidation of 1 mol of NADH and attributed not reaching the theoretical value of 3.0 to lack of sensitivity of the equipment and to imperfect mitochondrial preparations. However, Hinkle and coworkers suggest that the actual yield of ATP generated by oxidative phosphorylation for every pair of electrons is 2.5. The generation of ATP can be accounted for by the following path of two electrons down the respiratory chain from NADH to water:

Coupling site	*Protons produced*	*ATP formed*
NADH to succinate	4	1
Succinate to ascorbate	2	0.5
Ascorbate to oxygen	4	1

For every four protons obtained, three are required for the synthesis of ATP by ATP synthase, and one proton is used in the transport of ADP and Pi across the mitochondrial membrane. Therefore, four protons account for 1 mol of ATP each at sites 1 and 3 and 0.5 at site 2. In addition, it becomes necessary to adjust the yield of ATP from the oxidation of 1 mol of glucose accordingly from the traditional value of 36 to 29.5 (Hinkle et al., 1991).

Oxidative phosphorylation can be inhibited directly by a number of chemical agents. One class of these agents, characterized by the antibiotic oligomycin, inhibits not only phosphorylation but respiration as well. A second class of agents, characterized by *m*-chlorocarbonylcyanide phenylhydrazone (CCP), causes inhibition of phosphorylation, a phenomenon known as *uncoupling*. A third class includes site-specific respiratory inhibitors such as rotenone and cyanide. The following experiments are designed to demonstrate (1) the extent of phosphorylation obtained with different substrates (P/O ratios), (2) the degree of respiratory control (RCR), and (3) the effect of various inhibitors and an uncoupler on the system.

Experimental Procedures

Calibrate the oxygraph according to the procedure given on page 244. (See Fig. 10.6 for arrangement of instrument and ice bucket with appropriate reagents.)

Determination of ADP/O Ratios

Saturate some respiratory control buffer ("RC mix") with air by stirring it vigorously at room temperature for about 20 min.

(The "RC mix" contains the following: 0.25 M sucrose; 2 mM EDTA; 4 mM $MgCl_2$; and 20 mM phosphate buffer, pH 7.4.)

1. Fill the reaction chamber with "RC mix" using 0.1 mL less than the volume of the cell.
2. Add 0.1 mL of the mitochondrial suspension. (The protein content will subsequently be determined.)
3. Start the oxygraph, and continue the tracing until a steady baseline is achieved.
4. Add one of the following substrates:
   a. 30 μL each of 0.5 M Na glutamate and 0.5 M Na malate (this is an NADH-linked substrate).
   b. 30 μL of 0.5 M Na succinate, pH 7.4.
   c. 30 μL each of 0.5 M Na ascorbate and cytochrome *c* (10 mg/mL).
5. Let the instrument record the tracing for a few minutes.
6. Add 20 μL of 10 mM ADP, pH 7.2, and allow the pen to continue the tracing until the ADP is used up.
7. Let the tracing proceed for another minute, and then add 10 μL of CCP (an uncoupler).
8. Remove the mixture from the chamber by aspiration. Rinse the chamber very well and fill it with deionized water. ***DO NOT LET IT DRY***. Then repeat this procedure with the next substrate.

Effect of Inhibitors on Respiration Rate

1. Fill the reaction chamber with "RC mix," 0.1 mL of the suspension of mitochondria, and substrate.
2. Add 10 μL of one of the following inhibitors: rotenone (1 mM); antimycin A (0.5 mg/mL); Na malonate (1.5 M); or KCN (0.1 M). Record the tracing for a few minutes, clean the chamber, and use another inhibitor.
3. Repeat until the effect of each inhibitor on each of the three substrates has been determined. Record your results (+ for "inhibition", − for "no inhibition") in Table 10.1.

Reconstruction of Enzyme Sequence of Respiratory Chain

1. From the data on the effect of inhibitors, try to reconstruct the chain in as few steps as possible. Remember that the

TABLE 10.1. Effect of Inhibitors

Substrate	Inhibitor			
	Rotenone	Malonate	Antimycin A	KCN
Glutamate + malate (NADH)				
Succinate				
Ascorbate + cyt *c*				

inhibition of one step can sometimes be overcome by addition of the substrate for the next.

2. Demonstrate the "branch point" for the entry of succinate into the chain by using the appropriate sequence of substrates and inhibitors.

Calculations and Data Analysis

Calculate P/O and RCR ratios for the mitochondria, and draw a diagram showing the sequence of steps used to establish the function of the respiratory chain.

D. Microsomal Enzyme Systems

The microsomes are liposomal artifacts created from cytomembranes by the processes involved in fractionating cellular components. Microsomes are not distinct organelles in intact cells, but rather they are a collection of fragments of endoplasmic reticulum that combine in circular fashion to form vesicles when the cells are ruptured. Because these vesicles retain the biochemical functions of the endoplasmic reticulum, the microsomes are convenient vehicles for study of the processes associated with them. The endoplasmic reticulum (ER), as discussed earlier, exists in smooth and rough forms. In this study, no attempt is made at further separation; therefore the microsomal fraction consists of a mixture of the two. To separate smooth from rough ER, a sucrose density gradient can be used and the two fractions separated by ultracentrifugation.

The microsomal fraction contains a variety of enzymes such as those involved in fatty acid and steroid biosynthesis and degradation and in peroxide reductions. These are activities associated with the normal metabolic functions of the cell. Of considerable interest is the role of some of these microsomal enzymes in metabolizing drugs, poisons, and other nonphysiological compounds. The reactions leading to desaturation of fatty acids by a series of transfers through an electron transport chain that contains cytochrome b_5, cytochrome b_5 reductase, and cytochrome P_{450} are sufficiently nonspecific to permit these nonphysiological substances to undergo similar transformations.

Most passively absorbed compounds are hydrophobic and must be made more polar to aid in their excretion into the bile or by the kidney. One way that the microsomes accomplish this conversion is by hydroxylation of the compound, which not only makes it more water soluble but also provides a site for the formation of even more soluble sulfate and/or glucuronide conjugates. This transformation is very useful for the elimination from the body of toxic substances and drugs, but this process unfortunately may sometimes convert a previously harmless compound to a toxic one.

The enzyme system generally associated with these reactions is known as the microsomal mixed-function oxygenase, monooxygenase, or microsomal cytochrome P_{450} system. Unlike most enzymes, which are specific for a particular substrate, the cytochrome P_{450} reductase is capable of acting on a number of compounds.

Hydroxylation reactions are important in the synthesis of cholesterol and the conversion of cholesterol into hormones and bile salts. The general equation for these reactions is as follows:

$$RH + O_2 + NADPH + H^+ \longrightarrow ROH + H_2O + NADP^+$$

In this reaction, one atom of the O_2 molecule goes into the substrate while the other is reduced to H_2O.

Cytochrome P_{450} is named for its absorption maximum at 450 nm. It is the terminal electron acceptor of an *electron transport chain* whose function is to hydroxylate hydrophobic compounds (see the diagram below). The hydroxylation reaction increases the solubility of otherwise insoluble molecules,

Microsomal Electron-Transport Chain

NADPH ⟶ b_5 reductase ⟶ cyt-b_5 ⟶ cyt-P_{450} - - - → (3) $R + O_2$ ⟶ $ROH + H_2O$

b_5 reductase - - (1) - → DCPIP

cyt-b_5 - - (2) - → cyt-*c*

thereby permitting their elimination through the kidneys. The NADPH-cytochrome P_{450} reductase system will be examined in this section of the module.

The dashed lines in this diagram indicate the steps where reducing equivalents may be diverted by the oxidants indicated, so that reactions 1, 2, and 3 can be followed by observing the changes occurring with time in these indicators.

R may represent any substrate to be hydroxylated. The compound 4-dimethylaminoantipyrine will be used in this study.

Determination of K_m *for NADPH-DCPIP Reductase Reaction*

$$\text{NADPH} \xrightarrow{e^-} \text{flavoprotein} \overset{e^-}{\dashrightarrow} \textit{dye}\ (\text{DCPIP})$$

The reduction of dichlorophenolindophenol (DCPIP) by the NADPH-reduced flavoprotein will be assayed by following the *decrease* in absorbance at 600 nm. The molar extinction coefficient of DCPIP at 600 nm is 5,100 M^{-1} cm^{-1}.

Reaction Mixture

1. Microsomes (amount is adjusted to reach ΔA/min of *ca* 0.1 with 200 μL NADPH in the reaction vessel).
2. 0.06 M Potassium phosphate buffer, pH 7.4.
3. 0.5 mM DCPIP.
4. 1 mM NADPH (or NADH may be substituted): 50, 75, 100, 150, and 200 μL.

Assay

Mix a stock solution containing 3 mL of 0.5 mM DCPIP and 27 mL of buffer, transfer 3.0 mL to a cuvette, and add a small amount of microsomes (start with 0.05 mL). Adjust the absorbance on the spectrophotometer to 0.5. Add 200 μL of NADPH, mix, and start timing. Read $A_{600\ nm}$ at 20-sec intervals for several minutes. If the ΔA is too great or too small, adjust the protein until ΔA is *ca* 0.1/min. When the proper amount of protein has been determined, add a sufficient amount for 10 assays to the stock solution.

Repeat the assay using the next concentration of NADPH. Use 50, 75, 100, and 150 μL. The 200-μL NADPH readings previously made can be used for this concentration.

Calculations

Plot $A_{600\ nm}$ *vs.* time for each NADPH concentration to determine the velocity of the enzyme. Convert absorbance to μmol of product reduced using the Beer-Lambert Law and the molar extinction coefficient given for DCPIP. Calculate K_m and V_{max} values from a Lineweaver-Burk plot.

Determination of K_m *for NADPH–Cytochrome c Reductase Reaction*

$$\text{NADPH} \xrightarrow{e^-} \text{flavoprotein} \xrightarrow{e^-} \text{cyt-}b_5 \xrightarrow{e^-} \textit{cyt-c}$$

The reduction of cytochrome *c* and NADPH and the indicated proteins will be assayed by following the *decrease* in absorbance at 550 nm. The molar extinction coefficient of reduced cytochrome *c* at 550 nm is 18,500 $M^{-1}\ cm^{-1}$.

Reaction Mixture

1. Microsomes (amount is adjusted to reach ΔA/min of *ca* 0.1 with 200 μL NADPH in the reaction vessel).
2. 0.06 M Potassium phosphate buffer, pH 7.4.
3. Cytochrome *c* (10 mg/mL).
4. 1 mM NADPH (or NADH): 50, 75, 100, 150, and 200μL.

Assay

See procedure for DCPIP reductase above, but substitute 3 mL of cytochrome *c* solution for DCPIP and read the absorbance at 550 nm instead of 600 nm.

Calculations

The same as for the DCPIP reductase reaction.

Hydroxylation of Substrate R

$$\text{NADPH} \xrightarrow{e^-} \text{flavoprotein} \xrightarrow{e^-} \text{cyt-}b_5 \xrightarrow{e^-} \text{cyt-P}_{450} \dashrightarrow (R + O_2 \rightarrow \textit{ROH} + H_2O)$$

The oxidation of NADPH in the presence of a substrate (R) and oxygen via the electron transport chain will be assayed by following the ΔA/min at 340 nm. The molar extinction coefficient for NADPH and NADH at 340 nm is 6200 M^{-1} cm^{-1}.

The reaction will be carried out in a 1-mL quartz cuvette in a spectrophotometer and repeated under aerobic and anaerobic conditions to demonstrate the role of oxygen in the hydroxylation reaction.

Reaction Mixture

0.06 M Potassium phosphate, pH 7.4	0.6 mL
Glucose	20 mg
1 mM NADPH (or NADH)	0.4 mL
Microsomes (1 to 2 mg/mL)	0.025 mL
50 mM substrate	0.05 mL (after equilibration)

Assay

1. *Aerobic conditions.* Mix all components except the substrate, and adjust the spectrophotometer to an absorbance of 0.5. Record the $A_{340\ nm}$ at 30-sec intervals until a steady rate is obtained (2 to 3 min). Add 50 μL of 50 mM of 4-dimethylaminoantipyrine, mix, and continue measuring $A_{340\ nm}$ for another 3 to 4 min.
2. *Anaerobic conditions.* Repeat the above, but use a solution of the reaction mixture bubbled with N_2 for 5 min. Add 1 mg of glucose oxidase (in 25μL H_2O). Incubate for 5 min at room temperature, and transfer the reaction mixture to a quartz cuvette. Start measuring $A_{340\ nm}$ as above after adding the substrate.

Calculations and Data Analysis

Plot two graphs of $A_{340\ nm}$ *vs.* time, one for each set of conditions. Explain the differences observed under aerobic and anaerobic conditions. Calculate the specific activity for cytochrome P_{450} reductase from the ΔA/min obtained and the protein concentration used in the assay (μmol/min NAD^+ formed/mg protein).

E. Analysis of Cell Components

The functions of a number of cell components have been established, but to confirm that a particular organelle is responsible for a reaction assigned to it, it is necessary to establish that the fraction containing those cell components is free of contamination. During the differential centrifugation procedure, the fractions were subjected to a washing step, but too much handling while resuspending the pellets and centrifuging the organelles could result in membrane breakage. Consequently, some contamination from neighboring fractions was unavoidable, and it is important to establish that a given function is properly assigned to the correct organelle.

The following experiments are designed to quantify the extent of cross-contamination between cell fractions. In each case, a unique enzymatic activity and the protein concentration are determined for each fraction, and the specific activity is calculated. The same assays are performed on a sample of each fraction, and the results are tabulated as shown in Table 10.2. The effectiveness of the fractionation procedure can then be evaluated by comparing the relative specific activities for each enzyme.

Determination of Protein by the Biuret Reaction

1. Set up a series of test tubes, and add the volumes of subcellular fractions given below and water to give a volume of 1.3 mL. Include tubes for a standard curve, if a biuret standard curve was not previously generated.
2. Add a set of tubes containing 0.02, 0.05, 0.10, 0.20, and 0.50 mL of sucrose-EDTA solution, and bring the volume in each tube to 1.3 mL with H_2O. These are sucrose blanks.
3. Include a tube containing a known amount of standard BSA and a water blank.
4. Add 0.2 mL of 5% DOC in 0.01 N KOH (deoxycholate, DOC, is a detergent used to disperse particulate material).
5. Add 1.5 mL of biuret reagent. Mix thoroughly on a Vortex mixer, and read $A_{540\ nm}$ after 15 min of incubation at 37°C.
6. Determine the amount of protein in the samples by reading the absorbance values against a standard curve. Subtract the values obtained for the sucrose blanks from the tubes containing protein and the same amounts of the sucrose-EDTA solution.

Subcellular Fraction Volumes to Assay (mL)

Homogenate	0.02 and 0.05
Nuclei	0.02 and 0.05
Microsomes	0.02 and 0.05
Mitochondria	0.05 and 0.10
Cytoplasm	0.20 and 0.50

TABLE 10.2. Summary Table for Cell Fractionation Data

	Homogenate (crude)	Nuclei	Mito-chondria	Micro-somes	Cyto-plasm
Volume (mL)					
Protein mg/mL: mg/g liver:					
DNA mg/g liver: % in fraction[a]:					
RNA mg/g liver: % in fraction[a]:					
Specific activities[b]					
Succinoxidase (nmol O/min·mg)					
Glucose-6-phosphatase (μmol P_i/min·mg)					
Acid phosphatase (μmol p-n-ϕ/min·mg)					
Lactate dehydrogenase (μmol NAD^+/min·mg)					

[a]Let homogenate = 100%.

[b]Evaluate the effectiveness of the fractionation procedure by comparing the relative specific activities for each enzyme. How well do these values agree with the actual localization of these enzymes within the cell?

Record the data in a summary table (Table 10.2).

Assay of Fractions for Mitochondrial Activity

To check for the presence of mitochondria in the fractions, oxidation of succinate by succinoxidase will be determined in the oxygraph, and a value for specific activity will be calculated.

1. Calibrate the oxygraph. Saturate some "RC mix" with oxygen, and fill the oxygraph chamber with the buffer, the fraction (see amounts indicated below), and 30 μL of 0.5 M Na succinate.
2. Use the following volumes of each fraction (mL): mitochondria, 0.1; homogenate, 0.1; nuclei, 0.1; microsomes, 0.2; cytoplasm, 0.2.
3. From the slopes of the oxygraph tracings, calculate the activity of succinoxidase in nmol of O utilized/min.
4. Calculate the specific activity of each fraction. Enter these data in a summary table (Table 10.2).

Assay of Fractions for Microsomal Activity

The presence of microsomal contamination will be determined by assaying the fractions for glucose-6-phosphatase activity. The enzyme catalyzes the following reaction:

$$\text{Glucose-6-phosphate} \longrightarrow \text{glucose} + P_i$$

The reaction will be carried out in one series of tubes, and portions of this reaction mixture will then be examined for P_i content.

1. Calculate the amount of each fraction that is required to obtain 0.5 and 1.0 mg of protein (use data from the biuret assay). These will be the enzyme samples assayed below.
2. Set up a series of 15 glass conical centrifuge tubes, and add the following reagents: 100 mM MES buffer, pH 6.5, 0.6 mL; fraction, 0.5 mg to 5 tubes (one for each fraction) and 1.0 mg to 10 tubes (5 of which will serve as controls); deionized water, to a final volume of 1.3 mL (MES buffer is 2[*N*-morpholino] ethanesulfonic acid, pK 6.8).
3. Preincubate the tubes for 5 min at 30°C, and then add 1.0 mL of 0.5 N TCA (trichloroacetic acid) to five of the tubes containing 1.0 mg of protein. These are control tubes and will be used to determine endogenous P_i.
4. At timed intervals, start the reaction by adding 0.2 mL of substrate (150 mM glucose-6-phosphate) to all tubes, and incubate the tubes for 10 min at 30°C.
5. Stop the reaction by adding 1 mL of 0.5 N TCA.

6. Spin the tubes for 10 min at top speed of a clinical centrifuge.
7. Pipette 1.0-mL aliquots carefully into a series of 15 test tubes. Add 1.0 mL of 0.5 N TCA to each, mix, and proceed with substep d below as for the standard curve tubes.

Determination of Inorganic Phosphate by a Modified Taussky-Shorr Method

a. Prepare seven tubes for a standard curve. Deliver amounts from 0 to 1.0 μmol of P_i by using varying volumes of 1 mM KH_2PO_4.

b. Add water to bring the volume to 1.0 mL, and mix on a Vortex mixer.

c. Add 1.0 mL of 0.5 N TCA, and mix vigorously.

d. Add 1.0 mL of molybdate reagent and mix again. (Prepare the molybdate reagent just prior to use. It contains 4 mL of 16% ammonium molybdate in 10 N H_2SO_4, 36 mL of H_2O, and 2 g of $FeSO_4 \cdot 7H_2O$).

e. Incubate for 5 min at room temperature. Read at 660 nm against a water blank.

8. Determine P_i in each fraction assayed. Then subtract the values obtained for the tubes in which endogenous phosphate was measured. Also account for the amount of sample assayed (1 mL of 2.5 mL of reaction mixture).
9. Calculate the specific activity of glucose-6-phosphatase in each fraction, and record these values in a summary table (Table 10.2).

Assay of Fractions for Lysosomal Acid Phosphatase

No attempt was made to isolate the lysosomes, because they are heterogeneous in size, and therefore during centrifugation they settle with other organelles. They are very important in the organization of the cell, as they are the sites of localization of many degradative enzymes.

The presence of lysosomes will be demonstrated by assaying the factions for acid phosphatase activity. Acid phosphatase splits phosphate ester linkages in a nonspecific fashion so that terminal phosphate groups can be cleaved from many compounds. The assay used here takes advantage of this fact by providing an artificial substrate, *p*-nitrophenyl phosphate, which is colorless and turns yellow when the phosphate group is removed:

$$\underset{\text{(colorless)}}{p\text{-nitrophenyl phosphate}} \longrightarrow \underset{\text{(yellow)}}{p\text{-nitrophenol}} + P_i$$

1. Set up a series of 10 tubes, two for each fraction, and add 0.1 mL of fraction to one and 0.1 mL of fraction diluted 10 fold to the other. Include a blank, substituting 0.1 mL of water for the protein.
2. Add 0.2 mL of 0.5% Na cholate. Mix the suspension well to achieve thorough dispersion of the proteins, and add 0.2 mL of 0.5 M acetate buffer, pH 5.0, to all the tubes.
3. Incubate at 37°C, and start the reaction by adding 0.5 mL of 10 mM *p*-nitrophenyl phosphate. Time for exactly 10 min, and stop the reaction by adding 3 mL of 0.5 N NaOH.
4. Read the absorbance at 400 nm against the water blank. Determine the μmol/min of product formed using the Beer-Lambert law and the molar extinction coefficient at 400 nm of *p*-nitrophenol of 21,000 M^{-1} cm^{-1}. Be sure to take the dilution factors into account when calculating the specific activity of each subcellular fraction. Record the data in a summary table (Table 10.2).

Assay of Fractions for Cytosol Activity

Lactate dehydrogenase activity is localized in the soluble cytosol. The enzyme catalyzes the following reaction:

$$\text{Pyruvate} + \text{NADH} \rightleftharpoons \text{lactate} + \text{NAD}^+$$

The fractions will be analyzed for the presence of this enzyme by measuring the change in absorbance at 340 nm with time following the addition of NADH.

1. In a cuvette, mix the following reagents: 0.20 mL of 0.01 M Na pyruvate; 0.1 mL of subcellular fraction (diluted as needed, see step 3 below); and 2.4 mL of 0.01 M Na phosphate buffer, pH 7.4.
2. Place the cuvette in the spectrophotometer set at 340 nm and adjust the absorbance to read 0.5. Add 0.3 mL of 1 mM NADH to initiate the reaction. Mix the contents of the cuvette, and start timing.

3. Record absorbance readings every 30 sec for 3 to 4 min. If the oxidation rate is too fast to record accurately, repeat the procedure with a more dilute sample of the fraction.
4. Repeat the procedure with each fraction. Plot $A_{340\ nm}$ *vs.* time and determine ΔA/min.
5. Convert ΔA/min to μmol/min NADH oxidized using the Beer-Lambert law (molar extinction coefficient at 340 nm of NADH is 6200 $M^{-1}cm^{-1}$). Calculate the specific activity for lactate dehydrogenase for each fraction and record the data in a summary table (Table 10.2).

Extraction and Quantitation of DNA and RNA

A sample of each of the fractions will be used to extract the nucleic acids. The extracts will then be analyzed for their DNA and RNA contents.

1. Proceed with the procedure outlined in the following flow chart and steps 2 and 3 below, being careful to discard the appropriate phases.

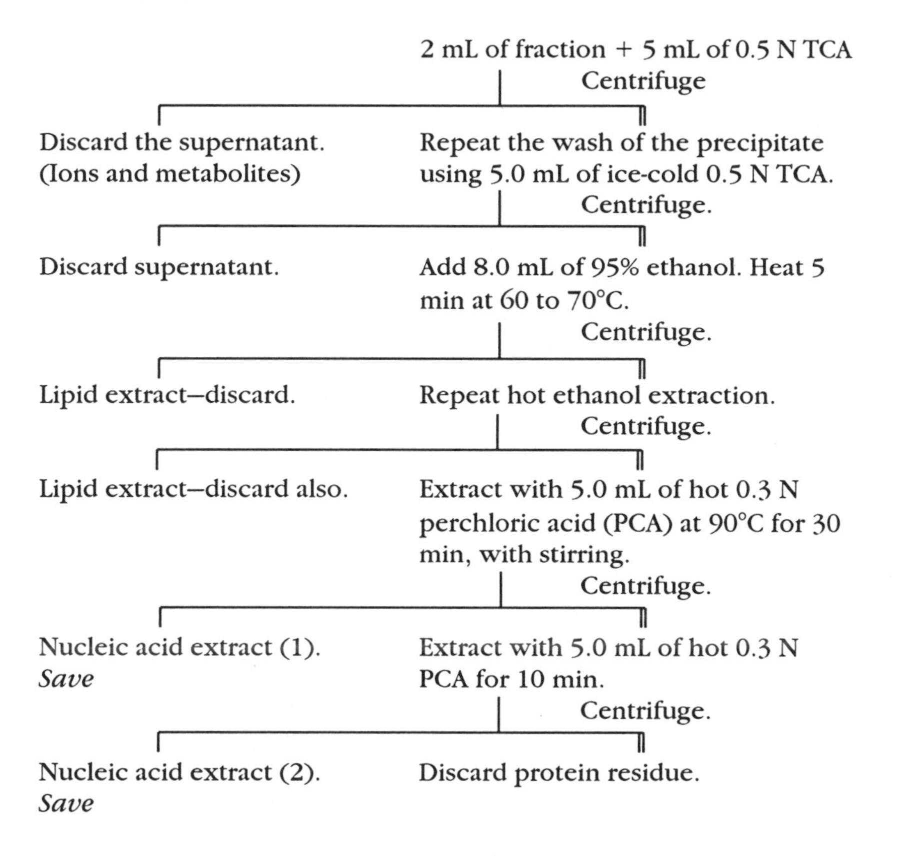

2. Use 2.0 mL of each of the fractions for the extraction. Transfer the aliquots into heavy-walled glass conical centrifuge tubes in an ice bath. Add 5.0 mL of ice-cold 0.5 N TCA and stir. (All spins are for 5 to 7 min in a table-top centrifuge).
3. Combine the "nucleic acid" extracts (1 and 2), and assay for DNA and RNA.

Diphenylamine Assay for DNA

4. The diphenylamine assay is used to measure the DNA content of the extracts.
   a. Use 1.0-mL and 2.0-mL aliquots of each extract for the assay. Construct a standard DNA curve using a solution containing 0.15 mg/mL. The series of standards should range from 0.015 to 0.30 mg of DNA per tube. Bring all of the tubes to a volume of 2.0 mL with 0.3 N PCA (perchloric acid). Prepare a blank containing 2.0 mL or of 0.3 N PCA.
   b. Add 2.0 mL of diphenylamine reagent to each tube (see Chapter 6, diphenylamine assay), and mix well. Cap the tubes with marbles, and heat the tubes for 15 min in a boiling water bath. Cool and read at 600 nm against the blank.
   c. Report your results in mg DNA/mg protein for each fraction and in mg DNA/g of liver. In making these calculations, the appropriate dilution factor for the sample volumes must be accounted for. Record the data in a summary table (Table 10.2).

Orcinol Assay for RNA

5. The orcinol assay is used to measure the RNA content of the extracts.
   a. Construct a standard curve for RNA using a solution containing 50 μg of RNA/mL. Set up a series of tubes containing from 10 to 100 μg of RNA, and bring the total volume to 2.0 mL with 5% TCA solution. Prepare a blank containing 2.0 mL of 5% TCA.
   b. Use the following volumes of sample extracts (mL):

Homogenate	0.3 and 0.6
Nuclear fraction	0.3 and 0.6
Mitochondrial fraction	0.7 and 1.4
Microsomal fraction	0.3 and 0.6
Supernatant fraction	0.7 and 1.4

c. Bring the total volume in each tube to 2.0 mL with 5% TCA.

d. Add 2.0 mL of orcinol reagent to each tube (see Chapter 6, orcinol assay), and mix well. Cap the tubes with marbles, and heat the tubes for 20 min in a boiling water bath. Cool and read at 640 nm against the blank.

e. Report your results in mg RNA/mg protein for each fraction and in mg RNA/g liver. Again, correct for dilution of the extract, and record the data in a summary table (Table 10.2).

F. Isolation and Characterization of Double-Stranded Rat Liver DNA

Double-stranded DNA can be denatured to single strands by heating the duplex at or above a characteristic temperature called the *melting point*. One of the properties of the denatured DNA is an enhanced absorbance of light at 260 nm (hyperchromism) that results from the transition of the UV light-absorbing bases from a stacked conformation to one similar to that of the free bases in solution. Since the melting temperature of double-stranded DNA is inversely proportional to the A-T base-pair content, a melting curve can be used to estimate the % G+C for rat DNA (see also Chapter 6 for further discussion).

Isolation of Duplex DNA

1. Combine all remaining samples of nuclear fraction, and record the volume of this suspension. Add 25 mL of urea containing buffer (8 M urea, 1 M $NaClO_4$, 0.2 M Na phosphate buffer, pH 7.0, 10 mM EDTA, and 1% SDS). Homogenize thoroughly in a hand-held glass homogenizer to disrupt the nuclei.
2. Add 25 mL of an isoamyl alcohol-chloroform (1:24) mixture. Stir vigorously for 15 min; then centrifuge 10 min at 10,000 rpm.
3. Remove the aqueous phase (upper layer) carefully, *record the volume,* and add 2 mL of hydroxyapatite (HAP) for each gram of liver in the fraction.* Centrifuge as before, and discard the supernatant.
4. Resuspend the HAP in 5 to 10 volumes (calculated from the volume of aqueous phase) of 0.2 M Na phosphate

buffer, pH 7.0, containing 8 M urea; centrifuge as before, and discard the upper layer.

5. Wash the gel with 10 volumes of 10 mM Na phosphate buffer, pH 7.0, to remove urea; centrifuge as before, and discard the upper layer.
6. Repeat the washing above.
7. Elute the duplex DNA with 5.0 mL of 0.4 M Na phosphate buffer, pH 7.0, by stirring the HAP suspended in the buffer for 5 min. Centrifuge as before, but save the supernatant.
8. Centrifuge the supernatant again to remove any residual hydroxyapatite, and discard the debris.

Melting Curve for Double-Stranded DNA

1. Read the absorbance of the DNA preparation against a 0.4 M Na phosphate buffer, pH 7.0 blank. Dilute with water to an $A_{260\ nm}$ of 0.2 to 0.3.
2. Use a sample of commercial DNA of known concentration for comparison. Prepare it as above, and treat it in the same manner as the unknown.
3. Follow the procedure for determing Tm for rat DNA and for the standard as described in Chapter 6.
4. Calculate the % G+C for rat liver DNA from the melting temperature data.

---

*Calculate the amount of liver in the nuclear fraction used in step 1, and multiply this quantity by 2 (mL) to determine the amount of HAP to use. For example, if 1 mL of fraction was used, the nuclear fraction volume was 5 mL, and the weight of the liver was 10 g, then 4 mL of HAP suspension should be used.

APPENDIX I

Equipment, Supplies, and Preparation of Reagents

CHAPTERS 2, 3, 4: BUFFERS, PROTEINS, AND THIN-LAYER CHROMATOGRAPHY

Day 1

Protein Determination by the Biuret and Warburg-Christian Methods

1. Biuret reagent.
2. 37°C H_2O bath.
3. Vortex mixers and test tube racks.
4. Protein solutions (1 tube/student).
   a. 2 mg/mL bovine serum albumin (BSA).
   b. 2 mg/mL lysozyme.
   c. 2 mg/mL gelatin.
   d. BSA unknowns.
5. Spectronic-20 below 625 nm and cuvettes.
6. Warburg-Christian—Set-up four cuvettes in the spectrophotometer as follows:
   a. H_2O—blank.
   b. BSA—1.00 mg/mL.
   c. RNA—0.03 mg/mL.
   d. Unknown (provided by students).

All reagents are made with deionized water unless otherwise stated.

Preparation of Reagents (for ca 40 students)

Biuret Method

1. *Biuret reagent.* Dissolve 4.5 g $CuSO_4 \cdot 5H_2O$ and 18.0 g K Na tartrate in 1500 mL H_2O. Add 900 mL 10% NaOH while stirring. Dilute to 3 L. **MAKE FRESH EACH SEMESTER.**
2. *2.0 mg/mL BSA.** Dissolve 1 g BSA in 500 mL H_2O.
3. *2.0 mg/mL lysozyme.** Dissolve 0.5 g lysozyme in 250 mL H_2O.
4. *2.0 mg/mL gelatin.** Dissolve 0.5 g gelatin in 250 mL H_2O. Warm gently to dissolve completely.
5. *BSA unknowns* (varied periodically)*
   a. *10.0 mg/mL*–dissolve 2.0 g BSA in 200 mL H_2O.
   b. *7.5 mg/mL*–dissolve 1.5 g BSA in 200 mL H_2O.
   c. *5.0 mg/mL*–dissolve 1.0 g BSA in 200 mL H_2O.
6. *10% NaOH*–100 g in 1L H_2O (Prepare in the fume hood.)

Warburg-Christian Method

Set up four cuvettes as follows:

1. *H_2O*–blank.
2. *BSA*–1.0 mg/mL (dilute 2 mg/mL 1:1 with H_2O).
3. *tRNA*–0.03 mg/mL (dissolve 3 mg bulk tRNA in 100 mL H_2O).
4. *Unknown*–students dilute their own to 1 mg/mL.

Day 2

Protein Determinations by the Lowry and Dye-Binding Methods

1. Lowry reagents:
   a. 1% $CuSO_4$.

---

*Freeze in 5-mL batches in plastic disposable centrifuge tubes and store at −20°C. (Solutions should be frozen while placed in a horizontal position to avoid breakage of tubes.)

b. 2% Na_2CO_3 in 0.1 N NaOH.
c. 2.7% K Na tartrate.
d. 1 N Folin-Ciocalteu reagent (in the hood).

2. Vortex mixers, stopwatches, test tubes, racks.
3. Protein solutions: 2.0 mg/mL BSA, BSA unknowns.
4. Spectronic-20 above 625 nm and cuvettes.
5. Curvette washing solution.
6. Dye-binding assay reagent.

---

1. *Lowry reagents.*
   a. *1% $CuSO_4$*—dissolve 5.0 g $CuSO_4 \cdot 5H_2O$ and dilute to 500 mL in H_2O (1 mL/student).
   b. *2.7% K Na tartrate*—dissolve 27.0 g K Na tartrate in H_2O and dilute to 1 L. **MAKE FRESH EACH SEMESTER.** Refrigerate.
   c. *2% Na_2CO_3 in 0.1 N NaOH*—dissolve 80.0 g Na_2CO_3 (anhydrous) and 16.0 g NaOH in H_2O and dilute to 4 L **MAKE FRESH EACH SEMESTER.** Store at 4°C.
   d. *Folin-Ciocalteu reagent*—purchased as 2 N, dilute to 1 N with H_2O. Store in a brown bottle.
2. *2.0 mg/mL BSA*. See day 1.
3. *BSA unknowns*. See day 1.
4. *Cuvette washing solution (50% EtOH, 5% HOAc)*. Mix 425 mL H_2O, 525 mL 95% EtOH, and 50 mL conc. HOAc. Dispense in wash bottles.
5. *Dye-binding assay*. Bio-Rad Protein Assay reagent purchased from Bio-Rad Laboratories, Richmond, CA.

Day 3

Titration of an Amino Acid

1. Five amino acid unknowns (*e.g.* glycine, histidine, glutamic acid, lysine, leucine).
2. Standardized KOH and HCl.
3. pH meters, burets, magnetic stirrers, stirring bars.
4. Buffer standards at pH 7 and pH 4.
5. 50-mL and 100-mL volumetric flasks (1 of each/group).
6. Beakers, various sizes.

Amino Acid Titration (100 mL/2 students)

1. *0.1 M lysine*. Dissolve 14.62 g lysine in H_2O and dilute to 1 L with H_2O (MW = 146.2).
2. *0.1 M glutamic acid*. Dissolve 16.91 g glutamic acid in H_2O. Dilute to 1 L with H_2O (MW monosodium salt = 169.1).
3. *0.1 M glycine*. Dissolve 17.51 g glycine in H_2O and dilute to 1 L with H_2O (MW = 175.1).
4. *0.1 M leucine*. Dissolve 13.12 g leucine in H_2O and dilute to 1 L with H_2O (MW = 131.2).
5. *0.1 M histidine*. Dissolve 15.52 g histidine in H_2O and dilute to 1 L with H_2O (MW = 155.2).
6. *6 N KOH*. Dissolve 168.3 g in H_2O. Dilute to 500 mL. (Prepare in the hood.)
7. *1 N KOH. Standardize with 1.000 N HCl acid.* Dilute 333.3 mL 6 N KOH to 2 L with H_2O. **MAKE FRESH EACH SEMESTER.**
8. *1 N HCl. Standardize with standard (KOH) made above.* Dilute 166.6 mL conc. HCl to 2 L with H_2O.

Acid and Base Titrations

1. *0.1% Bromthymol-blue*. Dissolve 0.025 g in 25 mL H_2O. (Use three to five drops to titrate a 25 mL solution. Yellow is acidic; blue is basic.)
2. Measure 25 mL of purchased standard 1 N HCl (in a volumetric flask) and transfer to a beaker.
   a. Fill a 25 mL buret with 1 N KOH and add three to four drops of indicator to HCl. Titrate until yellow solution turns blue.
   b. Calculate the normality of KOH:

      N × volume acid = N × volume base.

      1 × 25 mL HCl = N × ? mL.

      Solve for N when the volume of base has been determined.
   c. Rinse the buret with H_2O, then rinse twice with acid of unknown normality and titrate 25 mL of standardized KOH to a yellow end point. Calculate normality of the acid as was done for the base.

Day 4

Preparation of a Buffer and Amino Acid Separation on TLC

1. Equipment and supplies for titrations.
   pH meters.
   Electrodes.
   Magnetic stirrers.
   Stirring bars.
   Burets.
   Buffer standards pHs 7 and 4.
   pH paper.
   Ring stands and clamps.
2. Conc. HCl and HOAc (store in the hood).
3. 0.1 N NaOH and 0.1 N HCl in dropper bottles.
4. 0.5 N HCl and 0.5 N NaOH.
5. 100-mL and 50-mL volumetric flasks.
6. Buffer reagents listed below.
7. Spatulas, weighing dishes.
8. Chromatography tanks–95% EtOH:H_2O (10:1) (50 mL solvent/1 L beaker). Cover beakers with aluminum foil.
9. TLC plates–cut to 10×9 cm and briefly dried in oven.
10. 2-μL microcaps.
11. Amino acids (*e.g.*, five standards: arginine, isoleucine, proline, glycine, trytophan). Three "unknown" mixtures of these.
12. Ninhydrin reagent in a spraying dispenser (store in the hood).
13. Spraying chamber (store in the hood).
14. Drying oven (110°C).
15. Assorted beakers.

---

1. *0.5 N HCl.* Dilute 1 L of 1 N HCl to 2 L with H_2O.
2. *0.1 N HCl.* Dilute 10 mL of IN HCl to 100 mL with H_2O. Dispense in dropper bottles.
3. *6 N NaOH.* Dissolve 240 g NaOH in H_2O and dilute to 1 L. Prepare in the hood. **MAKE FRESH EACH SEMESTER**.

4. *1 N NaOH*. Dilute 167 mL of 6 N NaOH to 1 L.
5. *0.5 N NaOH*. Dilute 84 mL of 6 N NaOH to 1 L.
6. *0.1 N NaOH*. Dilute 10 mL of 1 N NaOH to 100 mL with H_2O. Dispense in dropper bottles.
7. *Buffer chemicals*.

   $NaC_2H_3O_2$(anhydrous).

   KH_2PO_4.

   K_2HPO_4.

   NaH_2PO_4.

   Na_2HPO_4.

   Na_2CO_3.

   $NaHCO_3$.

   Tris (hydroxymethyl) aminomethane.
8. *Standard amino acids*.

   Arginine–2.5 mg/mL.

   Isoleucine–2.5 mg/mL.

   Proline–2.5 mg/mL.

   Glycine–2.5 mg/mL.

   Tryptophan–2.5 mg/mL.

   Prepare *stock solutions* of 10 mg/mL in H_2O. Dilute to 2.5 mg/mL as needed, and freeze in 0.05-mL batches in 0.5-mL microfuge tubes.
9. *Unknown amino acid mixtures (vary mixtures periodically)*.
   a. No. 1 mix: Arginine, isoleucine, proline.
   b. No. 2 mix: Arginine, isoleucine, glycine.
   c. No. 3 mix: Arginine, glycine, tryptophan.*
   d. To make unknown mixtures: Use the three amino acids from the *stock solutions* in equal quantities. Freeze in 0.05-mL batches in 0.5-mL microfuge tubes.
10. *Developing tanks*. Set up the morning of use. Pour 50 mL of 95% EtOH:H_2O (10:1) solution in 1 L beakers. Cover with aluminum foil.

***NOTE:** Isoleucine and tryptophan have similar R_f values and colors. Do not combine in same unknown mixture.

11. *Ninhydrin spraying reagent.* Dissolve 6 g ninhydrin in 150 mL of 95% EtOH. Pour into a spraying apparatus. (Prepare and store in the hood.) **MAKE FRESH EACH SEMESTER.**

CHAPTER 5: ENZYMOLOGY

Day 1

Extraction and Partial Purification of Invertase.

1. Yeast cake (purchased from a bakery, 1 lb).
2. Toluene (store in the hood).
3. 1 N HOAc in dropper bottles.
4. pH meters, standard buffers, beakers.
5. 95% EtOH (*ca* 10 mL/student; store in the freezer).
6. 0.2 M acetate buffer, pH 4.5.
7. Clinistix test strips (Miles Laboratories, Elkhart, IN).
8. Glass beads.
9. 0.5 M sucrose (*ca* 0.5 mL/student).
10. Mortars and pestles.
11. 50°C H_2O bath.
12. Magnetic stirrers and stirring bars (1/student).
13. Plastic centrifuge tubes (50-mL size).
14. Top loading balances.
15. Safety glasses.
16. Spot plates.
17. Refrigerated Sorvall centrifuge with SS-34 rotor.
18. Rubber policemen.

All reagents are made with deionized H_2O unless otherwise stated.

Preparation of Reagents (for ca 40 students)

1. *1 N HOAc.* Dilute 58 mL conc. HOAc to 1 L H_2O.
2. *0.2 M Na acetate buffer, pH 4.5* Dissolve 21.64 g Na acetate (anhydrous) in H_2O, and add approximately 34 mL conc.

HOAc. Adjust the pH with 6 N NaOH to pH 4.5. Dilute to 4 L. Dispense in 200 mL portions and freeze.

3. *0.5 M sucrose.* Dissolve 171.15 g sucrose in 700 mL H_2O, and dilute to 1 L with H_2O. Store in 5-mL batches in the freezer.

Day 2

DEAE Cellulose Column Chromatography

1. DEAE cellulose.
2. 0.05 M Tris buffer, pH 7.3.
3. 0.05 M Tris + 100 mM NaCl buffer.
4. Clinistix reagent strips.
5. 0.2 M acetate buffer, pH 4.5.
6. Sand, glass wool.
7. Conductivity meter.
8. Plastic tubes for storing fractions.
9. 0.5 M sucrose (1 tube/4 students).
10. Column set-up.
    Ring stands.
    Clamps for stands.
    Magnetic stirrers.
    5-mL disposable serological pipettes.
    Rubber tubing.
    Corks–size 00.
    Syringe needles–20-gauge.
    Gradient makers and stirring bars.
    Plastic clamps for rubber tubing.
    Syringes with rubber adapter.
11. Rubber policemen.
12. Refrigerated Sorvall centrifuge with SS-34 rotor.
13. Glass tubes for electrophoresis gel preparation.
14. Top loading balances.

Gradient Maker Set-Up With Needle Adapter and Tubing.

Melt parafin wax in a beaker. As it is melting, push a syringe needle (20 gauge) through a 00 cork, and push out any cork left in the needle with a small thin wire. With a Pasteur pipette, cover the cork with melted wax while holding the cork and needle horizontally and rotating it to avoid drips. Once the corks are sealed completely, allow them to dry, and then attach the needle tops to the tubing of the gradient makers.

1. *DEAE 23.* 40 g dry DEAE cellulose is a convenient amount to prepare. It gives *ca* 250 mL wet "settled" volume. (Allow *ca* 1.5 g/student of dry DEAE). DEAE 23 is purchased from Whatman, Hillsboro, OR.
   a. To the dry DEAE add 600 mL of 0.5 N HCl (per 40 g), allow the suspension to stir for *ca* 1 hr, and let the DEAE resin settle for 15 min. Remove the upper aqueous layer by aspiration. Add 1 to 2 L of H_2O, stir the resin for 1 to 2 min, and allow it to settle. Repeat this procedure until effluent is at *ca* pH 4. (This will take *ca* 8 L of H_2O.) This method adequately removes fine particles (defines) from the resin.
   b. Add 600 mL of 0.5 N NaOH to the DEAE, stir for *ca* 1 hr, filter, and wash as before until the effluent is at pH 8. The DEAE should be stored in H_2O at 4°C after this step until the day it is to be used.
   c. The DEAE resin should be equilibrated the day it is to be used and allowed to come to room temperature. To equilibrate the DEAE, allow it to sit for 1 hr in 1 L 10× Tris buffer and filter on a Buchner funnel, rinse with 1× Tris until the pH is exactly 7.3 (*ca* 1 to 3 L of 1× Tris). Remove fine particles by aspiration and resuspend in *ca* 500 mL of 1× Tris. Distribute 25-mL DEAE portions in 50-mL Erlenmeyer flasks.
   d. Just before use, "de-fine" once more.
2. *10× Tris buffer.* 0.5 M Tris, pH 7.3. Dissolve 121.1 g Tris in 1.5 L H_2O and titrate to pH 7.3 with 4 N HCl (*ca* 250 mL). Dilute to 2 L with H_2O. Refrigerate.
3. *1× Tris buffer.* 0.05 M Tris, pH 7.3. Dilute 100 mL 0.5 M Tris to 1 L with H_2O (check pH). Refrigerate.
4. *0.05 M Tris, pH 7.3 + 100 mM NaCl.* Dissolve 5.58 g NaCl in H_2O, add 100 mL 10 × 0.5 M Tris, adjust pH 7.3, and dilute to 1 L with H_2O.
5. *0.2 M acetate buffer, pH 4.5.* See day 1.

6. *0.5 M sucrose*. See day 1.

Day 3

Nelson's Procedure for Analysis of Reducing Sugar

1. 4.0 mM glucose (1 tube/student).
2. 4.0 mM fructose (1 tube/2 students).
3. 4.0 mM sucrose (1 tube/2 students).
4. Nelson's A (keep warm as it easily comes out of solution).
5. Nelson's B.
6. Arsenomolybdate reagent.
7. 0.2 M Na acetate buffer, pH 4.5.
8. 0.5 M sucrose (1 tube/student).
9. Marbles.
10. Vortex mixers.
11. Boiling H_2O baths.
12. Stopwatches.
13. Spectronic-20 below 625 nm, cuvettes.
14. Dispenser filled with H_2O, adjusted to deliver 7.0 mL.

---

1. *Nelson's A.*

Na_2CO_3(anhydrous)	100 g
K Na tartrate	100 g
$NaHCO_3$	80 g
Na_2SO_4 (anhydrous)	800 g

Dissolve these salts slowly in 3 L H_2O while stirring and heating, and dilute to 4 L with H_2O. (Store in a *Pyrex* reagent bottle–heating is required periodically). Store in 1-L amounts.

2. *Nelson's B*. Dissolve 150.0 g $CuSO_4 \cdot 5H_2O$ in 900 mL H_2O. Add 4 drops of conc. H_2SO_4 and dilute to 1 L with H_2O.
3. *Arsenomolybdate reagent*. The solution should be yellow; if it turns green it should be discarded.
   a. Dissolve, 200.0 g ammonium molybdate in 3.5 L H_2O.
   b. In the hood, add 168 mL conc. H_2SO_4 while stirring.

c. Add 24.0 g Na arsenate to the solution. Stir thoroughly, and dilute to 4 L with H_2O.

d. Incubate at 37°C for 24 hr.

e. Store at room temperature in brown plastic 1-L bottles.

4. *4.0 mM glucose*. Dissolve 0.721 g glucose in water, and dilute to 1 L with H_2O. Freeze in 5-mL batches.
5. *4.0 mM sucrose*. Dissolve 0.274 g enzyme-grade sucrose in 200 mL H_2O. Freeze in 5-mL batches.
6. *4.0 mM fructose*. Dissolve 0.144 g fructose in 200 mL H_2O. Freeze in 5-mL batches.
7. *0.2 M acetate buffer, pH 4.5*. See day 1.
8. *0.5 M sucrose*. See day 1.

Day 4

Protein Determinations by the Lowry Method

1. 2.0 mg/ml BSA (1 tube/2 students).
2. 0.05 M Tris buffer, pH 7.3 (2 mL/student).
3. Lowry reagents.
   a. 2% Na_2CO_3 in 0.1 N NaOH (*ca* 60 mL/student).
   b. 2.7% K Na tartrate (1 mL/student).
   c. 1% $CuSO_4$ (1 mL/student).
   d. 1 N Folin-Ciocalteu phenol reagent (store in the hood).
4. Vortex mixers.
5. Stopwatches.
6. Spectronic-20 above 625 nm, cuvettes.

---

1. *0.05 M Tris buffer, pH 7.3*. See day 2.
2. *2.0 mg/mL BSA*. See Chapter 3, day 1.
3. *Lowry reagents*. See Chapter 3, day 2.

Day 5

Assay of Column Fraction for Invertase Activity and Electrophoresis of the Fractions

1. 0.2 M Na acetate pH 4.5
2. Nelson's A (keep warm)
3. Nelson's B

4. Arsenomolybdate reagent
5. 4.0 mM glucose (1 tube/student)
6. 0.5 M sucrose (1 tube/2 students)
7. Electrophoresis gel solutions A, B, C (store in the hood)
8. Vortex mixers, stopwatches
9. Boiling H_2O baths
10. 1X-Upper buffer A
11. *n*-butanol in dropper bottles (store in the hood)
12. Gel forms and racks for preparing gels
13. "Ultra-spin" triacetate microfuge tubes with cut off size of 30,000 MW (Cole-Palmer, Chicago, IL)
14. 50% glycerol
15. Spectronic-20 below 625nm, cuvettes
16. Dispenser filled with H_2O and adjusted to deliver 7 mL
17. Automatic pipettors and tips
18. 0.5-mL microfuge tubes
19. 0.1% Bromophenol blue
20. Microfuge at 4°C

---

1. *Nelson's reagents*. See day 3.
2. *4.0 mM glucose*. See day 3.
3. *0.2 M Na acetate, pH 4.5*. See day 1.
4. *0.5 M sucrose*. See day 1.
5. *Resolving gel, stock solution A*. Dissolve 64.0 g acrylamide* and 2.0 g bisacrylamide* in *ca* 50 mL H_2O and dilute to 200 mL with H_2O. Wrap aluminum foil around the bottle and refrigerate.
6. *Resolving gel, stock solution B*. Dissolve 18.15 g of Tris in *ca* 25 mL H_2O. Titrate with 6 N HCl to pH 9.1; add 0.25 mL of TEMED. (Prepare in the hood.) Dilute to 100 mL with H_2O. Wrap aluminum foil around the bottle and refrigerate. (Stable for 6 weeks.)

*This is a neurotoxin and must be handled with care. (Wear a respirator, wear gloves, and work in the hood). Students are to wear gloves and use "bench" paper when casting gels. Excess solution is discarded in an "acrylamide waste" container.

7. *Resolving gel, stock solution C.* Dissolve 0.2 g ammonium persulfate (ammonium peroxydisulfate) in H_2O and dilute to 100 mL with H_2O. **PREPARE FRESH EACH DAY**.
8. *10×-Upper buffer A.* Dissolve 63.2 g Tris and 39.4 g glycine in H_2O. Dilute to 1 L (pH 8.7 to 9.1). Refrigerate.
9. *1×-Upper buffer A.* Dilute 100 mL of 10× upper buffer A to 1 L with H_2O. Refrigerate.

Day 6

Electrophoresis and Product Formation *vs.* Time

1. 0.2 M Na acetate, pH 4.5.
2. Nelson's A (keep warm).
3. Nelson's B.
4. Arsenomolybdate reagent.
5. 4.0 mM glucose (1 tube/student).
6. 0.5 M sucrose (1 tube/student).
7. Electrophoresis apparatus.
8. Upper buffer A (1× – *ca* 1.5 L).
9. Lower buffer B (1× – *ca* 1.5 L).
10. 12.5% TCA.
11. Coomassie blue stain.
12. Destaining solution.
13. Destaining station with aspirator.
14. Syringes to remove gels from tubes.
15. Automatic pipettors and tips.
16. Vortex mixers and stopwatches.
17. Spectronic-20 below 625 nm, cuvettes.
18. Boiling H_2O baths.
19. Dispenser filled with H_2O and adjusted to deliver 7 mL.
20. Clinistic reagent strips.
21. Spot plates.
22. Razor blades.
23. Glass plates (*ca* 6 × 12 cm) to slice gels.

---

1. *50% glycerol.* Add 100 mL glycerol to 100 mL H_2O.

2. *0.1% bromophenol-blue*. Dissolve 50.0 mg bromophenol blue in 50 mL H_2O.
3. *1×-Upper buffer A*. See day 5 for 10× solution. Dilute 100 mL 10X upper buffer A to 1L. Refrigerate.
4. *1×-Lower buffer* B. Dissolve 121.0 g Tris in *ca* 400 mL H_2O. Add 65 mL 4 N HCl and adjust to pH 7.9 to 8.3. Dilute to 1 L with H_2O. Refrigerate.
5. *50% TCA*. Prepare the full jar (500 g), in the hood. Add 300 mL of H_2O and stir until dissolved. Transfer the solution to a 1-L graduated cylinder, and add water slowly to bring the volume to 1 L with H_2O. Pour into a glass reagent bottle and store in the hood. (Take proper precautions when preparing this solution.)
6. *12.5% TCA*. Dilute 250 mL of 50% TCA to 1 L with H_2O. (Store in the hood.)
7. *Coomassie blue staining solution (Brilliant blue R)*. Dissolve 2.5 g Coomassie blue in a mixture of 454 mL MeOH, 454 mL H_2O, and 92 mL HOAc.
8. *Destaining solution*. Mix 300 mL conc. HOAc with 200 mL MeOH and 3.5 L H_2O.
9. *0.2 M Na acetate, pH 4.5*. See day 1.
10. *Nelson's reagents*. See day 3.
11. *4.0 mM glucose*. See day 3.
12. *0.5 M sucrose*. See day 1.

Day 7

Effect of Substrate Concentration and Inhibition by Urea

1. 0.2 M acetate buffer, pH 4.5.
2. 0.5 M sucrose (1 tube/student).
3. 4.0 mM glucose (1 tube/2 students).
4. Nelson's A (keep warm).
5. Nelson's B.
6. Arsenomolybdate reagent.
7. Vortex mixers, stopwatches, marbles.
8. Boiling H_2O baths.
9. Clinistix reagent strips.
10. Polaroid camera and film (SX-70 film).

11. Light-box for visualizing gels.
12. 4 M urea.
13. Spectronic-20 below 625 nm, cuvettes.
14. Dispenser filled with H_2O and adjusted to deliver 7 mL.
15. Acrylamide waste container for disposal of gels.

---

1. *4 M urea*. Dissolve 156 g urea in 100 mL H_2O (heating gently). Dilute to 500 mL with H_2O. **MAKE FRESH EACH DAY.** (Keep warm in the hood.)
2. *0.2 M acetate buffer, pH 4.5*. See day 1.
3. *0.5 M sucrose*. See day 1.
4. *4.0 mM glucose*. See day 3.
5. *Nelson's reagents*. See day 3.

Day 8

Glucose Oxidase Method and Inhibition by Raffinose and Fructose

1. 4.0 mM glucose (1 tube/2 students).
2. 5 N HCl.
3. 0.2 M acetate buffer, pH 4.5.
4. Glucostat reagent I.
5. Glucostat reagent II.
6. 0.5 M sucrose (1 tube/2 students).
7. 0.5 M fructose.
8. 0.5 M raffinose.
9. Hot plates with beakers and boiling chips or boiling H_2O baths.
10. Vortex mixers, stopwatches, marbles.
11. Spectronic-20 below 625 nm, cuvettes.

---

1. *0.5 M raffinose*. Dissolve 29.7 g raffinose in H_2O. Heat gently to help dissolve. Dilute to 100 mL with H_2O. **MAKE FRESH EACH DAY**. (Keep warm.)
2. *0.5 M fructose*. Dissolve 9.0 g fructose in H_2O and dilute to 100 mL with H_2O. **MAKE FRESH EACH DAY**. Refrigerate.

3. *5 N HCl*. Dilute 208.3 mL conc. HCl to 500 mL with H_2O.
4. *0.1 M Na phosphate buffer, pH 7.0*. Dissolve 41.1 g Na $H_2PO_4 \cdot H_2O$ in 2.4 L H_2O. Adjust pH to 7.0 with 6 N NaOH. Dilute to 3 L with H_2O. (For use with glucostat reagent I, sufficient for 20 students.) Refrigerate.
5. *Glucostat reagent I*. Dissolve 40 mg of peroxidase per 1 L of 0.1 M Na phosphate buffer. Add *ca* 1500 units of glucose oxidase (amount will vary with batch purchased). (Dispense 150 mL/student in 250-mL Erlenmeyer flasks.) Use at room temperature. **MAKE FRESH EACH DAY**.
6. *Glucostat reagent II* (6.6 mg/mL). Dissolve 0.66 g *O*-dianiside dye in 100 mL H_2O. Store in a brown bottle in the refrigerator. (Wear gloves–it is a carcinogen.)
7. *0.2 M acetate buffer, pH 4.5*. See day 1.
8. *4.0 mM glucose*. See day 3.
9. *0.5 M sucrose*. See day 1.

Day 9

STUDENTS MUST WEAR GLOVES, LAB COATS AND WORK ON "BENCH" PAPER.

Transferase Activity of Invertase and Use of Radioactive Materials

1. 0.2 M Na acetate, pH 4.5.
2. Anion exchange resin, soaked in borate buffer.
3. 15 M MeOH.
4. Scintillation fluid (store in the hood).
5. 1.0 μCi/mL ^3H sucrose.
6. Scintillation vials.
7. Automatic pipettors and tips.
8. Ring stands and clamps.
9. Pasteur pipettes; glass wool.
10. Reagents for SDS electrophoresis gels (for experiment done on day 10).

    Gel A–acrylamide.

    Gel B–SDS buffer.

    Gel C–ammonium persulfate.

n-butanol in dropper bottles.

11. Hot plates, boiling chips.
12. Gel forms and racks for preparing gels.
13. Racks for scintillation vials.
14. 13 × 100 mm disposable tubes.
15. Roll of "bench" paper (plastic-backed absorbent paper).
16. 0.04% bromthymol blue in 95% EtOH.
17. *Rad-Con*—radioactive decontaminant spray (purchased from Nuclear Assoc., Carle Place, NY).

---

1. *Saturated Na borate.* Add approximately 60 g Na borate to 1 L H_2O. Stir for 1 hr, add more Na borate until saturated.
2. *Anion exchange resin.* AG-1-X2, 100 to 200 mesh (purchased from Bio-Rad Laboratories, Richmond, CA). **MAKE FRESH EACH DAY**.
   a. Allow 3 g of resin per student, and add 2 volumes of saturated sodium borate. Stir for 1 hr at room temperature. Allow resin to settle and decant, and discard the buffer.
   b. Add 300 mL H_2O to the resin, stir 2 min, and take a pH reading. Allow the resin to settle and decant, and discard the liquid.
   c. Repeat step b until pH 7 is reached (May require six or more H_2O washes.)
   d. Monitor the pH, and add saturated borate solution dropwise to maintain the pH. Dispense the resin as a slurry in 50-mL Erlenmyer flasks.
3. *15 M MeOH.* Dilute 604 mL MeOH to 1 L with H_2O. (Store in the hood.)
4. *Opti-fluor scintillation fluid*—(Purchased from Packard Instrument Co., Downers Grove, IL.) Pour into a dispenser adjusted to deliver 8 mL. (Store in the hood.)
5. *0.5 M sucrose.* Dissolve 42.8 g sucrose in 100 mL H_2O. Dilute to 250 mL.
6. *1.0 μCi/mL ^{3}H sucrose (0.322 μmole/mL).* Add 250 μCi^{3}H sucrose to 250 mL of 0.5 M sucrose. Add 0.75 g benzoic acid (does not go into solution-preservative).

Store in 5-mL batches in the freezer. Label tubes with radioactive labeling tape.

7. *0.04% Bromthymol blue in 95% EtOH.* Dissolve 40 mg bromthymol blue in 100 mL 95% EtOH. Store in brown bottle.
8. *SDS Buffer, pH 7.0.* Dissolve 15.6 g $NaH_2PO_4 \cdot H_2O$ and 40.0 g Na_2HPO_4 (anhydrous) in 800 mL H_2O. Dilute to 2 L, warm the solution, and slowly add 4.0 g sodium dodecylsulfate (SDS). (Dilute 1:1 for use in electrophoresis in both upper and lower chambers). If it comes out of solution, heat gently.
9. *Gel A.* See day 5.
10. *Gel B.* Use undiluted SDS buffer as prepared above. (Store in the hood.)
11. *Gel C.* Dissolve 0.8 g ammonium persulfate in 100 mL H_2O, and add 0.13 mL TEMED. **MAKE FRESH EACH DAY.** (Store in the hood.)

Day 10

Molecular Weight Determinations

A. Molecular-Sieve Chromatography

1. Agarose beads. Bio-Gel A-1.5 m, 100-200 mesh (Bio-Rad Laboratories, Richmond, CA).
2. 0.05 M Tris, pH 7.3.
3. 40 mg/mL glucose.
4. 2.0 mg/mL blue dextran.
5. Columns, sand, glass wool.
6. Column reservoirs, ring stands, clamps.
7. Tubing to connect columns to fraction collectors.
8. Fraction collectors.
9. Molecular weight standards for molecular-sieve chromatography.
10. Clinistic reagent strips.
11. Spot plates.
12. 0.5 M sucrose.
13. 0.2 M Na acetate, pH 4.5.

1. *Agarose beads.* Mix in a suction flask *ca* 30 mL suspended beads with an equal volume of 0.05 M Tris, pH 7.3. Degas by pulling a vacuum on the flask 10 min before use.
2. *0.05 M Tris, pH 7.3.* See day 2.
3. *40.0 mg/mL Glucose.* Dissolve 4.0 g glucose in 100 mL H_2O. **MAKE FRESH AS NEEDED.** Refrigerate.
4. *2.0 mg/mL blue dextran.* Dissolve 0.2 g blue dextran in 100 mL H_2O. **MAKE FRESH EACH SEMESTER.** Refrigerate.
5. *Molecular-sieve chromatography standards.* Prepare 4 mg/mL of the following proteins in 0.05 Tris buffer, pH 7.3:

   Cytochrome *c*.

   Hemoglobin.

   Alcohol dehydrogenase.

   Catalase.

   Ferritin (Refrigerate).

   Dispense in 0.5 mL batches in 1.5-mL microfuge tubes and freeze.
6. *0.5 M sucrose.* See day 1.
7. *0.2 M Na acetate, pH 4.5.* See day 1.

B. SDS Gel Electrophoresis

1. Buffer solution for MW determinations containing:

   50% Glycerol, 5% SDS.

   0.1 M Na phosphate, pH 7.0.

   0.1% Bromophenol blue.

   Mercaptoethanol.

   (Store in the hood.)
2. Electrophoresis equipment: power supply, gel electrophoresis chambers.
3. Automatic pipettors and tips.
4. Coomassie blue stain.
5. Destaining solution.
6. 12.5% TCA.
7. Syringes for removal of gels from glass tubes.

8. Molecular-weight standard mixture SDS-6H (Sigma Chemical Co., St. Louis, MO).

---

1. *5% SDS.* Dissolve 5 g sodium dodecylsulfate in 100 mL H_2O. Store at room temperature.
2. *50% Glycerol.* Dilute commercial solution 1:1 with H_2O.
3. *0.1 M Na phosphate,* pH 7.0. See day 8.
4. *0.1% Bromophenol-blue.* Dissolve 0.1 g bromophenol blue in 100 mL H_2O. Store in a brown bottle. **MAKE FRESH EACH SEMESTER.**
5. *Coomassie blue stain.* See day 6.
6. *Destaining solution.* See day 6.
7. *12.5% TCA.* See day 6.
8. *SDS molecular weight standards.* Prepare according to Sigma Chemical Co. kit instructions. See text for composition of mixture of proteins and their molecular weights.
9. Buffer solution for MW determinations. Prepare the following mixture:

5% SDS	2.0 mL
50% Glycerol	2.0 mL
0.1 M Na phosphate buffer	1.0 mL
0.1% Bromophenol blue	1.0 mL
Mercaptoethanol	0.5 mL

Dispense in 0.5-mL microfuge tubes in 65-μL batches. Store in the freezer.

Day 11

Effect of pH on Enzyme Activity

1. 0.5 M sucrose.
2. 4.0 mM glucose.
3. Nelson's A (keep warm).
4. Nelson's B.
5. Arsenomolybdate reagent.
6. Vortex mixers, stopwatches, marbles.
7. Boiling water baths.

8. Buffers.
   a. 0.2 M Na succinate, pHs 2.5, 3.0, 3.5, 4.5, 5.5, 6.5.
   b. 0.2 M Na acetate, pHs 3.5, 4.0, 4.5, 5.0, 5.5, 6.0.
   c. 0.2 M K phosphate, pHs 6.0, 6.5, 7.0.
9. Spectronic-20 below 625 nm, cuvettes.
10. Dispenser filled with H_2O and adjusted to deliver 7 mL.

---

1. *0.2 M Na succinate, pHs 2.5, 3.0, 3.5, 4.5, 5.5, 6.5.* Dissolve 5.4 g Na succinate in 50 mL H_2O. Adjust to pH desired with 4N HC1. Dilute to 100 mL with H_2O.
2. *0.2 M Na acetate buffer.* Dissolve salt in 20 mL of H_2O and add HOAc as indicated below. Dilute to 100 mL with H_2O.

pH	*Na acetate, anhydrous (g)*	*1 M HOAc (mL)*
3.5	0.08	19.1
4.0	0.20	17.3
4.5	0.55	13.3
5.0	1.00	7.8
5.5	1.37	3.3
6.0	1.54	1.2

3. *0.2 M potassium phosphate buffer, pHs 6.0, 6.5, 7.0.* Dissolve 2.72 g KH_2PO_4 (monobasic) in 50 mL H_2O. Adjust to desired pH with 1 M KOH. Dilute to 100 mL with H_2O.
4. *Nelson's reagents.* See day 3.
5. *0.5 M sucrose.* See day 1.
6. *4.0 mM glucose.* See day 3.

Day 12

Effect of Temperature on Enzyme Activity

1. 0.2 M acetate buffer, pH 4.5.
2. 0.5 M sucrose.
3. 4.0 mM glucose.
4. Nelson's A (keep warm).
5. Nelson's B.
6. Arsenomolybdate reagent.

7. Vortex mixers, stopwatches.
8. Constant temperature baths (as shown below).
9. Spectronic-20 below 625 nm, cuvettes.
10. Dispenser filled with H_2O and adjusted to deliver 7 mL.
11. Thermometers.

---

1. *0.2 M acetate buffer, pH 4.5.* See day 1.
2. *0.5 M sucrose.* See day 1.
3. *4.0 mM glucose.* See day 3.
4. *Nelson's reagents.* See day 3.
5. Constant temperature water baths:

   0°C–ice.

   14°C–fill a pan with water and maintain temperature by adding ice as needed.

   ca 25°C–room temperature.

   37°C, 50°C, 60°C, 70°C, 80°C, 90°C.
   100°C–boiling water bath.

CHAPTER 6: NUCLEIC ACIDS

Day 1

Isolation of Leu-tRNA Synthetase

1. 0.05 M Tris, pH 7.3 + 0.1 mM DTT.
2. Ammonium sulfate, enzyme grade.
3. Blue dextran (2 mg/mL).
4. Burets, sand, glass wool.
5. Ring stands, clamps.
6. G-25 Sephadex resin.
7. *E. coli* cell paste (0.5 g/2 students).
8. Bio-Rad protein assay reagent.
9. Plastic cuvettes.
10. Cuvette washing solution.
11. BSA solution (2 mg/mL).
12. 50-mL centrifuge tubes.

13. Magnetic stirrers and stirring bars.
14. Plastic spatulas.
15. Vortex mixers.
16. Refrigerated Sorvall centrifuge with SS-34 rotor.
17. Rubber policemen.
18. 30-mL beakers.
19. Spectrophotometer, quartz cuvettes.

Preparation of Reagents (for *ca* 16 students)

1. *0.5 M Tris,* pH 7.3. See Chapter 5, day 2. Dilute 1:10 to prepare buffer.
2. *1 M DTT.* Dissolve 308.6 mg dithiothreitol (DTT) in 2 mL H_2O. Store in the freezer.
3. *0.05 M Tris + 0.1 mM DTT.* Add 5.0 μL 1 M DTT to 500 mL of 0.05M Tris buffer, pH 7.3.
4. *1.0 mg/mL blue dextran.* Dissolve 0.1 g blue dextran in 100 mL H_2O. **MAKE FRESH EACH SEMESTER.** Refrigerate.
5. *G-25 Sephadex.* Use 2 to 3 g per group of 2 students. Weigh resin, and soak beads overnight in 10 mL H_2O/g.
6. *Cuvette washing solution.* See Chapter 3, day 2.
7. *2 mg/mL BSA.* See Chapter 3, day 1.
8. *Protein assay reagent.* Purchased from Bio-Rad Laboratories, Richmond, CA.
9. E. coli *cell paste.*

Instructions for Making *E. coli* Cells

(8 L of cell culture yields *ca* 16 g of cell paste)

Supplies Needed

1. 4 3-L flasks, sterile.
2. 8 L minimal medium (MM).
3. 400 mL 4% glucose.
4. Extra flasks, various sizes, sterile.
5. I L, 0.05 M Tris buffer, pH 7.3.
6. 0.9% NaCl solution.
7. 5 mL nutrient broth (NB).

Use aseptic technique.

1. Autoclave the following solutions separately in flasks with cotton plugs.
   a. 4 × 1.9 L MM in 3-L flask.
   b. 4 × 100 mL 4% glucose in 1-L flask.
   c. 1 L of 0.05 M Tris buffer in 2-L flask.
   d. 500 mL 0.9% NaC1 in 1-L flask.
2. Transfer a culture to 5 mL NB and grow at 37°C overnight with shaking (cultures may be stored at − 20°C in NB with 20% glycerol or on nutrient agar slants).
3. Harvest the cells by centrifuging the overnight culture (*ca* 2000 rpm) for 10 min, decant, and discard the supernatant. Resuspend the cells in 3 mL saline solution.
4. Add 160 mL MM and 40 mL 4% glucose solution to a sterile 500-mL Erlenmeyer flask and inoculate with the 3 mL of cell suspension prepared above.
5. Incubate at 37°C overnight with shaking.
6. To each of four 3-L flasks containing 1.9 L MM, add 100 mL of 4% glucose and 50 mL of the overnight culture.
7. Incubate at 37°C with shaking, and begin monitoring the absorbance at 450 nm after 90 min. Aseptically remove 1 mL from one of the flasks and read against a MM blank.
8. Continue incubation with shaking until an $A_{450\ nm}$ of 0.75 to 0.90 is reached. Withdraw a sample and read its absorbance every 30 to 45 min.
9. To harvest the cells, use 250-mL sterile centrifuge bottles in the GSA rotor. Fill each bottle with 200 mL of cell suspension, and centrifuge at 8000 rpm for 15 min in a Sorvall centrifuge.
10. After centrifuging handle each bottle as follows.
    a. Discard supernatant.
    b. Resuspend the cells in 20 mL sterile 0.05 M Tris buffer, pH 7.3.
    c. Pipette the cell suspension into a 50-mL centrifuge tube. Centrifuge the cells at 8000 rpm for 15 min in an SS-34 rotor. Discard the supernatant and cover with parafilm. Store the pellets in the freezer.

1. Prepare minimal medium (MM) as follows:

Reagent	*Per 1900 mL*
K_2HPO_4	21.0 g
KH_2PO_4	9.0 g
$(NH_4)_2 SO_4$	2.0 g
Na citrate (dihydrate)	1.94 g
Mg SO_4	0.1 g

   Autoclave at 15-lb pressure, 110°C, 15 min.

2. *4% Glucose.* Dissolve 4 g in H_2O and bring volume to of 100 mL with H_2O. (Use 100 mL of 4% glucose per 1900 mL MM.)
3. *0.05 M Tris.* See Chapter 5, day 2.
4. *0.9% saline.* Dissolve 0.9 g NaCl in 100 mL H_2O.
5. *Nutrient Broth (NB).* NB, 0.8 g (commercial); NaCl, 0.4 g; H_2O, 100 mL. Filter sterilize or autoclave and dispense in 5-mL batches in 15-mL screw-capped tubes and freeze.

Day 2

Charging of Leu-tRNA With ^{14}C Leucine

1. ^{14}C leucine solution.
2. "Charging" mix.
3. 1.2 mg/mL bulk tRNA.
4. Scintillation fluid, scintillation vials.
5. 0.05 Tris pH 7.3 with 0.1 mM DTT.
6. Vortex mixers, stopwatches.
7. 13 × 100 mL disposable test tubes.
8. 37°C waterbath.
9. 8% TCA solution.
10. 5% TCA solution.
11. Automatic pipettors and tips.
12. 110°C drying oven.
13. "Radioactive use" station.
14. "Bench" paper, plastic gloves.

---

1. *1 mM leucine.* Dissolve 26.2 mg leucine in 200 mL H_2O. Freeze in 5-mL batches.

2. *0.5 μCi/mL ^{14}C leucine (10 μCi/μmol).* To 94.7 mL H_2O, add 4.84 mL of 1 mM leucine and 0.5 mL ^{14}C leucine (as obtained from manufacturer). Label tubes with radioactive labeling tape. Freeze in 5-mL batches.
3. *"Charging" mix.* Mix together 25 mL of each of the following reagents.
   a. *1.0 M Tris, pH 7.3.* Dissolve 3.03 g Tris in 17 mL H_2O. Adjust pH to 7.3 with 4 N HCl. Dilute to 25 mL with H_2O.
   b. *0.2 M $MgCl_2$ + 30.0 mg DTT.* Dissolve 1.02 g $MgCl_2$ with 30.0 mg DTT in H_2O and dilute to 25 mL with H_2O.
   c. *0.01 M EDTA.* Dilute 2.5 mL 0.1 M EDTA (see day 3) to 25 mL of H_2O.
   d. *0.05 M ATP, pH 7.0.* Dissolve 0.64 g ATP in 20 mL H_2O. Adjust pH to 7.0 with 1 N NaOH and dilute to 25 mL. Freeze in 5-mL batches.
4. *1.2 mg/mL bulk tRNA.* Dissolve 0.12 g tRNA in 100 mL H_2O. Freeze in 5-mL batches.
5. *Scintillation fluid.* Dissolve 4.0 g omnifluor in 1 L toluene in a dispenser bottle adjusted to deliver 8 mL. Store in the hood.
6. *0.05 M Tris, pH 7.3 + 0.1 mM DTT.* See day 1.
7. *50% TCA.* Dissolve 500 g TCA in 1 L H_2O.
8. *8% TCA.* Dilute 160 mL of 50% TCA to 1 L with H_2O.
9. *5% TCA.* Dilute 100 mL 50% TCA to 1 L with H_2O and dispense in wash bottles.

"Radioactive Use" Station

STUDENTS WEAR LAB COATS AND GLOVES. THEY SHOULD NOT LEAVE THE WORK AREA DURING THIS PROCEDURE.

1. The working area is covered with a large sheet of "bench" paper (plastic side down).
2. Supplies provided per group of 2 students.
   1 Ring stand.
   1 Clamp.
   1 1-L suction flask.
   1 Millipore apparatus.

Tweezers.

Millipore filter pads.

Rad-Con radioactive decontaminant spray.

3. A suction flask is attached to a ring stand with a clamp, and a Millipore filtering apparatus is inserted.
4. The suction flask is connected to a vacuum pump or a faucet with an aspirator attachment.
5. Effluent in the suction flask is discarded in an appropriate bottle labeled "**RADIOACTIVE WASTE**" and stored in a bucket in the hood. Flasks are then rinsed and the wash added to the waste bottle.
6. Each student should wipe down the "use" area and equipment touched with decontamination foam (Rad-Con, a commercial product).
7. Scintillation vials should be wiped down with Rad-Con before they are loaded in the counter.
8. Students should wash hands before leaving. Gloves, disposable tubes, and used pipettor tips are tightly wrapped in the sheet of "bench" paper that covered the work area, and the package is discarded in an appropriate solid waste disposal container.

Day 3

DNA Isolation From Calf Thymus

1. CS buffer.
2. 0.15 M Na citrate buffer.
3. 20% SDS solution.
4. NaCl, solid.
5. Chloroform: isoamyl alcohol (24:1).
6. 95% EtOH (cold).
7. 55°C H_2O bath.
8. 250-mL separatory funnels.
9. Magnetic stirrers at 4°C and stirring bars.
10. Blenders at 4°C.
11. Glass stirring rods.
12. Acetone (AR grade).
13. 0.1 M EDTA solution.

14. 1 mM EDTA solution.
15. Razor blades.
16. Refrigerated Sorvall centrifuge with SS-34 rotor.
17. Calf thymus tissue (*ca* 15 g/2 students).
18. 50-mL centrifuge tubes.
19. 0.1 M Na phosphate buffer with 1.0 mM EDTA.
20. Plastic spatulas.

---

1. *CS buffer.* Dissolve 5.88 g Na citrate and 18.0 g NaCl in 2 L H_2O. Adjust the pH to 7.0 with dropwise addition of 0.5 N HCl. Refrigerate.
2. *0.15 Na citrate, pH 7.0.* Dissolve 44.1 g sodium citrate in 800 mL H_2O. Adjust the pH to 7.0 with addition of 4 N HCl. Dilute to 1 L with H_2O. Refrigerate.
3. *20% SDS.* Dissolve 40.0 g SDS in 200 mL H_2O. **DO NOT REFRIGERATE.**
4. *Chloroform: isoamyl alcohol (24:1).* Mix 1920 mL chloroform with 80 mL isoamyl Isoamyl alcohol (isopentyl alcohol). Store in the hood.
5. *0.1 M EDTA.* Dissolve 37.2 g EDTA (disodium salt) in H_2O. Dilute to 1 L with H_2O. Refrigerate.
6. *1.0 mM EDTA.* Dilute 10 mL 0.1 M EDTA to 1 L with H_2O. Refrigerate.
7. *0.1 M Na phosphate, pH7.0.* Dissolve 13.7 g $NaH_2PO_4 . H_2O$ in 800 mL H_2O. Adjust pH to 7 with 6 N NaOH. Dilute to 1 L. Refrigerate.
8. *0.1 M Na phosphate + 1 mM EDTA, pH 7.0.* Add 2 mL of 0.1 M EDTA to 198 mL of 0.1 M Na phosphate. Refrigerate.

Day 4

Determination of Base Composition of Calf Thymus DNA

1. 0.1 M Na phosphate buffer, pH 7.0.
2. 0.33 N HCl.
3. 3.6 N KOH.
4. G, C, A, and T standards (5.0 mg/mL).
5. pH paper.

6. Microcaps, 2μL, 10μL, and 20 μL.
7. Chromatography tanks with 150 mL H_2O saturated butanol.
8. Rods and clips for hanging chromatograms in tanks.
9. Chromatography paper.
10. 0.15 mg/mL DNA standard solution.
11. Nelson's A (keep warm).
12. Nelson's B.
13. Arsenomolybdate reagent.
14. H_2O in automatic dispenser adjusted to deliver 7 mL.
15. 4 mM glucose.
16. Spectronic-20 below 625 nm, cuvettes.
17. Boiling H_2O bath, marbles.
18. Spectrophotometer with water jacketted cuvette holder, quartz cuvettes.
19. Refrigerated Sorvall centrifuge with SS-34 rotor.

---

1. *0.1 M Na phosphate buffer, pH 7.0.* See day 3.
2. *0.33 N HCl.* Dilute 8.3 mL of 4 N HCl to 100 mL with H_2O.
3. *3.6 N KOH.* Dissolve 20.2 g KOH in H_2O and dilute to 100 mL with H_2O. Prepare in the hood.
4. *5.0 mg/mL DNA bases standards.* Dissolve 50.0 mg of each of the four bases (adenine, thymine, cytosine, and guanine) in 5 mL 0.02 N KOH. Add 3.6 N KOH dropwise to help dissolve; dilute to 10 mL with H_2O. Freeze in 0.5-mL batches.
5. *0.02N KOH.* Dilute 0.55 mL of 3.6 N KOH to 100 mL with H_2O.
6. *Developing tanks.* Use 150 mL H_2O-saturated butanol per tank. To prepare saturated butanol mix *ca* 130 mL butanol and 50 mL H_2O in a separatory funnel. Discard the aqueous layer, and transfer the butanol to a covered chromatography tank at least 1 hr before needed. After use, discard solvent in an appropriate waste bottle in the hood.
7. *0.05 N HCl.* Dilute 10 mL of 0.5 N HCl to 100 mL with H_2O.

Determination of Tm for Calf Thymus DNA

1. *0.15 mg/mL calf thymus DNA standard solution.* Dissolve 15 mg DNA in 100 mL 0.1 M Na phosphate buffer. Stir overnight at 4°C or until DNA is dissolved.

Polysaccharide Determination:

See Chapter 5, day 3, for the preparation of the following:

1. *Nelson's A.*
2. *Nelson's B.*
3. *Arsenomolybdate reagent.*
4. *4 mM glucose.*

Day 5

Isolation of Plasmids by the "Alkaline Miniprep" Method

1. Automatic pipettors, various sizes.
2. Sterile pipettor tips.
3. Sterile GET solution.
4. 0.4 M NaOH.
5. 2% SDS solution.
6. K acetate "working solution."
7. 0.5 mL and 1.5 mL microfuge tubes, sterile.
8. Leder phenol solution.
9. 100 % Ethanol.
10. Vacuum pump and desiccator.
11. 70% Ethanol.
12. TE buffer, sterile.
13. RNase T1.
14. Lysozyme.
15. 10× Enzyme buffer.
16. Sterile H_2O.
17. *Hind* III, *PVU* II restriction enzymes.
18. 5× tracking dye.
19. 37°C H_2O in microfuge tubes.
20. 1-mL and 10-mL sterile pipettes.
21. Overnight *E. coli* HB 101/pBR322 culture in LB medium.

---

1. For items 1 to 14 above, see Chapter 7, day 1.
2. *10× Enzyme buffer.* Mix 2.5 mL of 2.5 M NaCl, 1.2 mL of 0.5 M Tris, pH 7.5, and 0.6 mL 1 M $MgCl_2$. Dilute to 10 mL with H_2O. Add 0.0154 g DTT. Freeze in 0.5-mL batches in 1.5-mL microfuge tubes.
3. *5× tracking dye.* See Chapter 7, day 5.
4. E.coli *HB 101/pBR322.* Grow this *E. coli* strain in 5 mL LB medium + 30 μL ampicillin at 37°C for 24 hr.
5. *LB medium.* See Chapter 7, Special procedures section.
6. *Ampicillin (10 mg/mL).* See Chapter 7, Special procedures section.

Day 6

A. Orcinol and Diphenylamine Assays for RNA and DNA

1. Orcinol (commercial).
2. Diphenylamine (commercial).
3. 1.6% acetaldehyde.
4. 5% TCA.
5. 0.05 mg/mL RNA standard solution.
6. 0.15 mg/mL DNA standard solution.
7. 0.1% $FeCl_3$ in conc. HCl.
8. Spectronic-20s below and above 625 nm curvettes.

B. Agarose Gel Electrophoresis

1. See Chapter 7, day 6.

---

1. *1.6% acetaldehyde.* Conc. acetaldehyde is stored at 4°C. Dilute 0.8 mL of acetaldehyde to 50 mL with H_2O. Store in well-stoppered reagent bottles. Refrigerate. **MAKE FRESH EACH WEEK.**
2. *5% TCA.* Dilute 100 mL of 50% TCA to 1 L with H_2O. Dispense in wash bottles. Store at room temperature.
3. *0.05 mg/mL RNA standard solution.* Dissolve 12.5 mg bulk tRNA in 250 mL of 0.1 M phosphate buffer. Refrigerate.

4. *0.15 mg/mL DNA standard solution.* Dissolve 15 mg of calf thymus DNA in 100 mL of 0.1 M Na phosphate buffer. Stir at 4°C until dissolved.
5. *0.1% $FeCl_3$ in conc. HCl.* Dissolve 0.5 g $FeCl_3 \cdot 6H_2O$ in 500 mL conc. HCl. Store in the hood.

CHAPTER 7: RECOMBINANT DNA METHODOLOGY

Day 1

Isolation of Plasmid DNA

Method 1—"Alkaline Miniprep." Method

1. *E. coli* JM 101 with pUC 9 plasmid.
2. 0.4 N NaOH.
3. 2% SDS solution.
4. 1.5-mL microfuge tubes, sterile.
5. Microfuge at 4°C.
6. K acetate working solution.
7. Leder phenol solution.
8. Automatic pipettors.
9. Sterile tips for pipettors.
10. 100% EtOH.
11. Vacuum pump and desiccator.
12. 70% EtOH.
13. TE buffer, sterile.
14. RNase T1.
15. 37°C water bath.
16. GET solution, sterile.

Preparation of Reagents (for ca 16 students)

Use aseptic technique.

1. *JM101 with pUC 9.* See Special Procedures Section below. Stock cultures are grown at 37°C for 24 hr and the cells are stored either as colonies on an agar plate (at 4°C) or for longer periods of storage, in liquid culture in LB medium

with glycerol (at −70°C). Add 0.3 mL 50% glycerol/0.7 mL LB broth culture.

To start a fresh culture, add 30 μL of ampicillin (for a final concentration of 60 μg/mL) to 5 mL of LB broth and inoculate with a loop from a single colony or transfer a loopful of liquid inoculum. Incubate at 37°C for 24 hr with shaking.

2. *0.4N NaOH.* Dilute 40 mL of 1 N NaOH to 100 mL with H_2O.
3. *2% SDS.* Dissolve 2 g sodium dodecyl sulfate in 100 mL H_2O. **DO NOT REFRIGERATE.**
4. *K acetate working solution.*
   a. Dissolve 49.07 g of K acetate in H_2O and bring volume to 100 mL (5 M K acetate stock solution).
   b. To 60 mL of stock solution, add 11.5 mL glacial HOAc and 28.5 mL H_2O.
5. *Leder phenol solution.* Mix the following in a reagent bottle: phenol (liquid):chloroform:isoamyl alcohol in the proportions of 24:24:1. Add 8-hydroxyquinoline powder to a final concentration of 0.1%. Then add a volume of TE buffer equal to the volume of the solvent mixture. Shake well and wrap the stopper of the bottle with parafilm. Store the solution at −20°C. Check the pH of the TE buffer before each use (the pH must be >8.). **(USE IN THE HOOD ONLY AND WEAR GLOVES TO HANDLE.)**
6. *70% EtOH.* Add 70 mL of 100% absolute alcohol to 30 mL H_2O.
7. *1 M Tris, pH 8.* Dissolve 12.1 g in 70 mL H_2O. Adjust pH to 8 with 6 N HCl. Dilute to 100 mL with H_2O. Refrigerate.
8. *0.1 M EDTA.* See Chapter 6, day 3.
9. *TE buffer (10 mM Tris, pH 8, 1 mM EDTA).* Mix 1.0 mL of 1 M Tris buffer, pH 8, and 1.0 mL of 0.1 M EDTA and dilute to 100 mL with H_2O. Use in part for Leder phenol (nonsterile) solution. Autoclave the remainder and dispense in 1-mL batches in sterile 1.5-mL microfuge tubes.
10. *GET solution (50 mM glucose, 10 mM EDTA, 25 mM Tris, pH 8).* Dissolve 0.9 g glucose in H_2O, add 10 mL of 0.1 M EDTA and 2.5 mL of 1 M Tris buffer, pH 8. Dilute to 100 mL with H_2O. Filter sterilize and dispense in 1-mL batches in sterile 1.5-mL microfuge tubes.

Method 2—CTAB Miniprep Method

1. *E. coli* JM101 with pUC 9 plasmid.
2. 2% SDS solution.
3. 0.4N NaOH.
4. K acetate working solution.
5. Automatic pipettors and tips.
6. TE buffer, sterile.
7. RNase T1.
8. Vacuum pump and desiccator.
9. 1.2 M NaCl.
10. 100% Ethanol.
11. 70% Ethanol.
12. 1.5-mL microfuge tubes, sterile.
13. Microfuge at 4°C.
14. 37°C water bath.
15. GET solution.
16. Isopropanol.
17. 5% CTAB solution.

---

1. See also Chapter 7, Method 1.
2. *5% CTAB.* Hexadecylmethyl ammonium bromide, purchased from Sigma Chemical Co., St. Louis, MO. Dissolve 5 g CTAB with gentle heating in 100 mL H_2O.
3. *1.2 M NaCl.* Dissolve 7.01 g NaCl in H_2O. Bring volume to 100 mL.

Day 2

Electrophoresis of DNA Digests on a Mini Agarose Gel

1. pUC 9 DNA (commercial).
2. *Eco*R1 enzyme and buffer (commercial).
3. H_2O, sterile.
4. Lambda DNA (commercial).
5. 37°C water bath.
6. Agarose (commercial).
7. 10× TBE buffer.

8. 1× TBE buffer.
9. Sterile microfuge tubes (0.5 mL).
10. Ethidium bromide solution (10 mg/mL).
11. "Minigel" apparatus.
12. Tracking dye.
13. 65°C water bath.
14. *Hind III* cut λ DNA marker (commercial).

---

1. *10× TBE buffer.* Dissolve 216 g Tris and 110 g boric acid in 1.2 L H_2O. Add 400 mL 0.1 M EDTA (see Chapter 6, day 3), and adjust the pH to 8.2 with Tris buffer or boric acid. Dilute to 2 L. Refrigerate.
2. *1× TBE buffer.* Dilute 100 mL of 10× TBE to 1 L. Refrigerate.
3. *Ethidium bromide (10 mg/mL).* Dissolve 20 mg ethidium bromide in 2 ml of water. **(Take proper precautions while handling-carcinogen.)** Freeze in 25-μL batches in 0.5-mL microfuge tubes.
4. *Tracking dye.* Dissolve 25 mg bromphenol-blue in 10 mL 50% glycerol. Freeze in 25-μL batches in 0.5-mL microfuge tubes.
5. *Water, sterile.* Sterilize H_2O as needed and dispense in 1-mL batches in sterile 1.5-mL microfuge tubes.
6. *Hind III λ DNA marker.* Mix 100 μL commercial preparation (*ca* 44 μg) with 300 μL TE buffer and 100 μL tracking dye. Freeze in 40 μL batches in 0.5 μL microfuge tubes.

Day 3

Ligation of pUC 9 and λDNA Fragments

1. Sterile 1.5-mL and 0.5-mL microfuge tubes.
2. Automatic pipettors and sterile tips.
3. H_2O, sterile.
4. 3 M NaOAc solution.
5. 100% EtOH.
6. Microfuge at 4°C.

7. 70% EtOH.
8. Vacuum pump and desiccator.
9. TE buffer, sterile.
10. 65°C H_2O bath.
11. 5× ligation buffer (commercial).
12. T_4 ligase (commercial).
13. 16° and 37°C water baths.

---

1. *3 M NaOAc solution.* Dissolve 24.6 g sodium acetate in 50 mL H_2O, adjust pH to 5, and bring volume to 100 mL. Sterilize and dispense in sterile 1.5-mL microfuge tubes as needed. Store at room temperature.
2. *70% EtOH.* See day 1, item 6.
3. *TE buffer.* See day 1, item 7.

Day 4

Transformation of Recombinant Plasmids Into Competent Cells

1. Competent cells.
2. 42°C H_2O bath.
3. Sterile 15-mL capped culture tubes.
4. LB medium with 20 mM glucose.
5. X-gal plates.
6. Glass spreaders.
7. Bunsen burners.
8. Alcohol to sterilize spreaders.
9. 37°C incubator.
10. 37°C water bath, shaking.

---

1. *Competent cells.* See Special Procedures Section, item 9, below.
2. *LB medium with 20 mM glucose.* See Special Procedures Section, items 2 and 6, below.
3. *X-gal plates.* See Special Procedures Section, item 1, below.

Day 5

Isolation of Plasmids by the Boiling Miniprep Method

1. 1.5-mL microfuge tubes, sterile.
2. Microfuge at 4°C.
3. Saline solution, sterile.
4. STET solution, sterile.
5. Lysozyme solution.
6. Boiling water bath.
7. Isopropanol.
8. 37°C water bath.
9. 3 M NaOAc solution.
10. 0.3 M NaCl solution.
11. 100% EtOH.
12. Vacuum pump and desiccator.
13. TE buffer, sterile.
14. λ DNA standard (commercial).
15. pUC 9 DNA standard (commercial).
16. RNase T1 (commercial).
17. *Eco*R1 restriction enzyme and appropriate buffer (commercial).
18. Tracking dye.

---

1. *Saline solution, sterile.* Dissolve 9 g NaCl in 1 L H_2O. Autoclave in small portions, and dispense in sterile 1.5-mL microfuge tubes.
2. *STET solution, sterile.* Dissolve 16 g sucrose, 10 mL Triton ×-100, 20 mL 0.1 M EDTA (see day 3), and 2.0 mL 1 M Tris, pH 8, in H_2O. Add 1.16 g NaCl, and bring volume to 200 mL with H_2O. Autoclave and dispense in sterile 1.5-mL microfuge tubes.
3. *Lysozyme solution* (10 mg/mL in 10 mM Tris, pH 8). Dilute 0.1 mL of 1 M Tris to 5 mL with water. Add 0.1 g lysozyme and bring volume to 10 mL with H_2O. Filter sterilize and dispense in sterile 1.5-mL microfuge tubes. Freeze.
4. *3 M NaOAc.* See day 3.

5. *0.3 M NaCl, sterile.* Dissolve 1.75 g NaCl in H_2O and bring volume to 100 mL with H_2O. Autoclave and dispense in microfuge tubes.
6. *TE buffer, sterile.* See day 1.
7. *Tracking dye.* See day 2.

Day 6

Separation of DNA Fragments by Electrophoresis

1. Agarose (commercial).
2. 9 × 13 inch pyrex dish, gel form and comb.
3. 10× TBE buffer.
4. 1× TBE buffer.
5. Ethidium bromide solution (10 mg/mL).
6. *Hind III* λ DNA marker.
7. Automatic pipettors and tips.
8. Power supply.

---

1. *10× TBE buffer.* See day 2.
2. *1× TBE buffer.* See day 2.
3. *10 mg/mL ethidium bromide.* See day 2.
4. *Hind III λ DNA marker.* See day 2.
5. *TE buffer.* See day 1.
6. *Tracking dye.* See day 2.

CHAPTER 7: SPECIAL PROCEDURES

Preparation of Solutions and Cultures

1. LB agar plates for *E. Coli* JM 101 with pUC 9.
   a. For 1 L LB medium, use the following.

      10 g bactotryptone.

      5 g bactoyeast extract.

      10g NaCl.

      Dissolve the ingredients in 500 mL H_2O, adjust the pH to 7.5 with NaOH, and bring volume to 1 L with H_2O.

b. Divide into four 250-mL portions in 500-mL flasks.

c. Add 3.75 g agar to each flask.

d. Cover with a cotton plug and autoclave.

e. At the time of use melt the agar, cool to 50°C, and add aseptically to each flask:

   i. 5.0 mL ampicillin.

   ii. 0.25 mL IPTG (isopropyl thio-β-galactoside).*

   iii. 0.25 mL X-gal (5-bromo-4-chloro-3-indolyl-β-D-galactoside).*

   Pour the warm agar into sterile petri dishes. Wrap with aluminum foil as X-gal is light sensitive, and store the cooled plates at 4°C.

2. *LB liquid medium.* The same reagents used in preparing the LB agar plates are used but the agar is omitted. Ampicillin (6 μL/mL medium) is added only when the culture carries a plasmid. Dispense as needed.

3. *Ampicillin.* Prepare a 10 mg/mL stock solution in H_2O, in 10-mL amounts, and filter sterilize. Use the sodium salt form of ampicillin. Remove 6 mL for the plates, and dispense the remainder into 0.5-mL batches. Freeze.

4. *X-gal.* Prepare 0.5 to 1 mL amounts of a 40 mg/mL solution in DMF (dimethylformamide). Filter sterilization is optional. Store in a glass tube in the freezer and protect from light. *X-gal* forms a blue precipitate when it is hydrolyzed, thereby affording a convenient marker for colonies having an intact β-galactosidase gene. **MAKE ON THE DAY OF USE.** Prepare in the hood and wear gloves.

5. *IPTG.* Prepare 1.5 to 2 mL of a 40 mg/mL solution in H_2O, and store in 0.25-mL batches. Filter sterilize and freeze. IPTG is an inducer of β-galactosidase activity. Stable at − 20°C.

6. *2 M glucose stock solution.* Dissolve 3.6 g in H_2O and dilute to 10 mL with H_2O, filter sterilize. Store at room temperature in sterile tubes. (Use 10 μL glucose stock solution/mL of LB medium for a final concentration of 20 mM.)

---

*IPTG and X-gal are purchased from Bethesda Research Laboratories, Gaitersburg, MD.

7. *CM buffer.* Dissolve 1.016 g $MgCl_2$ and 0.735 g $CaCl_2$ in H_2O, add 10 mL glycerol, and dilute to 100 mL with H_2O.
8. E. coli *JM 101 culture (for preparation of competent cells).* Inoculate 5 mL of LB medium with JM 101 cells from either a broth culture or an agar plate. Incubate 24 hr at 37°C with shaking.

Preparation of Competent Cells

Equipment, Supplies, and Solutions

1. *E. coli* JM 101 fully grown culture.
2. 37°C shaking H_2O bath.
3. Spectrophotometer, cuvettes.
4. LB medium.
5. Refrigerated Sorvall centrifuge with SS-34 rotor.
6. Sterile 50-mL capped centrifuge tubes, sterile.
7. CM buffer, sterile.
8. Sterile 1.5-mL microfuge tubes.
9. Sterile pipettes.
10. Bucket with ice.
11. DMSO (dimethylsulfoxide) reagent.
12. Freezer at −70°C.
13. pUC 9 (commercial).
14. X-gal plates.
15. Glass spreaders and 95% EtOH in a beaker.
16. Bunsen burner.
17. 37°C incubator.

Procedure (allow *ca* 4 hr to complete and carry out in duplicate)

Use aseptic technique.

1. Grow *E. coli* JM 101 overnight in 5 mL medium.
2. Inoculate 1 mL of the overnight culture into 100-mL sterile LB medium, prewarmed to 37°C. Incubate at 37°C with shaking.
3. Monitor cell growth at $A_{600\ nm}$ until $A = 0.425$ is reached. (Start monitoring at 1 hr.)

4. Cool CM buffer on ice.
5. Transfer the cells to sterile 50-mL tubes, and incubate on ice for 10 min.
6. Centrifuge the cells in a refrigerated centrifuge at 6000 rpm for 5 min.
7. Resuspend the cell pellets from 100 mL of culture in a total of 7.5 mL cold CM buffer.
8. Incubate on ice for 15 min.
9. Centrifuge the cells again for 10 min at 6000 rpm.
10. Resuspend in 3.5 mL of cold CM buffer.
11. Incubate on ice for 5 min.
12. Add 125 μL DMSO reagent and incubate on ice for 5 min.
13. Add another 125 μL DMSO and again incubate on ice for 5 min.
14. Dispense in 200-μL batches in 0.5-mL microfuge tubes and store in the freezer at −70°C.

Procedure for Determining the "Competency" of the Cells

1. Proceed with one tube containing 200 mL of cell suspension.
2. Add 50 ng of commercial pUC 9 (usually 1 μL of a 1:10 dilution).
3. Incubate on ice for 30 min.
4. "Heat shock" by incubating the cellc at 42°C for 90 sec.
5. Transfer to an ice bath for 2 min.
6. Add 0.8 mL of prewarmed LB medium with 20 mM glucose.
7. Incubate at 37°C for 1 hr with shaking. Make a 1:10 dilution in LB medium.
8. Transfer 50 and 100 μL of the diluted cell suspension to X-gal plates, and spread the liquid evenly over the surface of each plate. The glass spreader used for this procedure is sterilized by dipping in alcohol, then flaming in a Bunsen burner. The spreader is allowed to cool before it is used by waiting a short time while holding it and then touching an area of the plate that has none of the cells.
9. Incubate the plates in an inverted position in an incubator at 37°C for 18 to 24 hr.

10. Calculate the cells' competency (see Chapter 7 in the text for this $CaCl_2$ method; a yield of 10^4 to 10^6 cells is expected).

CHAPTER 8: LIPIDS

Day 1

Extraction of Lipids From a Natural Source

1. Ethyl ether, methanol (store in the hood).
2. 55°C H_2O bath (store in the hood).
3. Mortars and pestles.
4. Clinical centrifuge with rotor for 50-mL size tubes.
5. 0.9% NaCl solution.
6. Filter paper, 9 to 11 cm (No.2 Whatman, fluted).
7. Glass stirring rods with flattened tip.
8. Watch glasses, 10-cm diameter.
9. Glass stoppered 50-mL test tubes (3/student).
10. Liver, egg yolk, or other lipid source.
11. 50-mL centrifuge tubes.
12. Plastic spatulas.
13. N_2 gas tank.
14. Aluminum weighing boats.
15. MeOH: ether (1:1) (store in the hood).
16. Top-loading balances.
17. Scissors, scalpels, and tweezers.
18. Aluminum foil.

Preparation of Reagents (for ca 16 students)

1. *0.9% NaCl.* Dissolve 9.0 g NaCl in 1 L H_2O. Refrigerate.
2. *MeOH:ether (1:1).* Mix equal parts of MeOH and ethyl ether. Store in the hood.

Day 2

Separation of Lipid Fractions and Gravimetric Analysis

1. Pressure bulbs.

2. 50- and 15-mL glass-stoppered tubes.
3. N_2 gas tank.
4. Aluminum foil.
5. Methanol.
6. Acetone.
7. Sep-Pak silica gel cartridges.
8. Ethyl ether:methanol (100:1; v/v) mixture.
9. Syringe barrels.
10. Acetone:acetic acid (100:1) mixture.
11. Ethyl ether:methanol (1:1) mixture.
12. Ring stands and clamps.
13. Analytical balances.
14. Aluminum foil.

---

1. *Ether:methanol (100:1).* Prepare 250 mL.
2. *Acetone:acetic acid (100:1).* Prepare 500 mL.
3. *Ether:methanol (1:1)* See day 1.

Day 3

Cholesterol Determination and Selection of Solvent Systems

1. 0.1 mg/mL cholesterol standard.
2. 5 μl microcaps.
3. Cholesterol reagent (store in the hood).
4. Spectronic-20 below 625 nm, cuvettes.
5. Automatic pipettors with tips.
6. Mix of neutral lipid standards (same as mix A, day 4).
7. Mix of polar lipid standards (same as mix B, day 4).
8. TLC silica gel plates (cut to 1 × 3 inch size).
9. Aluminum foil.
10. $CHCl_3$, MeOH, hexane, ethyl ether, HOAc (store in the hood).
11. Mix of glycolipid standards (optional).
12. Iodine crystals (store in the hood).

13. N_2 gas tank.
14. Absolute EtOH (store in the hood).
15. 250-mL beakers for chromatography procedure.
16. Watch glasses.
17. Tweezers.

---

1. *1.0 mg/mL cholesterol stock standard.* Dissolve 0.1 g cholesterol in 100 mL absolute EtOH. Refrigerate.
2. *0.1 mg/mL cholesterol standard.* Dilute 5.0 mL stock solution to 50 mL in absolute EtOH. Refrigerate.
3. *Iron stock reagent.* Dissolve 2.5 g $FeCl_3 \cdot 6H_2O$ in 100 mL conc. H_3PO_4. **MAKE FRESH EACH SEMESTER.** (Store in the hood).
4. *Cholesterol reagent.* Add 40 mL iron stock reagent to 460 mL conc. H_2SO_4. (Store in the hood).

Day 4

Thin-Layer Chromatography Procedure for Lipid Fractions

1. 2 and 5 μL microcaps.
2. TLC chambers with lids.
3. Lipid standards.
4. Ninhydrin spray and 10% H_2SO_4 spray (store in the hood).
5. Spraying chamber lined with paper (store in the hood).
6. N_2 gas tank.
7. TLC silica gel plates.
8. Drying oven (110°C).
9. $CHCl_3$, MeOH, hexane, ethyl ether, HOAc (store in the hood).
10. $CHCl_3$:MeOH (1:1) mixture (store in the hood).
11. Chromatography paper to line TLC chambers.
12. Tracing paper.
13. Colored pencils.
14. Iodine vapor chromatography tank (store in the hood).
15. 0.2 mM KH_2PO_4 in milli-Q. H_2O.
16. 10 × 75 mm test tubes–**NEW**.

17. Automatic pipettors and tips.

Lipids standards

1. The standards are stored in the freezer on arrival.
2. If suspended in a solvent, dilute to 20 to 25 mg/mL with the same solvent. If the standard is 10 to 25 mg/mL, use as is. If crystalline, make a solution of 20 mg/mL in a mixture of 1:1 $CHCl_3$:MeOH.
3. Store the solutions of standards in small glass test tubes with screw caps. Before sealing, *store under* N_2 to prevent *lipid oxidization*. Store in the freezer.
4. The standards are dispensed to students in 0.05 to 0.10-mL batches in 0.5-mL microcentrifuge tubes. **(BE CAREFUL TO AVOID CONTAMINATION!)**
   a. Polar lipid standards.
      i. Phosphatidylserine
         25 mg/mL in $CHCl_3$:MeOH (1:1)
      ii. Phosphatidylcholine
         25 mg/mL in $CHCl_3$:MeOH:EtOH (1:1:2)
      iii. Phosphatidylethanolamine
         25 mg/mL in $CHCl_3$:MeOH (1:1)
      iv. Sphingomyelin
         25 mg/mL in $CHCl_3$:MeOH (1:1)
      v. Cardiolipin
         5.5 mg/mL in absolute ethanol
      vi. *Mixture A:* 1 to 4 (1:1:1:1)
         10 μL/student
      *vii. Gangliosides
         25 mg/mL in MeOH
      *viii. Cerebrosides
         25 mg/mL in MeOH
      *ix. Sphingosine
         25 mg/mL in MeOH

---

*Standards 7 to 9 are used for brain tissue only.

b. Neutral lipid standards.

i. Cholesterol

25 mg/mL in $CHCl_3$:MeOH (1:1)

ii. Cholesterol esters, cholesterol stearate, or cholesterol oleate

20 mg/mL in $CHCl_3$:MeOH (4:1)

iii. Monoglycerides; 1-monoolein
25 mg/mL in $CHCl_3$:MeOH, (1:1)

iv. Diglycerides; 1,3-diolein
10 mg/mL in $CHCl_3$:MeOH, (1:1) (Use in 2X amount.)

v. Triglycerides; triolein
25 mg/mL in $CHCl_3$: MeOH (1:1)

vi. *Mixture B:* 1 to 5 (1:1:1:1:1)

c. *Glycolipids* (if lipids were extracted from plant material).

i. Monogalactosyldiglyceride

10 mg/mL in MeOH

ii. Digalactosyldiglyceride

10 mg/mL in MeOH

iii. *Mixture C:* 1 and 2 (2:1)

Detection Reagents

1. *I_2 vapor.* Crystals in a covered dry tank. Prepare in the hood. Wear gloves.
2. *10% H_2SO_4.* Dilute 10 mL conc. H_2SO_4 to 100 mL with H_2O. Dispense in a spraying apparatus. Store in the hood.
3. *Ninhydrin.* See Chapter 4, day 4.

Phosphate Standard Curve

1. *0.2 mM KH_2PO_4.* Dissolve 6.8 mg KH_2PO_4 in 250 mL Milli-Q H_2O. Refrigerate.

Day 5

Determination of Phosphate in the Fractions

1. Heating blocks at 150°C, thermometers.
2. 10% H_2SO_4 in Milli-Q H_2O.
3. 5% Ammonium molybdate in Milli-Q H_2O.

4. 30% H_2O_2.
5. 50°C H_2O bath.
6. Automatic pipettors and tips.
7. $CHCl_3$:MeOH mixture (1:1) (store in the hood).
8. **NEW** disposable 10 × 75 mm tubes.
9. Milli-Q H_2O.
10. Ascorbic acid.
11. Spectrophotometer and **NEW** plastic cuvettes.

---

1. *5% ammonium molybdate.* Dissolve 5.0 g ammonium molybdate in 100 mL Milli-Q H_2O. **MAKE FRESH EACH DAY.**
2. *10% H_2SO_4.* Dilute 10 mL conc. H_2SO_4 to 100 mL with Milli-Q H_2O.
3. *30% H_2O_2.* Commercial preparation.
4. *15% ascorbic acid.* Made with Milli-Q H_2O by the students (1.5 g ascorbic acid in 10 mL Milli-Q H_2O). Prepare in a new disposable tube and wrap in aluminum foil to protect from light.

Day 6

Extraction and Identification of Fatty Acids

1. 0.5 N KOH in EtOH (store in the hood).
2. 55°C H_2O.
3. Conc. HCl (store in the hood).
4. Petroleum ether (store in the hood).
5. Glass wool.
6. Sodium sulfate, anhydrous.
7. Diethyl ether (store in the hood).
8. Hexane (store in the hood).
9. Vortex mixers.
10. Glass funnels.
11. 5% PCA in MeOH (store in the hood).
12. N_2 gas tank.
13. Glass-stoppered 50-mL test tubes.

14. 10-μL Microcaps.
15. Drying oven (110°C).
16. TLC chambers with lids.
17. Ring stands and rings to attach funnels.
18. Tracing paper and colored pencils.
19. 0.2 M KH_2PO_4 in Milli-Q H_2O (see day 4).
20. Silver-impregnated silica gel plates.
21. Standards for fatty acid identification.
22. Spraying reagent: 0.025% dichlorofluorescein (store in the hood).
23. Diethyl ether:hexane mixture (5:95).

Extraction of Fatty Acids

1. *0.5 N KOH in 95% EtOH.* Dissolve 7.0 g KOH in 95% EtOH and dilute to 250 mL with 95% EtOH. Store in the hood.
2. *5% PCA in MeOH.* Dilute 36 mL of 70% PCA (conc.) to 500 mL with MeOH. Store in the hood.

Fatty Acid Separation by Argentation Chromatography

1. *Silver-impregnated silica gel plates.* Activate the plates by drying for 15 min at 110°C prior to use (purchased from Altech, Deerfield, IL).
2. *Fatty acid standards:* Methyl esters (m.) (20 to 25 mg/mL in hexanes).
   a. m.arachidate (C20:0).
   b. m.caprylate (C8:0).
   c. m.oleate (C18:1).
   d. m.linoleate (C.18:2).
   e. m.stearate (C18:1).
   f. m.palmitoleate (C16:1).
   g. m.linolenate (C18:3).
   h. m.palmitate (C15:0).
   i. Free fatty acids.

   Dispensed as needed in 0.5-mL microfuge tubes (10 μL/student).

3. *Spraying reagent.* Dilute 10 mL of 0.25% dichlorofluorescein to 100 mL with 95% EtOH. **MAKE FRESH EACH SEMESTER**. Store in the hood in a spraying apparatus. (Purchased from Sigma Chem. Co., St. Louis, MO)

CHAPTER 9: CLINICAL/NUTRITIONAL BIOCHEMISTRY

Day 1

Blood Analysis

1. 10% Na tungstate.
2. 0.67 N H_2SO_4.
3. Alkaline copper reagent.
4. Phosphomolybdic acid.
5. 0.2 mg/mL glucose standard.
6. K ferricyanide solution.
7. Urease solution.
8. 0.2 mg/mL urea standard.
9. Phenol color reagent.
10. Alkaline hypochlorite reagent.
11. Fluted filter paper (Whatman No. 2).
12. Hematocrit tubes and sealing clay.
13. Automatic pipettors and tips.
14. Boiling H_2O bath.
15. 37°C H_2O bath.
16. Spectronic-20, below 625 nm, cuvettes.
17. Autoclavable bag (for disposal of blood waste materials).
18. Plastic centrifuge tubes, 15-mL size.
19. Antibody solutions for blood typing.
20. Clinical centrifuge with rotor for small tubes.
21. Clinical centrifuge for hematocrit.
22. 13 × 100 mm test tubes.
23. Disposable latex gloves.

Preparation of Reagents (for ca 16 students)

Glucose Content in Blood

1. *10% Na tungstate.* Dissolve 100.0 g Na tungstate in H_2O and dilute to 1 L.
2. *0.67 N H_2SO_4.* Add 18.6 mL conc. H_2SO_4 to H_2O and dilute to 1 L.
3. *10.0 mg/mL glucose stock standard.* Dissolve 1.0 g glucose in 100 mL of H_2O. Freeze in 2.5-mL batches.
4. *0.2 mg/mL glucose working standard.* Dilute 2 mL stock glucose (10.0 mg/mL) standard to 100 mL with H_2O. Refrigerate. **MAKE FRESH EACH DAY**.
5. *Alkaline copper solution.* Dissolve 40.0 g anhydrous Na_2CO_3 in 800 mL H_2O, add 7.5 g tartaric acid, and stir until dissolved. Then add 4.5 g $CuSO_4 \cdot 5H_2O$ and continue stirring. Dilute to 1 L with H_2O.
6. *Phosphomolybdic acid solution.* Dissolve 70.0 g molybdic acid, 10.0 g Na tungstate, and 40.0 g of NaOH in 650 mL H_2O. Slowly add 250 mL conc. H_3PO_4. Dilute to 1 L with H_2O. Prepare this solution on ice in the hood. Add acid slowly while stirring gently.

Determination of Blood Urea Nitrogen

1. *EDTA buffer.* Dissolve 7.5 g EDTA-free acid in 400 mL H_2O. Bring pH to 6.5 with 10% NaOH. Dilute to 750 mL with H_2O. Refrigerate.
2. *2.5% Na nitroferricyanide stock solution.* Dissolve 2.5 g Na nitroferricyanide in 100 mL H_2O. Store in a brown bottle. Refrigerate.
3. *10.0 mg/mL urea stock standard.* Dissolve 2.14 g urea in 100 mL H_2O. Freeze in 2.5-mL batches.
4. *0.2 mg/mL urea working standard.* Add 2.0 mL stock solution to 98 mL H_2O. Refrigerate. **MAKE FRESH EACH DAY**.
5. *Urease solution.* Deliver 5 mL urease-glycerol extract to a 100-mL volumetric flask. Dilute to 100 mL with EDTA buffer. Store in a brown bottle. Refrigerate. **MAKE FRESH EACH DAY**.
6. *Alkaline hypochlorite reagent.* Mix 25 mL 2.5 N NaOH (10%) and 4.0 mL Clorox. Dilute to 100 mL with H_2O. Store in a brown bottle. Refrigerate. **MAKE FRESH EACH DAY**.

7. *Phenol color reagent.* Deliver 5 mL of liquefied phenol to a 100 mL volumetric flask. Add 80 mL H_2O and 1.0 mL Na nitroferricyanide stock solution, and bring to volume with H_2O. Refrigerate in a brown bottle. **MAKE FRESH EACH DAY**. (Prepare in the hood and wear gloves).

Hemoglobin Content of Blood

1. *K ferricyanide solution.* Dissolve the following in 1 L H_2O:

   1.0 g $NaHCO_3$.

   50.0 mg KCN.

   200.0 mg $K_3Fe(CN)_6$.

   Store in a brown bottle. Refrigerate. (Prepare in the hood and wear respirator and gloves.)

Day 2

1. Cholesterol stock solution.
2. Cholesterol working solution.
3. Iron stock solution.
4. Color reagent for cholesterol determinations.
5. Ponceau S in 5% HOAc.
6. Destaining solution and staining/destaining dish.
7. ALT reagent (Technicon Instruments Corp., Tarrytown, NY).
8. Electrophoresis kit for serum protein separations (Helena Laboratories, Beaumont, TX), electrophoresis chambers, power supply.
9. Triglyceride kit (Seradyn Inc., Indianapolis, IN).
10. (HR) buffer for serum electrophoresis.
11. 37°C H_2O bath.
12. 50-mL Erlenmeyer flasks.
13. Timers.
14. Refractometer (American Optical Corp., Buffalo, NY).
15. Spectrophotometer and quartz cuvettes.
16. Absolute ethanol.
17. Screw-capped tubes (15 × 125 mm).

Determination of Cholesterol in Serum

1. *Cholesterol stock solution*. See Chapter 8, day 3.
2. *Cholesterol working solution*. Add 2 mL cholesterol stock standard to 98 mL absolute ethanol. **MAKE FRESH EACH DAY**. Refrigerate.
3. *Iron stock solution*. Dissolve 5.0 g $FeCl_3 \cdot 6H_2O$ in 200 mL conc. H_3PO_4. Store in the hood.
4. *Color reagent*. Dilute 40 mL iron stock solution to 500 mL with conc. H_2SO_4. Transfer to an automatic dispenser set to deliver 2 mL. Store in the hood.

Electrophoresis of Serum Proteins

1. *Ponceau S*. Dissolve 1.25 g in 500 mL of 5% HOAc. May need filtration before use if a precipitate forms. Stable for *ca* 2 months at room temperature.
2. *5% Acetic acid*. Dilute 50 mL conc. HOAc to 1 L with H_2O.
3. *Destaining solution*. Mix 250 mL MeOH with 50 mL conc. HOAc and 250 mL H_2O.
4. *HR buffer*. Dissolve one packet of buffer (provided by vendor) in 750 mL H_2O with stirring. Refrigerate. Stable for *ca* 6 weeks.

Day 3

Urine Analysis

1. Cliniscan 2 densitometer.
2. 2.0 mg/mL BSA.
3. 50% TCA.
4. 3% NaOH.
5. Biuret reagent.
6. 50°C H_2O bath.
7. 37°C H_2O bath.
8. 0.1 M Na phosphate buffer, pH 6.5.
9. 6 N NaOH.
10. 5.0 M Na acetate, pH 4.8.
11. 1.5 N HCl.
12. 6 N HCl.
13. 1% Na acetate.

14. 0.05 μg/mL coproporphyrin standard.
15. Ethyl acetate (store in the hood).
16. Ring stands, ring attachments.
17. Separatory funnels.
18. Refractometer.
19. Ames Multistix (Miles Laboratories, Elkhart, IN).
20. 2-L graduated cylinders.
21. Vortex mixers.
22. Spectronic-20, below 625 nm, cuvettes.
23. Fluorimeters, borosilicate tubes (13 × 100 mm).

Determination of Protein in Urine

1. *2.0 mg/mL BSA*. See Chapter 3, day 1.
2. *50% TCA*. See Chapter 6, day 2.
3. *3% NaOH*. Dissolve 30g NaOH in 1 L H_2O.
4. *Biuret reagent*. See Chapter 3, day 1.

Determination of Coproporphyrins in Urine

1. *5.0 M Na acetate, pH 4.8*. Slowly dissolve 115.0 g Na acetate in 125 mL concentrated HOAc and 125 mL H_2O. Adjust pH to 4.8. Bring volume to 500 mL with H_2O. Refrigerate.
2. *1.5 N HC1*. Mix 125 mL conc. HC1 with 875 mL H_2O.
3. *1% Na acetate*. Dissolve 10.0 g Na acetate (anhydrous) in 1 L H_2O. Refrigerate.
4. *5.0 μg/mL coproporphyrin stock standard*. Dissolve 1.0 mg coproporphyrin (tetramethyl ester) in 200 mL 1.5 N HCl. Store in a brown bottle. Refrigerate.
5. *0.05 μg/mL coproporphyrin working standard*. Dilute 1.0 mL stock solution to 100 mL with 1.5 N HC1. Store in a brown bottle. **MAKE FRESH EACH DAY**.

Day 4

Tay-Sachs Assay and Catecholamines in Urine

1. Alumina washed and dried (Sigma Chemical Co., St. Louis, MO).
2. 1 N NaOH, 5 N NaOH.

3. 0.2 N HOAc.
4. 0.04% bromthymol blue.
5. 2 M K_2CO_3.
6. 0.2 M Na phosphate buffer, pH 6.5.
7. 0.25% $ZnSO_4$.
8. 0.25% K ferricyanide.
9. Ascorbic acid, solid.
10. O.1 N HCl.
11. 0.25 μg/mL norepinephrine standard.
12. 0.25 μg/mL epinephrine standard.
13. EDTA, solid.
14. 1.0 mM methylumbelliferyl-*N*-acetyl-β-D-glucosaminide.
15. 0.2 mg/mL 4-methylumbelliferone standard.
16. 0.17 M glycine-carbonate buffer, pH 9.9.
17. 37°C H_2O bath.
18. Fluorimeters borosilicate tubes (13× 100 mm).
19. Vortex mixers.
20. 50-mL capped centrifuge tubes.
21. 0.04 M Na citrate-phosphate buffer, pH 4.4.

Assay for Tay-Sachs Disease

1. *0.04 M Na citrate–phosphate buffer, pH 4.4.* Dissolve 5.88 g Na citrate and 2.76 g NaH_2PO_4 in 800 mL H_2O. Adjust to pH 4.4 with 6 N HC1. Dilute to 1 L. **MAKE FRESH EACH MODULE**. Refrigerate.
2. *0.17 M glycine-carbonate buffer*, pH 9.9. Add 12.76 g glycine and 18.02 g Na carbonate to 1600 mL H_2O. Adjust pH to 9.9 with 1 N NaOH. Dilute to 2 L with H_2O.
3. *1.0 mM methylumbelliferyl-N-acetyl-*β-D-*glucosaminide.* Dissolve 38.0 mg methylumbelliferyl-*N*-acetyl-β-D-glucosaminide in 100 mL citrate-phosphate buffer. Warm gently to dissolve. Store in a brown bottle. Label "Substrate for Tay-Sachs Assay." **MAKE FRESH EACH DAY**. Refrigerate.
4. *0.2 mg/mL 4-methylumbelliferone standard.* Dissolve 20.0 mg methylumbelliferone in 100 mL glycine-carbonate

buffer. Store in brown bottle. **MAKE FRESH EACH DAY**. Refrigerate.

Determination of Catecholamines in Urine

1. *1 N NaOH*. Dissolve 40.0 g NaOH in 1 L H_2O.
2. *0.2 N HOAc*. Add 12 mL conc. HOAc to H_2O and dilute to 1 L with H_2O. Store in the hood.
3. *0.04% Bromthymol blue*. Dissolve 0.04 g bromthymol blue in 100 mL 95% EtOH. Store in a brown bottle. Refrigerate.
4. *2N K carbonate*. Dissolve 138.0 g K_2CO_3 in H_2O. Dilute to 1 L with H_2O.
5. *0.2 M Na phosphate buffer, pH 6.5*. Dissolve 13.8 g NaH_2PO_4 in 400 mL H_2O. Adjust pH to 6.5 with 10% NaOH. Dilute to 500 mL with H_2O. Refrigerate.
6. *0.25% $ZnSO_4$*. Dissolve 0.25 g $ZnSO_4 \cdot 7H_2O$ in 100 mL H_2O. Refrigerate.
7. *0.25% K ferricyanide*. Dissolve 0.25 g $K_2Fe(CN)_6$ in 100 mL H_2O. Refrigerate.
8. *5 N NaOH*. Dissolve 200.0 g NaOH in H_2O and dilute to 1 L with H_2O. Prepare in the hood.
9. *6 N HCl*. Dilute 500 mL conc. HC1 to 1 L with H_2O.
10. *0.1 N HCl*. Dilute 16.7 mL 6 N HC1 to 1 L with H_2O.
11. *50 μg/mL norepinephrine stock standard*. Dissolve 10 mg L-arterenol bitartrate in 200 mL 0.1 N HC1. Store in a brown bottle. Freeze.
12. *0.25 μg/mL norepinephrine working standard*. Dilute 0.5 mL stock standard to 100 mL with 0.1 N HCl. **MAKE FRESH EACH DAY**.
13. *50 μg/mL epinephrine stock standard*. Dissolve 10 mg L-epinephrine bitartrate in 200 mL 0.1 N HCl. Store in a brown bottle. Freeze.
14. *0.25 μg/mL epinephrine working standard*. Dilute 0.5 mL stock standard to 100 mL with 0.1 N HCl. **MAKE FRESH EACH DAY**.
15. *Alumina (1 g student)*. Commercial product; chromatographic grade (Sigma Chemical Co., St. Louis, MO).

Day 5

Analysis of Foods for Nutritional Value and Electrophoresis of Lipoproteins in Serum

1. Carotene stock standard.
2. Carotene working standard.
3. Hexanes.
4. Spectronic-20 below 625 nm and cuvettes.
5. 50% $(NH_4)_2 SO_4$ solution.
6. Acetone:hexane (1:1) solution with 0.1% quinol solution.
7. 10% Acetone in hexane.
8. Sep-pak plus cartridges-silica gel (Millipore Corp., Bedford, MA).
9. Spinach or other green leafy vegetable.
10. Sand.
11. Mortars and pestles, glass homogenizers.
12. Separatory funnels.
13. Syringe barrels, ring stands, and clamps.
14. Pressure bulbs.
15. Egg whites for analysis.
16. Cheese for analysis.
17. pH meter.
18. Ampules, Bunsen burner, and aluminum foil.
19. 0.1 M Na phosphate buffer, pH 6.5.
20. 6 N NaOH.
21. 110°C drying oven.
22. (LDL) buffer for lipoprotein electrophoresis.
23. Electrophoresis supplies for lipoprotein separations (Helena Laboratories, Beaumont, TX).
24. Automatic pipettors and tips.

Carotene Standard Curve and Determination of Carotenes in Serum

1. *0.5 mg/mL carotene stock standard.* Dissolve 12.5 mg α-carotene in 25 mL $CHC1_3$. Store in a brown bottle under N_2. Freeze.

2. *8.0 μg/mL carotene working standard*. Bring 1.6 mL of stock standard to 100 mL with hexanes in a volumetric flask. Store in a brown bottle under N_2.

Carotene Content of a Vegetable

1. *50% $(NH_4)_2SO_4$*. Prepare saturated $(NH_4)_2SO_4$ by dissolving *ca* 500 g in 625 mL H_2O. Dilute 500 mL saturated $(NH_4)_2SO_4$ to 1 L with H_2O. Store at room temperature.
2. *Acetone:hexane (1:1) with 0.1% quinol*. Mix 500 mL acetone (AR grade) with 500 mL hexane. Add 1.0 g hydroquinone and mix. Store in the hood.
3. *10% Acetone in hexane*. Mix 900 mL hexane with 100 mL acetone (AR grade). Store in the hood.
4. *0.1 M Na phosphate buffer*. Dilute 0.2M Na phosphate buffer 1:1 with H_2O.
5. *6 N NaOH*. Dissolve 120 g NaOH in H_2O and bring volume to 500 mL with H_2O.
6. *LDL buffer*. Dissolve one packet of buffer in 1500 mL H_2O. Refrigerate.

Day 6

Determination of Protein Content in a Food and Carotene Determinations in Serum

A. Method 1. Phenylalanine Content by Fluorimetry

1. 60°C H_2O bath.
2. Fluorimeters and borosilicate tubes (13 × 100 mm).
3. 6 N HCl.
4. Microfuge at 4°C.
5. Succinate buffer, pH 5.8.
6. 30 mM ninhydrin solution.
7. 5 mM L-leucyl-L-alanine.
8. Copper reagent for phenylalanine assay.
9. 0.3 N TCA.
10. 0.6 N TCA.
11. 0.05 mg/mL phenylalanine standard.
12. Cliniscan 2 densitometer.
13. 1.5-mL microfuge tubes.

14. Screw-capped tubes (15 × 125 mm).
15. 95% EtOH.
16. Petroleum ether.
17. Spectronic-20 below 625nm, cuvettes.
18. Fat red stain.
19. Lipoprotein destaining solution.

Assay of Protein Extract for Phenylalanine Content

1. *6 N HCl.* Dilute conc. HC1 (1:1) with H_2O. Prepare and store in the hood.
2. *0.3 M Succinate buffer, pH 5.8.* Dissolve 11.88 g disodium succinate dihydrate ($Na_2C_4H_4O_4 \cdot 2\,H_2O$) in 150 mL of H_2O and adjust to pH 5.8 with 1 N HC1. Dilute to 200 mL with H_2O. Refrigerate.
3. *30 mM Ninhydrin solution.* Dissolve 1.068 g in H_2O and dilute to 200 mL. Store in a brown bottle. Refrigerate.
4. *5 mM Leu-ala solution.* Dissolve 101 mg of L-leucyl-L-alanine in H_2O and dilute to 100 mL. Dispense in 2-mL batches. Freeze.
5. *Copper reagent.* Dissolve the following in *ca* 300 mL. Mix in the order given, and dilute to 1 L with H_2O. Stir between additions.

Sodium carbonate, anhydrous	1.60 g
Potassium sodium tartrate	0.10 g
Copper sulfate ($CuSO_4 \cdot 5H_2O$)	0.06 g

   Store at room temperature. Discard if cloudy.

6. *0.6 N TCA.* Dilute to 19.5 mL of 50% TCA to 100 mL with H_2O.
7. *0.3 N TCA.* Dilute 0.6 N TCA (1:1) with H_2O.
8. *0.5 mg/mL Phenylalanine standard stock solution.* Dissolve 50 mg L-phenylalanine in 0.3 N TCA and dilute to 100 mL. Freeze in 5-mL batches. For working standard dilute 5 mL stock solution to 50 mL with 0.3 N TCA.
9. *0.05 mg/mL Phenylalanine working solution.* Dilute 5 mL of stock solution to 50 mL with 0.3 N TCA. **MAKE FRESH EACH DAY.**

Lipoprotein Separation by Electrophoresis

1. *Fat Red 7B stain*. Dissolve 0.5 g of Fat Red 7B in MeOH and dilute to 500 mL with MeOH. Stir; filter next day. Store at room temperature. Discard if precipitate develops.
2. *Lipoprotein destaining solution*. Mix 750 mL methanol and 250 mL H_2O.

B. Method 2. Tryptophan Content by a Microbiological Assay

1. 2.5 mg/mL tryptophan stock standard.
2. 0.2 mg/mL tryptophan working solution.
3. Saline, sterile.
4. Minimal medium, sterile.
5. 4% Glucose, sterile.
6. Nutrient broth, sterile.
7. Test tubes, 16 × 150 mm.
8. Foam plugs for test tubes.
9. 37°C H_2O bath, shaking.
10. Food hydrolysates (prepared on day 5, same as for Method 1).
11. *E. coli* 514 culture grown in minimal medium with tryptophan.

Assay of Protein Extract for Tryptophan Content by a Microbiological Method

1. *2.5 mg/mL Tryptophan standard stock solution*. Dissolve 250 mg L-tryptophan in 100 mL H_2O. Freeze.
2. *0.2 mg/mL Tryptophan working solution*. Dilute 8.0 mL stock solution to 100 mL H_2O. **MAKE FRESH EACH DAY.**
3. *Saline, sterile*. Dissolve 1.8 g NaCl in 200 mL H_2O. Dispense 10-mL batches in screw-capped tubes. Autoclave.
4. *Minimal medium (MM)*. Dissolve the following salts in 1 L H_2O:

K_2HPO_4	10.5g
KH_2PO_4	4.5g

$(NH_4)_2SO_4$	1.0g
Na citrate (dihydrate)	0.97g
$MgSO_4$ (anhydrous)	0.05g

Add 5 mg tryptophan to 100 mL of MM and dispense in 5-mL batches in screw-capped tubes. Dispense the remaining minimal medium in 200-mL batches in 1-L flasks. Sterilize the flasks and screw-capped tubes in an autoclave.

5. *4% Glucose, sterile*. Dissolve 2 g glucose in 50-mL H_2O. Dispense in 5-mL batches in screw-capped tubes and sterilize.
6. *Nutrient broth, sterile*. Dissolve 0.8 g nutrient broth powder and 0.4 g NaC1 in 100 mL H_2O. Dispense in 5-mL batches in screw-capped tubes and sterilize.
7. E. coli *514 culture*. Every 2 to 3 months, streak a loopful of culture on nutrient agar plates. Pick some cells from a colony, and inoculate into nutrient broth when needed. Incubate overnight at 37°C with shaking. Centrifuge cells in a clinical centrifuge, and resuspend in 5 mL of 0.9% saline. Inoculate two tubes of MM + tryptophan with 0.05 mL of cells. Incubate at 37°C with shaking until culture has fully grown, usually overnight.
8. *Nutrient Agar Plates*. Dissolve 0.8 g nutrient broth, 0.4 g NaCl, and 1.5 g agar in 100 mL H_2O. Sterilize the solution and pour plates when the broth has cooled. Plates can be stored in the refrigerator for a few days.

CHAPTER 10: CELL COMPONENTS MODULE

Day 1

Fractionation of Rat Liver Cells and Oxidative Phosphorylation

1. Rats and guillotine.
2. Homogenizers.
3. Cheese cloth.
4. Scalpels, tweezers.
5. 0.25 M sucrose+1.0 mM EDTA, pH 7.5.
6. 0.34 M sucrose+1.0 mM EDTA, pH 7.5.

7. 0.25 M sucrose+3.0 mM $CaCl_2$, pH 7.5.
8. 5% Ficoll-400.
9. RC buffer mixture.
10. 10.0 mM ADP.
11. 10.0 mg/mL cytochrome *c*.
12. 0.5 M Na succinate.
13. 0.5 M Na glutamate.
14. 0.5 M Na malate.
15. 1.0 M Na ascorbate.
16. Inhibitors: CCP, antimycin A, rotenone, and Na malonate.
17. Oxygraphs, ice buckets, and automatic pipettors with tips.
18. Autoclavable bag for disposal of rats.
19. Plastic centrifuge tubes for storing fractions.
20. Refrigerated Sorvall centrifuge and ultracentrifuge with rotors.
21. 50-mL centrifuge tubes, 30-mL tubes for ultracentrifuge.

All reagents are made with deionized H_2O unless otherwise stated.

Preparation of Reagents (for ca 6 students for the isolation procedure, the other solutions for 16 students)

Solutions for Fractionation

1. *1.0 M sucrose.* Dissolve 342.3 g sucrose in 600 mL of H_2O. Dilute to 1 L with H_2O. Refrigerate.
2. *0.2 M EDTA (disodium salt).* Dissolve 7.44 g EDTA in H_2O. Dilute to 100 mL with H_2O. Refrigerate.
3. *0.25 M sucrose with 1.0 mM EDTA, pH 7.5.* Mix 500 mL 1.0 M sucrose and 10 mL 0.2 M EDTA with 1.5 L H_2O. Adjust pH to 7.5 with 1 N NaOH and dilute to 2 L with H_2O. **MAKE FRESH EACH DAY.** Refrigerate.
4. *0.34 M sucrose with 1.0 mM EDTA, pH 7.5.* Mix 170 mL 1.0 M sucrose and 2.5 mL 0.2 M EDTA with 300 mL H_2O. Adjust pH to 7.5. Dilute to 500 mL. **MAKE FRESH EACH DAY.** Refrigerate.

5. *0.25 M sucrose with 3.0 mM $CaCl_2$, pH 7.5*. Mix 62.5 mL 1.0 M sucrose and 0.11 g $CaCl_2$ with 100 mL with H_2O. Adjust pH to 7.5 (with 1 N NaOH). Dilute to 250 mL. **MAKE FRESH EACH DAY**. Refrigerate.
6. 5% Ficoll-400 (Pharmacia LKB Biotechnology, Piscataway, NJ). Dissolve 5g Ficoll-400 in 100 mL H_2O. Refrigerate.

Mitochondrial Oxidation Studies

1. *0.2 M K phosphate buffer, pH 7.4* Dissolve 55.68 g K_2HPO_4 and 10.88 g KH_2PO_4 in 1.6 L H_2O. Adjust pH to 7.4 (with 1 NaOH). Dilute to 2 L with H_2O.
2. *0.1 M $MgCl_2$.* Dissolve 2.03 g $MgCl_2 \cdot 6H_2O$ in H_2O and dilute to 100 mL with H_2O.
3. *RC Mixture*

Final concentration	*For 500 mL solution*
0.25 M sucrose	25 mL of 1 M sucrose
2 mM EDTA	5 mL of 0.2 M EDTA
20 mM KPi, pH 7.4	50 mL 0.2 M KPi, pH 7.4
4 mM $MgCl_2$	20 mL 0.1 M $MgCl_2$

Dilute to volume with H_2O. Freeze in 100-mL batches.

4. *10 mM ADP.* Dissolve 52.52 mg ADP in 5 mL H_2O. Adjust pH to 6.8 (with 1 N NaOH). Dilute to 10 mL with H_2O. **PREPARE FRESH EACH DAY.**

Substrates for Oxidative Phosphorylation Study

1. *0.5 M Na succinate.* Dissolve 10.04 g in 30 mL H_2O. Adjust pH to 7 to 7.4 (with 1 N HCl). Dilute to 75 mL with H_2O.
2. *0.5 M Na glutamate.* Dissolve 6.34 g in 30 mL H_2O. Adjust pH to 7 to 7.4 (with 1 N HCl). Dilute to 75 mL with H_2O. (L-glutamic acid, sodium salt.)
3. *0.5 M Na malate.* Dissolve 5.85 g in 30 mL H_2O. Adjust pH to 7 to 7.4 (with 1 N HCl). Dilute to 75 mL with H_2O. (L-malic acid, monosodium salt.)
4. *1.0 M Na ascorbate.* Dissolve 13.2 g ascorbic acid in 10 to 20 mL H_2O. Add 3 N NaOH *dropwise* to a pH of 7.1 to 7.2. Dilute to 75 mL with H_2O.
5. *20.0 mg/mL cytochrome c (C-2506 Sigma type III).* Dissolve 500 mg cytochrome *c* in 10 to 20 mL H_2O. Adjust pH to *ca* 7.2. Dilute to 25 mL with H_2O. Wrap in aluminum foil.

Store all substrates frozen in 2- to 3-ml batches.

Inhibitors of Oxidative Phosphorylation

1. *0.1 mM CCP in EtOH.* Dissolve 2.0 mg of *m*-chlorocarbonyl-cyanide phenylhydrazone in 100 mL of 95% EtOH.
2. *0.5 mg/mL Antimycin A in EtOH.* Dissolve 5.0 mg of antimycin A in 10 mL of 95% EtOH.
3. *1 mM Rotenone in EtOH.* Dissolve 10.0 mg of rotenone in 25 mL of 95% EtOH.
4. *1.5 M Na Malonate, pH* ca *7.0.* Dissolve 0.76 g of Na malonate in 20 mL of H_2O. Adjust to pH 7.0 to 7.4 (with 1 N HCl). Dilute to 25 mL with H_2O.

   Store all inhibitors frozen in 1-mL batches. The inhibitors are toxic. Use care in handling.

Day 2

Microsomal Activity Studies

1. 0.06 M K phosphate buffer, pH 7.4.
2. 0.5 mM DCPIP.
3. Automatic pipettors and tips.
4. 10.0 mg/mL cytochrome *c*.
5. 50.0 mM 4-dimethylaminoantipyrine.
6. 40.0 mg/mL glucose oxidase.
7. Glucose, solid.
8. Spectrophotometer and quartz cuvettes.
9. 1.0 mM NADH.
10. Stopwatches.

Determination of K_m NADH–DCPIP Reductase

1. *0.06 M K phosphate, pH7.4.* Dilute 300 mL 0.2 M K phosphate, pH 7.4, to 1 L with H_2O (see day 1). Refrigerate.
2. *0.5 mM DCPIP.* Dissolve 14.5 mg dichlorophenol-indophenol (DCPIP) in H_2O and dilute to 100 mL with H_2O. Store in a brown bottle. **MAKE FRESH EACH DAY.** Refrigerate.
3. *1.0 mM NADH.* Dissolve 7.5 mg NADH in 10 mL H_2O (MW, 750.6). **MAKE FRESH EACH DAY.** Refrigerate.

Determination of K_m NADH for NADH-Cytochrome-*c* Reductase

1. *0.06 M K phosphate, pH 7.4.* See above.
2. *10.0 mg/mL Cytochrome* c. See day 1.
3. *1.0 mM NADH.* See above.

Determination of Cytochrome P_{450} Reductase Activity

1. *0.06 M K phosphate, pH 7.4.* See above.
2. *20 mg Glucose.* Provide glucose and weighing paper.
3. *1.0 m M NADH.* See above.
4. *50.0 mM 4-dimethylaminoantipyrine (DMAAP).* Dissolve 0.578 g DMAAP in H_2O and dilute to 50 mL with H_2O (MW, 231.3). Store in a brown bottle. Refrigerate.
5. *40.0 mg/mL Glucose oxidase.* Dissolve 80 mg glucose oxidase in 2 mL H_2O. **MAKE FRESH EACH DAY.** Refrigerate.

Day 3

Protein Content of Liver Cell Fractions and Reconstruction of the Electron Transport Chain

1. RC buffer mixture.
2. 0.1 N KCN.
3. 5% DOC in 0.01N KOH.
4. 37°C H_2O bath.
5. Biuret reagents.
6. Oxygraphs.
7. 0.25 M sucrose–1.0 mM EDTA, pH 7.5.
8. Inhibitors (same as day 1 but omit CCP).
9. Substrates (see day 1).
10. BSA 2 mg/mL.
11. Automatic pipettors and tips.

Determination of Protein by Biuret Method

1. *5% DOC in 0.01 N KOH.* Dissolve 5.0 g deoxycholic acid in 80 mL H_2O. Add 1 mL 1N KOH. Dilute to 100 mL with H_2O. Refrigerate.
2. *Biuret reagent.* See Chapter 3, day 1.

Reconstruction of the Electron Transport Chain

1. *0.1 M KCN.* Dissolve 32.0 mg KCN in 5 mL H_2O. **MAKE FRESH EACH DAY. TOXIC REAGENT, HANDLE WITH CARE.** Refrigerate.
2. *RC buffer mixture.* See day 1.
3. *0.25 M sucrose with 1.0 mM EDTA, pH 7.5.* See day 1.

Day 4

Assay of Fractions for Various Enzymatic Activities

1. 30°C and 37°C H_2O baths.
2. 100 mM 2(*N*-morpholino) ethane sulfonic acid (MES), pH 6.5.
3. 150 mM glucose-6-phosphate, pH 6.5.
4. 1 mM KH_2PO_4.
5. 0.5 N TCA (8%).
6. 1.6% Ammonium molybdate in 1.0 N H_2SO_4.
7. $FeSO_4 \cdot 7H_2O$, solid.
8. 0.01 M Na pyruvate.
9. 1 mM NADH.
10. 0.01 M Na phosphate, pH 7.4.
11. 0.5% Na cholate.
12. 0.5 M acetate buffer, pH 5.0.
13. 10 mM *p*-nitrophenyl phosphate.
14. 0.5 M NaOH.
15. Spectronic-20 below 625 nm, cuvettes.
16. Clinical centrifuge and 15-mL glass conical centrifuge tubes.

Assay for Lysosomal Acid Phosphatase Activity

1. *0.5% Na cholate.* Dissolve 0.50 g Na cholate in H_2O and dilute to 100 mL. Refrigerate.
2. *0.5 M Na acetate buffer, pH 5.0.* Dissolve 8.20 g Na acetate in 100 mL H_2O. Adjust pH to 5.0. Dilute to 200 mL with H_2O. Freeze.

3. *10 mM p-nitrophenyl phosphate.* Dissolve 0.186 g *p*-nitrophenyl phosphate (Sigma 104) in 50 mL H_2O (MW 371.1). Freeze in 10-mL batches.

Assay for Glucose-6-Phosphatase Activity

1. *100 mM MES, pH 6.5.* Dissolve 1.95 g MES in 75 mL H_2O. Adjust pH to 6.5 with 1 N NaOH. Dilute to 100 mL with H_2O. Refrigerate.
2. *150 mM glucose-6-phosphate, pH 6.5.* Dissolve 4.23 g in 75 mL H_2O. Adjust pH to 6.5 with 1 N NaOH. Dilute to 100 mL with H_2O. Freeze in 5-mL batches.
3. *0.5 N TCA (8%).* Dilute 80 mL 50% TCA to 500 mL with H_2O. Refrigerate.

Determination of Inorganic Phosphate

1. *1 mM KH_2PO_4.* Dissolve 0.068 g KH_2PO_4 in H_2O and dilute to 500 mL with H_2O or dilute 2.5 mL of 0.2 M K phosphate to 500 mL with H_2O. Refrigerate.
2. *0.5 N TCA (8%).* See above.
3. *16% Ammonium molybdate in 10 N H_2SO_4.* Dissolve 16 g ammonium molybdate in 50 mL H_2O, add 28 mL conc. H_2SO_4, and dilute to 100 mL with H_2O. Refrigerate.

Assay for Lactate Dehdyrogenase

1. *0.01 M Na pyruvate.* Dissolve 0.11 g Na pyruvate in H_2O, and dilute to 100 mL with H_2O. Refrigerate.
2. *0.01 M Na phosphate, pH 7.4.* Add 12.5 mL 0.4 M Na phosphate, pH 7.0, to 400 mL H_2O; adjust pH to 7.4 with 1 N NaOH. Dilute to 500 mL with H_2O. Refrigerate.
3. *1 mM NADH.* Dissolve 7.5 mg NADH in 10 mL H_2O. **MAKE FRESH EACH DAY.**

Day 5

Isolation of Nucleic Acids

1. 8% TCA, 5% TCA.
2. 95% EtOH.
3. 2% PCA.
4. 0.4 M Na phosphate, pH 7.0.
5. 10.0 mM Na phosphate, pH 7.0.
6. 0.2 M Na phosphate with 8 M urea, pH 7.0.

7. Hydroxyapatite (HAP) (Sigma Chemical Co., St. Louis, MO).
8. Urea containing solution for DNA suspension.
9. Isoamyl alcohol: chloroform (1:24) (store in the hood).
10. 60° to 70°C H_2O bath.
11. 90°C and boiling H_2O bath.
12. Spectronic-20s above and below 625 nm, cuvettes.
13. Homogenizers.
14. Glass stirring rods (extra thin).
15. Centrifuge tubes (heavy and regular walled), 15 mL.
16. Refrigerated Sorvall centrifuge and SS-34 rotor.

Extraction of DNA and RNA From the Cell Fractions

1. *8% TCA.* See day 4.
2. *5% TCA.* See day 4.
3. *95% EtOH.* Provide 95% stock solution.
4. *2% PCA (0.3 N).* Mix 14 mL 70% PCA with H_2O and dilute to 500 mL with H_2O.
5. *0.4 M Na phosphate, pH 7.0.* Dissolve 54.84 g $NaH_2PO_4{\cdot}H_2O$ in 800 mL H_2O. Adjust pH to 7.0 and dilute to 1 L with H_2O. Refrigerate.
6. *10.0 mM Na phosphate, pH 7.* Dilute 25 mL 0.4 M Na phosphate to 1 L with H_2O. Refrigerate.
7. *0.2 M Na phosphate with 8 M urea, pH 7.* Dissolve 96.0 g urea in 100 mL 0.4 M Na phosphate, pH 7. Dilute to 200 mL with H_2O.

Isolation of Duplex DNA

1. *1 M $NaClO_4$, pH 7.0.* Add 4.0 g NaOH pellets to 50 mL H_2O. Add 12.2 mL 70% perchloric acid. Adjust pH to 7.0 with 6 N NaOH. Dilute to 100 mL with H_2O. Refrigerate.
2. *Urea containing solution for DNA suspension.* Add the following to 50 mL of 1M $NaClO_4$:

   48.0 g urea

   1.70 g NaH_2PO_4

   1.08 g Na_2HPO_4

0.03 g EDTA

1.00 g SDS

Dissolve slowly with stirring, and dilute to 100 mL with 1M $NaClO_4$.

3. *$CHCl_3$:isoamyl alcohol (24:1).* See Chapter 6, day 3.

Day 6

Determination of Base Composition by UV Spectrophotometric Analysis

1. 2% PCA.
2. 0.15 mg/mL DNA standard solution.
3. Diphenylamine, solid.
4. Glacial acetic acid (store in the hood).
5. Conc. H_2SO_4 (store in the hood).
6. 1.6% acetaldehyde.
7. 0.05 mg/mL RNA standard solution.
8. Orcinol, solid.
9. 0.1% $FeCl_3$ in concentrated HCl (store in the hood).
10. Spectrophotometer with water-jacketted cuvette holders, quartz cuvettes.
11. Boiling H_2O bath.
12. Water bath attached to a spectrophotometer.

---

1. *0.4 M Na phosphate, pH 7.0.* See day 5.

Assays for RNA and DNA Contents of the Cell Fractions

1. *2% PCA.* See day 5.
2. *0.15 mg/mL DNA.* See Chapter 6, day 6.
3. *1.6% Acetaldehyde.* See Chapter 6, day 6.
4. *0.05 mg/mL RNA standard.* See Chapter 6, day 6.
5. *0.1 M $FeCl_3$ in conc. HCl.* See Chapter 6, day 6.

APPENDIX II

Problem Sets

CHAPTER 2: BUFFER PROBLEMS

2.1. What volume of glacial acetic acid (17.6 N) and what weight of sodium acetate (MW = 82) would be required to make 100 mL of 0.2 M buffer, pH 3.9 ($pK_a = 4.8$)?

2.2. What weights of monobasic potassium phosphate (KH_2PO_4; MW = 136) and dibasic potassium phosphate (K_2HPO_4; MW = 174) are required to make 100 mL of 0.2 M buffer, pH 6.5 ($pK_a = 6.8$)?

2.3. What weights of monobasic potassium phosphate (KH_2PO_4; MW = 136) and dibasic sodium phosphate (Na_2HPO_4; MW = 142) are required to make 100 mL of 0.2 M buffer, pH 7.1 ($pK_a = 6.8$)?

2.4. What weights of sodium carbonate (Na_2CO_3; MW = 106) and sodium bicarbonate ($NaHCO_3$; MW = 84) are required to make 100 mL of 0.2 M buffer, pH 10.2 ($pK_a = 9.8$)?

2.5. What volume of hydrochloric acid (11.7 N) and what weight of Tris ([hydroxymethyl] aminomethane) base (MW = 121) would be required to make 100 mL of 0.2 M buffer, pH 8.5 ($pK_a = 8.0$)?

2.6. How would you prepare 100 mL of 0.01 M acetate buffer pH 4.5, given a solution of 0.2 M buffer at pH 4.5?

2.7. What is the pH of a solution containing 0.15 M potassium acetate and 0.30 M acetic acid ($pK_a = 4.8$)?

2.8. a. Write the reactions for the stepwise titration of H_3PO_4 with NaOH.

b. Draw the titration curve, label the axes, and indicate the pK_a values on the graph.

c. Calculate the pH of a solution prepared by mixing 125 mL of 0.1 M H_3PO_4 with 125 mL of 0.1 M NaOH ($pK_{a1} = 2.1$, $pK_{a2} = 7.2$, $pK_{a3} = 12.3$).

CHAPTER 3: DILUTION PROBLEMS

3.1. How would you prepare 500 mL of a 0.5 N HCl solution from a 2 N HCl solution?

3.2. Show by serial dilution how you would prepare a 1:800 dilution of a protein solution. Assume the largest piece of measuring equipment is a 10-mL pipette.

3.3. How would you prepare a protein solution of 0.2 mg/mL from one that has a concentration of 1 mg/mL? Make a total of 10 mL.

3.4. Calculate how you would prepare 300 mL of 0.5 M buffer from a 1.5 M buffer solution.

3.5. Use the Warburg-Christian nomograph to answer the following questions:

a. A solution composed of protein and nucleic acids gave an absorbance at 280 nm of 0.96 and at 260 nm of 0.81.

How much protein in mg/mL is present?

How much nucleic acid in mg/mL is present?

b. If you dilute the solution 1:2

What will the absorbance be at 280 nm?

What will the absorbance be at 260 nm?

What will the new concentration of protein be?

What will the new concentration of nucleic acids be?

CHAPTER 5: ENZYMOLOGY PROBLEMS

5.1. The following data were obtained for the fractionation of yeast extract during the purification of the enzyme alcohol dehydrogenase:

Fraction	*Total volume per fraction (mL)*	*Protein (mg/mL)*	*Units/ mL*
Yeast extract (crude)	320	10.9	28,000
0–60% $(NH_4)_2SO_4$	10	22.1	670,000
60–80% $(NH_4)_2SO_4$	20	13.4	34,400
80% supernatant	500	3.5	1,300

a. For each fraction calculate the following values and tabulate the results:
   i. The specific activity.
   ii. The percent recovery (sometimes termed *yield*).
   iii. The relative purification, assuming a value of 1.0 for the crude fraction.

b. What conclusions can be reached from these calculations?

c. Briefly discuss the following:
   i. What does $(NH_4)_2SO_4$ fractionation accomplish?
   ii. What is the basis for the separation?
   iii. What other substances can be used for the same purpose?

5.2. Determine the dilution of a protein solution to assay for enzyme activity. Use the data provided on the graph to calculate the dilution of enzyme that is needed to get an activity of 1 μmol/10 min in 0.2 mL. The fraction contains 0.025 mg/mL of protein.

5.3. Determination of K_m and V_{max} values for invertase.

mL of 0.3M sucrose	*Sucrose concentration [S]*	*ΔA/10 min*	*Velocity (v) (μmol reducing sugar/min)*	*1/[S]*	*1/v*
.05		.243			
.10		.345			
.20		.436			
.40		.516			

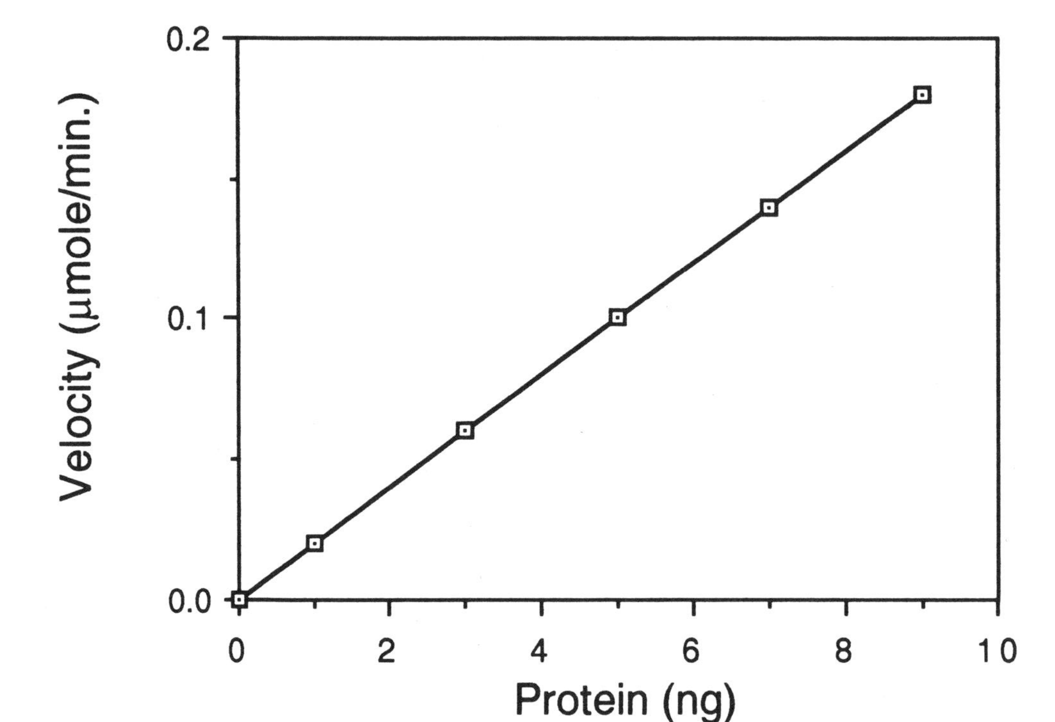

a. Calculate the sucrose concentration in each sample.

b. Using a standard curve for the Nelson's assay, calculate the velocity at each sucrose concentration. (The slope of a typical curve is *ca* 1.0 absorbance unit/µmol sucrose.)

c. Plot 1/[S] vs. 1/v and calculate values for K_m and V_{max}. Be sure to report these values in the proper units.

5.4. Effect of pH on the active site of invertase.

a. The following data were obtained from an enzyme catalyzed reaction.

Velocity (µmol/min)	*pH*
16.7	3
40.0	5
50.0	7
40.0	9
16.7	11

i. Draw the pH optimum curve on graph paper.
ii. Determine the pK values from the graph.
iii. What are the possible amino acids involved in this titration?

b. The following Lineweaver-Burk plot was obtained using the same enzyme from an experiment in which substrate concentration was varied at three different pH values.
   i. What kind of inhibition is shown?
   ii. How do the curves prove this?
   iii. What do these data and the pK values determined from the pH optimum curve tell about the active site of the enzyme?

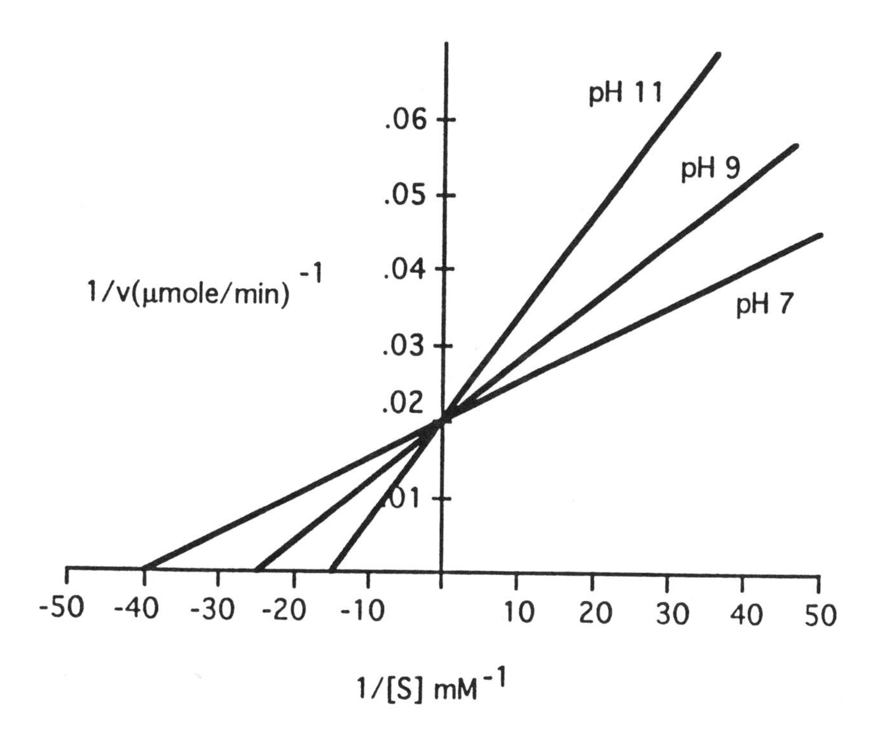

5.5 For determination of the energy of activation (E_a) for an enzyme, the following data were obtained when a protease was assayed for its activity at different temperatures.

T°C	*Velocity (v)* *(μmol/min)*
5	0.0417
25	0.1250
35	0.2000
40	0.2500

a. Graph 1/T *vs.* log (*v*) and determine the energy of activation for this enzyme in Kcal/mol from the slope of this curve and the Arrhenius equation given in Chapter 5 of the text.

CHAPTER 6: NUCLEIC ACIDS PROBLEM

6.1. A solution containing 10 mg/mL of protein and 0.6 mg/mL of nucleic acids was too concentrated to use in an experiment. Dilute this solution to an absorbance of 0.70 at 280 nm and 0.84 at 260 nm.(Give an approximate dilution.)

[Hint: Determine the concentrations of protein and nucleic acids at the absorbances given.]

CHAPTER 7: RECOMBINANT DNA METHODOLOGY PROBLEM

7.1. The following data were obtained from an agarose gel electrophoresis separation.

Hind *III ladder*		*Recombinant digests from*	
No. of base pairs (kb)	*Distance (mm)*	Eco*R1* *(mm)*	Bam *H1* *(mm)*
23.1	13.0	19	22.5
9.4	15.0	26	29.0
6.6	17.5		
4.4	21.0		
2.3	27.0		
2.0	29.0		

a. Construct a standard curve by graphing log kb *vs.* distance (mm) using the data above for the *Hind* III ladder.

b. Calculate the sizes of the fragments of the *Eco*R1 and *Bam* H1 digests.

i. Referring to the λ restriction map (Fig. 7.2), identify the fragment inserted into the recombinant plasmid. (The site on the map is identified by the size of the insert.)

ii. When the recombinant plasmid is digested with *Bam* H1, what were the sizes of the resulting fragments?

c. Show the two possible orientations for the insertion.

i. Which one is correct?

ii. Why?

CHAPTER 8: LIPIDS PROBLEMS

8.1. Lipids were extracted from 10.0 g of liver. The total volume of the lipid extract was 25.0 mL. A 5.0-mL aliquot of this extract was then applied to a silicic acid column for separation into lipid classes. A neutral lipid fraction and a polar lipid fraction, each having a volume of 35.0 mL, were obtained.

A 2.0-mL aliquot from each of these three fractions was subjected to gravimetric analysis and the following data obtained.

Total lipids	
Weight of weighing boat	1.50713 g
Weight of weighing boat + sample	1.51977 g
Neutral lipids	
Weight of weighing boat	1.51432 g
Weight of weighing boat + sample	1.51497 g
Polar lipids	
Weight of weighing boat	1.51018 g
Weight of weighing boat + sample	1.51115 g

a. How many mg of lipid were extracted from the 10.0 g of liver? Calculate the number of mg lipid per gram of liver.

b. How much polar lipid was obtained from the column? How much neutral lipid? Calculate the mg polar lipid and mg neutral lipid per 10.0 g of tissue and per gram of tissue.

c. What was the percent recovery from the column?

8.2. Lipids were extracted from 2.00 g of egg yolk. The volume of the extract was 25.0 mL. A 1.00-mL portion was separated on a silicic acid column. Neutral and polar fractions were collected, each of which had a volume of 40.0 mL. These two fractions were then evaporated down to 5.0 mL each.

The cholesterol assay was performed on the total, polar, and neutral lipids. The total lipid extract was used without further concentration, but the neutral and polar lipid fractions were concentrated as follows: 1.0 mL of the neutral fraction and 2.0 mL of the polar fraction were evaporated to dryness. A 1.0 mL aliquot of total lipid fraction was also evaporated to dryness and each fraction was then resuspended in 0.2 mL of EtOH.

A standard curve for the assay was generated using aliquots of a 0.1 mg/mL cholesterol standard solution. This curve is shown on the following graph.

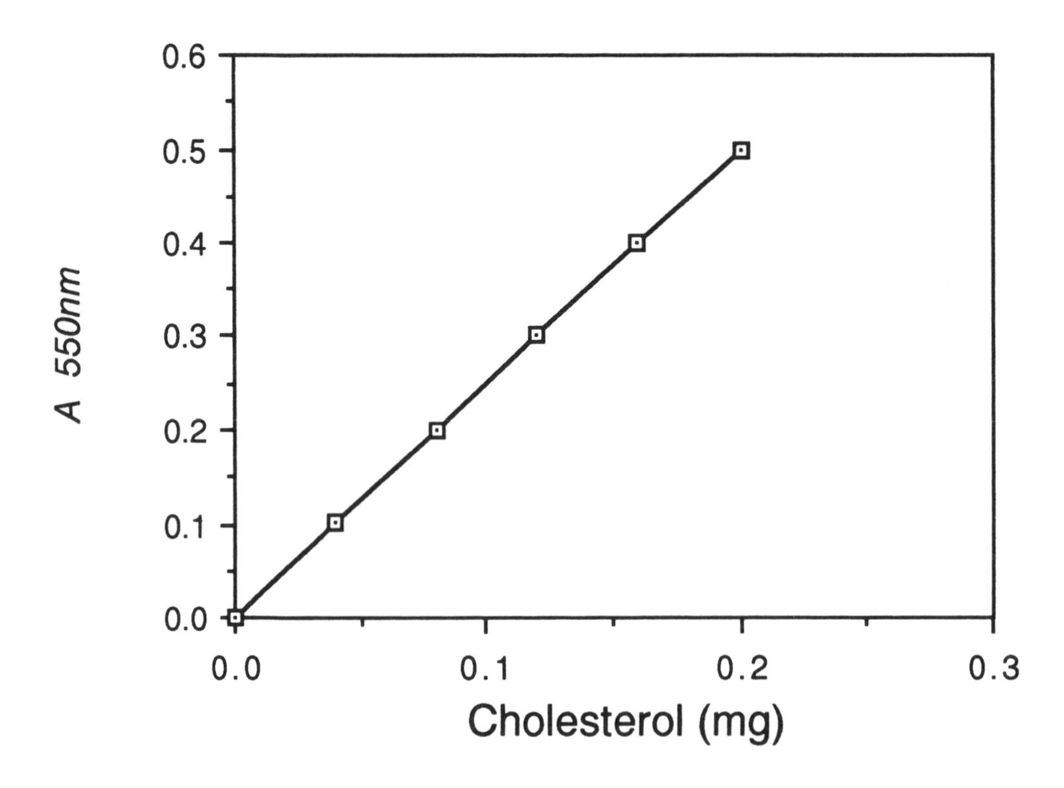

The polar, neutral, and total lipid fractions were tested for cholesterol in the same assay using 0.02-, 0.05-, and 0.10-mL aliquots of each. The absorbance measurements at 550 nm are shown in the table below.

Fraction	*Volume (mL)*		
	0.02	0.05	0.10
		$A_{550\ nm}$	
Polar	.020	.055	.113
Neutral	.101	.250	.505
Total	.180	.450	.910

a. Using the standard curve, determine the milligrams of cholesterol in each sample.

b. Calculate the mg per mL in each fraction.

   i. Averaging together the samples that fall within the range of the standard curve, calculate the mg/mL of cholesterol in each of the lipid fractions before they were concentrated for the assay.

   ii. Calculate the mg cholesterol in the original "total lipids" extract.

c. Calculate the mg/g of cholesterol in egg yolk.

d. What is the percentage of cholesterol in egg yolk?

e. Calculate the percent yield of cholesterol from the column.

CHAPTER 9: CLINICAL/NUTRITIONAL BIOCHEMISTRY PROBLEM

9.1. Determination of tryptophan in foodstuffs.

Protein extracts of egg whites and cheddar cheese were hydrolyzed and the hydrolysates were assayed by measuring the growth of a culture of *E. coli* 514, a tryptophan requiring mutant. The following data were recorded.

Standard tryptophan (μg)	$A_{540\ nm}$
0	0
10	.13
20	.34
30	.50

40	.55
50	.60
Cheese extract (7.5 mg/mL), 1 mL	.06
Egg white extract (10 mg/mL), 1 mL	.14

a. Construct a standard curve by graphing $A_{540\ nm}$ *vs.* tryptophan (μg).

b. Determine the amount of tryptophan in the extract (remember that a racemic mixture is obtained and *E. coli* can only use the L-trp form of the amino acid).

c. Calculate the percent tryptophan in each of the two foods analyzed.

Index

3